For a complete listing of titles in the
Artech House Radar Library,
turn to the back of this book.

Signal Processing
Passive Bistatic Ra

Signal Processing for Passive Bistatic Radar

Mateusz Malanowski

ARTECH HOUSE
BOSTON | LONDON
artechhouse.com

Library of Congress Cataloging-in-Publication Data
A catalog record for this book is available from the U.S. Library of Congress.

British Library Cataloguing in Publication Data
A catalogue record for this book is available from the British Library.

Cover design by John Gomes

ISBN 13: 978-1-63081-662-9

685 Canton Street
Norwood, MA 02062

10 9 8 7 6 5 4 3 2 1

Contents

Preface

My adventure with passive radar started around 2005, at the beginning of my Ph.D. studies at the Warsaw University of Technology, Warsaw, Poland. My supervisor, Professor Krzysztof Kulpa, who worked in a NATO (North Atlantic Treaty Organization) task group focusing on passive radar, inspired me to pursue this research topic.

Passive radar is a unique area in radar technology from the point of view of research, especially research conducted at the university. It does not require elaborate infrastructure or high-tech equipment, such as advanced microwave transmit/receive modules and rotary joints, or permit for electromagnetic energy emission. In fact, a simple passive radar can be built using COTS (commercial off-the-shelf) components, the most advanced of which is a software-defined radio receiver. The cost of such a radar can be as low as several hundred dollars. As the processing is realized in the digital domain, the capabilities of the constructed radar depend heavily on the software. As is commonly known, the software written by Ph.D. students worldwide is (almost) cost-free. Therefore, a passive radar can be built at a relatively low cost, but it provides great potential in terms of detection capabilities as well as a research topic.

As a result of research leading to my Ph.D. degree, a passive radar was constructed called *PaRaDe* – Passive Radar Demonstrator. It was an FM (frequency modulation) radio-based passive radar, equipped with an 8-element antenna array, performing detection and tracking in real time. It has been used in numerous trials, during which various types of targets were observed. The radar has also been installed onboard a car and an aircraft, providing valuable insight into passive radar on moving platforms.

Passive radar was also the topic of my habilitation, a degree also referred to as D.Sc. (Doctor of Science), awarded in Poland after the Ph.D. The habilitation dissertation, which included a description of the complete processing chain in passive radar, was the starting point for this book. Several years passed since the defense of the habilitation in 2013, but finally I was able to write the book.

Building the *PaRaDe* system was very rewarding and allowed me to conduct a vast amount of research, but it was still only a demonstrator. For a long time, I had an urge to be involved in the development of an operational passive radar. This happened at the end of 2012, when a project was initiated by the Polish Ministry of Defense, whose aim was to construct a military passive location system. The system is to be a combination of two technologies: passive coherent location (PCL) and passive emitter tracking (PET). The PCL part, which is the topic of the book, is based on the reflection of the signals transmitted by the illuminators of opportunity (nonradar transmitters, such as radio or television) by the target. The PET system relies on the transmissions generated by the target, such as radar, communications or navigation signals. The consortium currently realizing the project consists of two Polish companies, PIT-Radwar and AM Technologies, alongside the Warsaw University of Technology. The PCL-PET system developed within the project has the chance to be one of the first operational, full-fledged military passive radars in the world.

This book would not be possible without the support of my colleagues from the Warsaw University of Technology. First of all, I would like to thank my supervisor and mentor, Professor Krzysztof Kulpa, who encouraged me to start work in the radar field. I would also like to express my gratitude to other colleagues, all of whom supported my work on the book: Marcin Bączyk, Grzegorz Krawczyk, Damian Gromek, Dr. Łukasz Maślikowski, Professor Jacek Misiurewicz, Marek Płotka, Dr. Stanisław Rzewuski, Professor Piotr Samczyński, Maciej Wielgo, and Marcin Żywek.

I am also very grateful to colleagues from other institutions who provided input or helped me in other ways with this book: Dr. Tadeusz Brenner from PIT-Radwar, Dr. Amerigo Capria from CNIT, Andrzej Chwastek from the Polish Rocket Society, Dr. Diego Cristallini from Fraunhofer FHR, Sébastien Faivre from Thales, Col. Dr. Tomasz Jakusz from the Polish MOD, Dr. Steffen Lutz from Hensoldt, Damian Mayer, Dr. Karl Erik Olsen from FFI, Dominique Poullin from ONERA, Dr. Vojtech Stejskal from Era, and Martin Ummenhofer from Fraunhofer FHR.

Last but not least, I would like to thank my wife, who endured endless evenings in which I spent writing the book.

Chapter 1

Introduction

In this book, signal processing algorithms, detection, and estimation methods for passive bistatic radar are analyzed. The type of radar under consideration has two characteristic features: it does not have its own transmitter, and it uses transmitters that are not radars. This type of radar is usually referred to as passive bistatic radar (PBR), passive covert radar (PCR), passive coherent location (PCL) radar, or simply passive radar. In this book, the last term will usually be used. There is also a class of bistatic radar that utilizes either cooperative or noncooperative active radars. This type of radar, however, will not be considered.

The term *bistatic* used above means that the transmitter and the receiver are spatially separated. In the case when multiple transmitter-receiver pairs are used, the system is called *multistatic*. The bistatic/multistatic configuration of the passive radar differs from the *monostatic* configuration of a classical active radar, where the transmitter and the receiver are in the same place (or very close to each other). The consequences of the bistatic/multistatic nature of passive radar will be apparent throughout the book.

As already mentioned, the passive radar under consideration in this book uses illuminators different than active radars. Typically, illuminators such as radio or television transmitters are used. This means that the signals utilized by the passive radar were not designed for the purpose of target detection; therefore, a suboptimal performance can be expected. This will be demonstrated in the analysis of different transmitter types in Chapter 2.

The question is: why do we need a passive radar using noncooperating transmitters when we have active radars that have been functioning very well for so many decades? As it turns out, passive radar can offer some benefits in comparison to active radar that makes it worth considering, or in some cases, a very desirable technology. First, because the radar does not have its own transmitter, its cost is substantially reduced when compared to an active radar. Moreover, the lack of moving parts operating with high-powered signals, such as a rotating antenna, reduces the cost even further.

Second, passive radar provides the possibility of creating an air picture without the need for transmitting high-powered signals which reveal the position of the radar. This is a very important feature in military applications. Another military-related feature is the possibility of detecting stealth targets due to the low operating frequencies (usually very high frequency (VHF) or ultrahigh frequency (UHF) band), where the RAM (radiation-absorbent material) is not as effective as in the microwave region.

Another advantage of the passive radar is connected with the congestion of the electromagnetic spectrum. As it is generally known, the ether is being filled with more emissions. It seems that telecommunications take priority regarding the allocation of frequency bands over other applications, such as radar, navigation or radio astronomy. In this situation, allocating a new frequency band for radar purposes, or even keeping a frequency band already allocated to radar, may become increasingly difficult. In this context, passive radar is an ideal solution as new telecommunication emissions can be used for radar purposes. For this reason, passive radar is sometimes referred to as "green" or "ecological" radar, as it "recycles" signals that are normally used for other purposes, without generating new electromagnetic pollution.

Nowadays the primary focus of passive radar designers is on improving and refining system performance. As only the receiver is under the control of the radar designer, the emphasis is placed on its improvement. Since passive radar is usually built using software-defined radio concept, most of the signal processing is carried out digitally. Therefore, a large portion of the improvements can be obtained by modifying the signal processing algorithms, which are the topic of this book.

1.1 BOOK CONTENT

It is assumed that the reader has a general knowledge of radar technology and signal processing. In some cases, slightly more specific knowledge on certain aspects is required, such as adaptive or Kalman filtering. Suitable references are provided for the interested reader.

The content of the book is obviously influenced by the author's involvement in certain areas of passive radar research and development. For this reason, these areas will be described in more detail and with greater focus. For example, certain types of illuminators of opportunity will be analyzed, while others will not. Nevertheless, the my intention is to give a relatively universal overview of the topic.

The book contains a description of all signal processing stages in passive radar, starting from the beamforming and ending with the target tracking. Some of the stages are very similar as in the case of classical active radar; however, their description is provided for the sake of the completeness of the book. Whenever applicable, the differences between active monostatic and passive bistatic radars will be highlighted.

This book is divided into 11 chapters. Chapter 2 introduces the basic concepts in passive radar, such as bistatic geometry or bistatic range equation. The influence of the transmitter azimuth and elevation patterns on the detection performance is analyzed. The ambiguity function is defined as a tool for the analysis of signals from different illuminators of opportunity. Next, the following signal types are analyzed: FM radio, DVB-T television, DAB radio, GSM cellular telephony, and WiFi wireless communication. At the end, different transmitters are compared from the point of view of the detection range and range resolution.

In Chapter 3 the subject of beamforming is investigated. The two most popular types of arrays (uniform linear array and uniform circular array) are analyzed and compared. The influence of interelement spacing, the number of array elements, and beam steering on the radiation pattern are investigated. The array tapering is analyzed. Methods for optimizing beamforming coefficients and the calibration of the mutual coupling of the circular array elements are proposed.

Chapter 4 deals with correlation processing, which is a basis for target detection and bistatic parameter estimation. As the calculation of the cross-ambiguity function from its definition would be very time-consuming, three practical cost-effective methods are introduced. The problem of the reduction of deterministic sidelobes by windowing in the time and frequency domains is analyzed. The coherent integration gain that is obtained by calculating the cross-ambiguity function is investigated next. As the classical cross-ambiguity function is based on some simplifying assumptions, a more complex processing algorithm is proposed that alleviates some of the limitations of the classical approach, namely, the range cell migration and the velocity cell migration.

In Chapter 5 clutter filtering is discussed. Methods from two classes of adaptive algorithms (i.e., iterative and block) are compared from the point of view of the convergence rate, frequency selectivity, and computational burden. The iterative methods that are analyzed include NLMS (Normalized Least Mean Squares), RLS (Recursive Least Squares) and LSL (Least Square Lattice) filters. In the case of block algorithms, the least squares matrix solution, lattice filter, modified lattice filter, and the CLEAN methods are investigated.

Chapter 6 discusses the problem of target detection in passive radar. First, the basics of radar detection theory are introduced. Next, the classical constant false alarm rate (CFAR) algorithms are applied to passive radar. The CFAR algorithms include Cell Averaging (CA), Greatest-of Cell Averaging (GOCA), Smallest-of Cell Averaging (SOCA), and Order Statistic (OS). At the end of the chapter, the problem of the estimation of the bistatic parameters of the target echo is investigated.

In Chapter 7, target tracking in bistatic coordinates based on the linear Kalman filter is considered. Bistatic tracking is an intermediate stage between target detection in the bistatic coordinates and target localization in the Cartesian coordinates. The influence of various parameters on tracking accuracy is investigated. A real-life example of accuracy analysis is provided.

Chapter 8 deals with target localization in Cartesian coordinates based on bistatic measurements. Two general approaches are considered, the first of which is the combination of bistatic measurements from different transmitter-receiver pairs. Two methods derived from time-difference-of-arrival (TDOA) systems are adopted for the passive radar scenario. In the second approach, the combination of the direction-of-arrival measurement with the bistatic range from a single transmitter-receiver pair is analyzed. In both approaches, the analysis of the localization accuracy is provided.

In Chapter 9, Cartesian target tracking based on the extended Kalman filter is analyzed. Two updating schemes are compared: parallel and sequential. A two-stage tracking algorithm is proposed based on bistatic and Cartesian trackers. As an example, the tracking of an experimental rocket in the Cartesian coordinates is shown.

Chapter 10 presents several examples of passive radar systems and their applications. These include solutions from Era (Czech Republic), Fraunhofer FHR (Germany), Hensoldt (Germany), PIT-Radwar (Poland), Thales and ONERA (France), the University of Pisa (Italy), and the Warsaw University of Technology (Poland). The examples also include topics that are beyond the scope of this book, such as passive radar imaging (both target and ground). This, however, shows the potential capabilities of the passive radar technology.

In Chapter 11, the final conclusions are presented.

1.2 HISTORICAL BACKGROUND

It is often considered that the history of radar started in 1904, when a German engineer, Christian Hülsmeyer, patented his invention: *the Telemobiloscope* [1]. It was a device that detected metal objects by means of electromagnetic waves. The device consisted of a spark-gap transmitter and a coherer receiver. When a metallic object was detected, an alarm was activated. The first version of the device was not able to calculate the range to the target, but indicated its direction. By today's standards, the Telemobiloscope is a simple form of radar. Despite its great potential, the device did not enjoy commercial success. The knowledge about the possibility of using electromagnetic waves for target detection seems not to have spread, as other inventors had to come up with the idea themselves.

Radar technology slowly developed in different forms over the following three decades. In 1922, for example, Albert Taylor and Leo Young from the U.S. Navy realized that ships passing through the beam of electromagnetic radiation used for communications caused interference, which could be used for target detection.

The next crucial event in the history of radar was the *Daventry Experiment* conducted in 1935 [2–4]. The experiment, whose aim was to detect an airborne target, was initiated by a request from the British Air Ministry directed to Robert Watson-Watt to calculate the required electromagnetic power able to heat up the body of an aircraft pilot, resulting in death. Robert Watson-Watt quickly concluded that the idea of the *death rays*, as the concept was called, was not feasible; however, the reflection of electromagnetic waves from an aircraft could be used for its detection. The idea was quickly turned into practice and an experiment was planned, where a 10-kW BBC radio transmitter in Daventry operating at the wavelength of 49m was used. The antenna system was set up so that the direct signal was canceled. When a target distorted the electromagnetic field, the change in the signal could be noticed and target detection could be declared. This simple setup did not allow the range nor azimuth of the target to be calculated. The experiment was successful, as a Heyford bomber was detected from the range of 8 miles. The Heyford bomber was chosen for this experiment because its wing span corresponded to the half of the wavelength used, thus resulting in a resonance.

The system used in the Daventry Experiment was the first passive bistatic radar using a noncooperative illuminator of opportunity. However, the experimental setup relied on the radio transmitter not because it was an advantageous solution, but rather out of necessity, as no suitable transmitter was available to Watson-Watt at that time. In the following years, dedicated transmitters based on magnetrons were created, which emitted high-power pulses at high frequency. Further radar development was focused mostly on active radars.

The Daventry Experiment started the British radar program, which resulted in the famous Chain Home radar system being built, which became operational just before the beginning of World War II [5]. Among a vast number of radar systems developed during World War II, one system is worth mentioning from the point of view of this book: Klein Heidelberg [2–4, 6–8]. This radar used the illumination of the Chain Home and was apparently the first operational bistatic radar with a noncooperative illuminator.

After World War II, several cooperative bistatic radar systems were developed, such as the American AN/FPS-23 early warning radar and the AN/FPS-133 Air Force Space Surveillance System. They were very specific systems used in niche applications. The mainstream of radar was, and still is, the active monostatic radar.

The modern era of passive radar started in the 1980s and 1990s when television transmitters were used for the detection and tracking of aircraft [9–11]. In 1998, possibly the first operational passive radar using nonradar illuminators, called the Silent Sentry, was developed by Lockheed Martin Corporation, USA. In 2005, a Special Issue of the *IEE Proceedings – Radar, Sonar and Navigation* (Volume 152, Issue 3, June 2005) was published, containing a series of fundamental papers on passive radar [12–26]. Later, passive radar gradually gained recognition and became a very active research area. Presently, sessions on passive radar are organized at most major radar conferences, and papers on this topic are regularly published in scientific journals. Over the years, several excellent books have also been published on passive and/or bistatic radar [27–30].

References

[1] H. Rohling, "From Huelsmeyer's Telemobiloskop to the Digital Radar," in *2014 11th European Radar Conference*, pp. 33–36, October 2014.

[2] H. Griffiths and N. Willis, "Klein Heidelberg – The First Modern Bistatic Radar System," *IEEE Trans. Aerospace and Electronic Systems*, vol. 46, pp. 1571–1588, October 2010.

[3] H. Kuschel, "Approaching 80 Years of Passive Radar," in *2013 International Conference on Radar*, pp. 213–217, September 2013.

[4] H. Griffiths, "Early History of Bistatic Radar," in *2016 European Radar Conference (EuRAD)*, pp. 253–257, October 2016.

[5] R. Buderi, *The Invention That Changed The World: How A Small Group of Radar Pioneers Won The Second World War and Launched A Technological Revolution.* New York, USA: Touchstone, 1997.

[6] R. Voles, "False Detections in Klein Heidelberg," *IET Radar, Sonar Navigation*, vol. 6, pp. 413–416, July 2012.

[7] H. Griffiths, "Klein Heidelberg: New Information and Insight," in *2015 IEEE Radar Conference*, pp. 527–532, October 2015.

[8] H. Griffiths, "Klein Heidelberg: New Information and Further Insight," *IET Radar, Sonar Navigation*, vol. 11, no. 6, pp. 903–908, 2017.

[9] H. D. Griffiths and N. R. W. Long, "Television-Based Bistatic Radar," *IEE Proceedings F – Communications, Radar and Signal Processing*, vol. 133, pp. 649–657, December 1986.

[10] P. E. Howland, "Passive Tracking of Airborne Targets Using Only Doppler and DOA Information," in *IEE Colloquium on Algorithms for Target Tracking*, pp. 37–39, May 1995.

[11] P. E. Howland, "Target Tracking Using Television-Based Bistatic Radar," *IEE Proceedings – Radar, Sonar and Navigation*, vol. 146, pp. 166–174, June 1999.

[12] P. E. Howland, D. Maksimiuk, and G. Reitsma, "FM Radio Based Bistatic Radar," *IEE Proceedings – Radar, Sonar and Navigation*, vol. 152, pp. 107–115, June 2005.

[13] D. K. P. Tan, H. Sun, Y. Lu, M. Lesturgie, and H. L. Chan, "Passive Radar Using Global System for Mobile Communication Signal: Theory, Implementation and Measurements," *IEE Proceedings – Radar, Sonar and Navigation*, vol. 152, pp. 116–123, June 2005.

[14] X. He, T. Zeng, and M. Cherniakov, "Signal Detectability in SS-BSAR with GNSS Non-cooperative Transmitter," *IEE Proceedings – Radar, Sonar and Navigation*, vol. 152, pp. 124–132, June 2005.

[15] R. Saini and M. Cherniakov, "DTV Signal Ambiguity Function Analysis for Radar Application," *IEE Proceedings – Radar, Sonar and Navigation*, vol. 152, pp. 133–142, June 2005.

[16] D. Poullin, "Passive Detection Using Digital Broadcasters (DAB, DVB) with COFDM Modulation," *IEE Proceedings – Radar, Sonar and Navigation*, vol. 152, pp. 143–152, June 2005.

[17] H. D. Griffiths and C. J. Baker, "Passive Coherent Location Radar Systems. Part 1: Performance Prediction," *IEE Proceedings – Radar, Sonar and Navigation*, vol. 152, pp. 153–159, June 2005.

[18] C. J. Baker, H. D. Griffiths, and I. Papoutsis, "Passive Coherent Location Radar Systems. Part 2: Waveform Properties," *IEE Proceedings – Radar, Sonar and Navigation*, vol. 152, pp. 160–168, June 2005.

[19] K. S. Kulpa, "Multi-Static Entirely Passive Detection of Moving Targets and its Limitations," *IEE Proceedings – Radar, Sonar and Navigation*, vol. 152, pp. 169–173, June 2005.

[20] K. S. Kulpa and Z. Czekała, "Masking Effect and Its Removal in PCL Radar," *IEE Proceedings – Radar, Sonar and Navigation*, vol. 152, pp. 174–178, June 2005.

[21] A. N. Morabito, M. G. Meyer, and J. D. Sahr, "Improved Computational Performance for Distributed Passive Radar Processing Through Channelised Data," *IEE Proceedings – Radar, Sonar and Navigation*, vol. 152, pp. 179–184, June 2005.

[22] M. Cetin and A. D. Lanterman, "Region-Enhanced Passive Radar Imaging," *IEE Proceedings – Radar, Sonar and Navigation*, vol. 152, pp. 185–194, June 2005.

[23] M. Tobias and A. D. Lanterman, "Probability Hypothesis Density-Based Multitarget Tracking with Bistatic Range and Doppler Observations," *IEE Proceedings – Radar, Sonar and Navigation*, vol. 152, pp. 195–205, June 2005.

[24] C. M. Hoyuela, A. J. Terzuoli, and R. P. Wasky, "Determining Possible Receiver Locations for Passive Radar," *IEE Proceedings – Radar, Sonar and Navigation*, vol. 152, pp. 206–214, June 2005.

[25] J. J. Xiu, Y. He, G. H. Wang, J. H. Xiu, and X. M. Tang, "Constellation of Multisensors in Bearing-Only Location System," *IEE Proceedings – Radar, Sonar and Navigation*, vol. 152, pp. 215–218, June 2005.

[26] W.-C. Li, P. Wei, and X.-C. Xiao, "TDOA and T^2/R Radar Based Target Location Method and Performance Analysis," *IEE Proceedings – Radar, Sonar and Navigation*, vol. 152, pp. 219–223, June 2005.

[27] N. J. Willis, *Bistatic Radar*. SciTech Publishing, Inc., 2005.

[28] M. Cherniakov, *Bistatic Radar: Principles and Practice*. John Wiley & Sons, Ltd., 2007.

[29] N. J. Willis and H. D. Griffiths, *Advances in Bistatic Radar*. SciTech Publishing, Inc., 2007.

[30] H. Griffiths and C. Baker, *An Introduction to Passive Radar*. Norwood, MA: Artech House, 2017.

Chapter 2

Passive Bistatic Radar Principles

2.1 INTRODUCTION

Passive radar, unlike the well-known active radar, is not equipped with its own transmitter, but relies on an external source of illumination, such as telecommunication transmitters. Target detection is based on a comparison of the direct signal from the transmitter with the echo signal reflected from a target [1, 2]. For this reason, passive radar is equipped with at least two receiving channels: reference and echo. The reference channel serves as the source of the original transmitted signal. This is obtained by pointing a directional antenna towards the transmitter, or by digitally creating a beam directed at the transmitter [3]. The other antenna, or the digitally formed beam, looks towards the intended area of target surveillance, and is the source of the echo signal. These two signals are processed to detect targets. The primary measurement performed by passive radar is the bistatic range, which is the difference in distance between the transmitter-target-receiver and transmitter-receiver paths. The locus of constant bistatic range (iso-range) points defines an ellipsoid with foci on the transmitter and receiver positions. The Doppler shift, also measured with passive radar, is proportional to the bistatic velocity, which is the rate at which the bistatic range changes. Target localization in Cartesian coordinates is carried out by calculating the intersection point of bistatic ellipsoids originating from different transmitter-receiver pairs. Alternatively, target position can be estimated by combining information on the bistatic range and direction of arrival, an approach similar to the classic monostatic radar.

In this chapter, basic definitions and concepts relevant to understanding passive radar will be introduced. These include bistatic geometry, range equation, and evaluation of illuminators of opportunity.

2.2 PASSIVE RADAR GEOMETRY

Let us consider the passive radar geometry in Cartesian coordinates depicted in Figure 2.1. A target is located at $(x(t), y(t), z(t))$, the transmitter is located at (x_t, y_t, z_t), and the receiver is located at (x_r, y_r, z_r). The angle between the target-transmitter and target-receiver segments, denoted with β, is called the *bistatic angle*. Transmitter-target range $R_1(t)$ and target-receiver range $R_2(t)$ can be calculated as follows:

$$R_1(t) = \sqrt{(x(t) - x_t)^2 + (y(t) - y_t)^2 + (z(t) - z_t)^2}, \tag{2.1}$$

$$R_2(t) = \sqrt{(x(t) - x_r)^2 + (y(t) - y_r)^2 + (z(t) - z_r)^2}. \tag{2.2}$$

Let us define the *baseline* length R_b, that is, the range from the transmitter to receiver:

$$R_b = \sqrt{(x_t - x_r)^2 + (y_t - y_r)^2 + (z_t - z_r)^2} \tag{2.3}$$

The difference between the indirect path $R_1(t) + R_2(t)$ and the direct path R_b:

$$R(t) = R_1(t) + R_2(t) - R_b, \tag{2.4}$$

is called the *bistatic range*. Depending on the context, the time dependency of $R(t)$ may be omitted in the rest of the book.

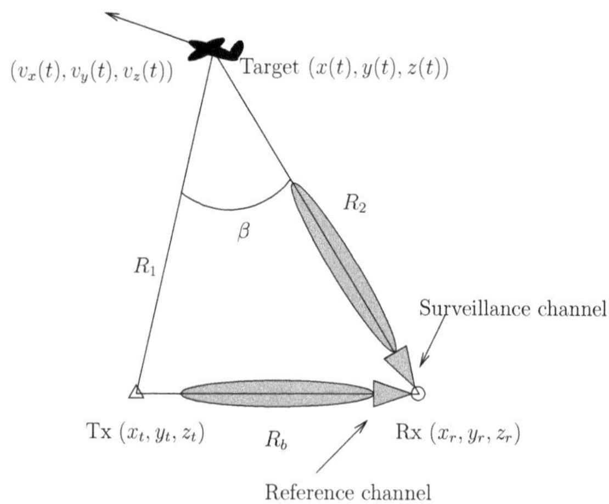

Figure 2.1 Geometry of passive radar.

The bistatic range is calculated from the measured delay τ between the echo and reference signals and the speed of light c (assumed to be 3×10^8 m/s throughout the book):

$$R = c\tau \tag{2.5}$$

The locus of constant bistatic ranges forms an ellipsoid in three-dimensional space, a *bistatic ellipsoid*. The foci of the ellipsoids are located at the transmitter and receiver positions. A two-dimensional cross-section of the bistatic ellipsoid with a plane containing the transmitter and receiver forms an ellipse, the *bistatic ellipse*. In Figure 2.2 an example of the bistatic ellipsoid is shown. The transmitter is located at $(-20, 0, 0)$ km and the receiver is located at $(+20, 0, 0)$ km. The ellipsoid corresponds to the bistatic range equal to 10 km. The plane intersects the bistatic ellipsoid, creating a bistatic ellipse.

In Figure 2.3 an example of a family of bistatic ellipses is shown. As in the previous example, the transmitter is located at $(-20, 0, 0)$ km and the receiver is located at $(+20, 0, 0)$ km. Different ellipses correspond to varying bistatic ranges, from 0 km to 40 km, every 5 km. The bistatic range equal to 0 km corresponds to a target located at the baseline (i.e., on the line between the transmitter and the receiver). An increasing bistatic range corresponds to larger bistatic ellipses.

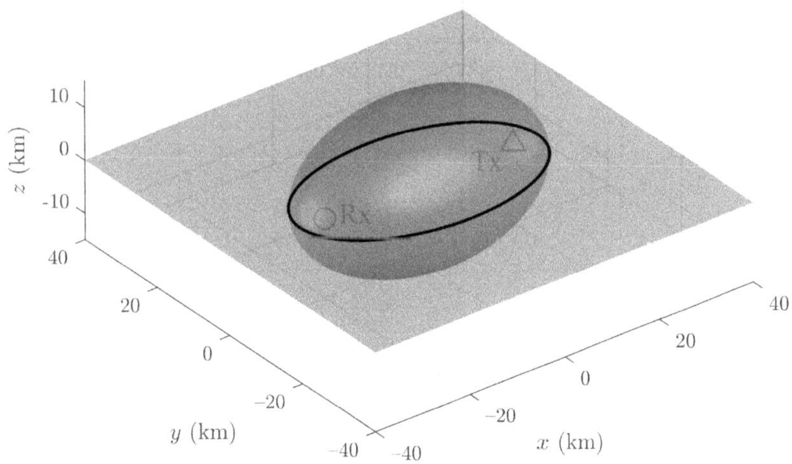

Figure 2.2 Example of a bistatic ellipsoid (locus of constant bistatic range, or iso-ranges) and corresponding bistatic ellipse (intersection of bistatic ellipsoid with the ground plane).

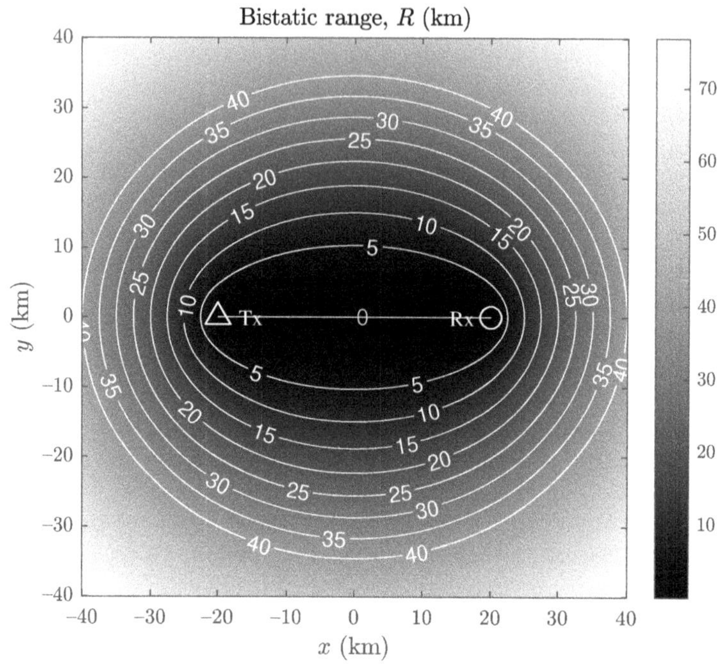

Figure 2.3 Example of bistatic ellipses (locus of constant bistatic range, or iso-ranges).

Apart from bistatic range, another parameter that is routinely measured with the passive radar is the *bistatic velocity*. It is defined as the time derivative of the bistatic range:

$$V(t) = \frac{\mathrm{d}R(t)}{\mathrm{d}t} = \frac{(x(t)-x_t)\,v_x(t) + (y(t)-y_t)\,v_y(t) + (z(t)-z_t)\,v_z(t)}{\sqrt{(x(t)-x_t)^2 + (y(t)-y_t)^2 + (z(t)-z_t)^2}} + \frac{(x(t)-x_r)\,v_x(t) + (y(t)-y_r)\,v_y(t) + (z(t)-z_r)\,v_z(t)}{\sqrt{(x(t)-x_r)^2 + (y(t)-y_r)^2 + (z(t)-z_r)^2}}, \tag{2.6}$$

The bistatic velocity is calculated from measured Doppler shift f_d between the reference and echo signals and the wavelength $\lambda = c/f_c$ (where f_c is the carrier frequency):

$$V = -\lambda f_d \tag{2.7}$$

The negative sign in (2.7) can be explained in the following way: the bistatic velocity is defined as the derivative of bistatic range, thus if the bistatic range is decreasing, the velocity is negative. This corresponds to positive Doppler frequency, hence the minus sign.

Figure 2.4 shows examples of calculated bistatic velocity depending on the target position on the xy plane. Unlike in the case of bistatic range, the bistatic velocity depends not only on the target position, but also on the target velocity vector. The three plots shown in the figure correspond to calculations performed for different azimuth angles of the velocity vector: $0°$, $45°$, and $90°$. In each case the magnitude of the velocity vector was 300 m/s. The white contours show lines of constant bistatic velocity (corresponding to iso-Doppler) with 200 m/s spacing.

If a target is moving along the bistatic ellipsoid, the bistatic range does not change, and the bistatic velocity is zero. If the target is moving in a direction perpendicular to the bistatic ellipsoid, maximum bistatic velocity is obtained. In addition to the velocity vector direction relative to the bistatic ellipsoid, target position with respect to the transmitter and receiver is also important. This effect is illustrated in two dimensions in Figure 2.5. A bistatic ellipse is shown with vectors of Cartesian velocity resulting in constant bistatic velocity (equal to 100 m/s). The velocity vectors in Cartesian coordinates are tangential to the bistatic ellipse, which means that the resulting bistatic velocity is maximum. As can be seen, in order to provide the same bistatic velocity, the magnitude of the Cartesian velocity vector has to change (from 50 to 110 m/s).

The bistatic range resolution depends on the ability to measure the relative delay between the echo and reference signals. The delay measurement resolution $\Delta\tau$ is inversely proportional to the signal bandwidth B; thus, $\Delta\tau = 1/B$. If the relationship between the delay and bistatic range (2.5) is taken into consideration, the resolution of the bistatic range can be expressed as:

$$\Delta R = c \cdot \Delta\tau = \frac{c}{B} \tag{2.8}$$

This is a simple yet important result. The bistatic range resolution depends on the signal bandwidth, which is not under radar constructor control. Therefore, passive radar performance depends heavily on the source of illumination. This aspect will be investigated when analyzing different types of illuminators of opportunity.

Similar reasoning can be applied to bistatic velocity measurement. The Doppler frequency resolution Δf_d is inversely proportional to the observation, or integration, time T; thus, $\Delta f_d = 1/T$. When the relation (2.7) is used, the expression for bistatic velocity resolution is obtained:

$$\Delta V = \lambda \cdot \Delta f_d = \frac{\lambda}{T} \tag{2.9}$$

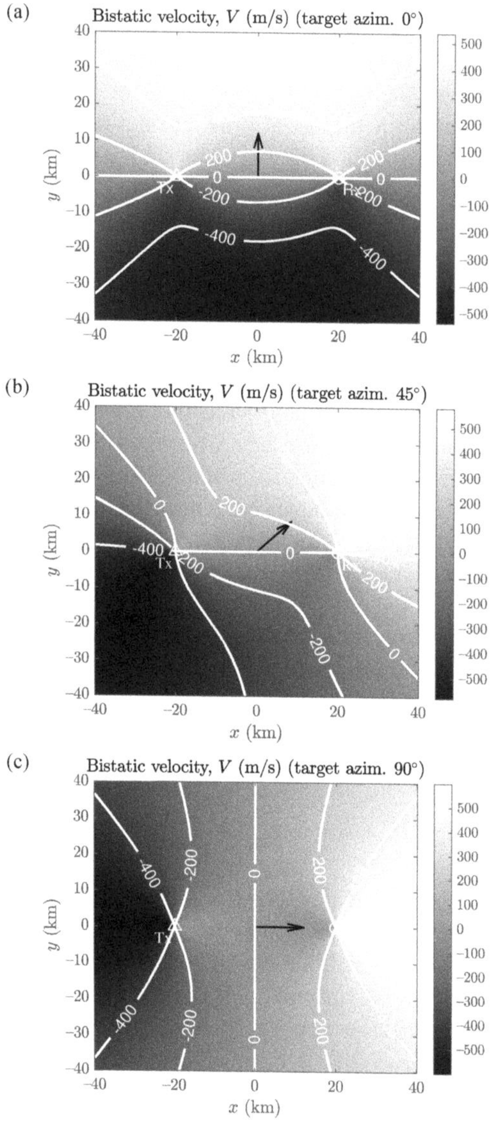

Figure 2.4 Bistatic velocity calculated for velocity vector azimuth equal to (a) 0°, (b) 45°, and (c) 90°. Velocity vector magnitude is 300 m/s in all three cases.

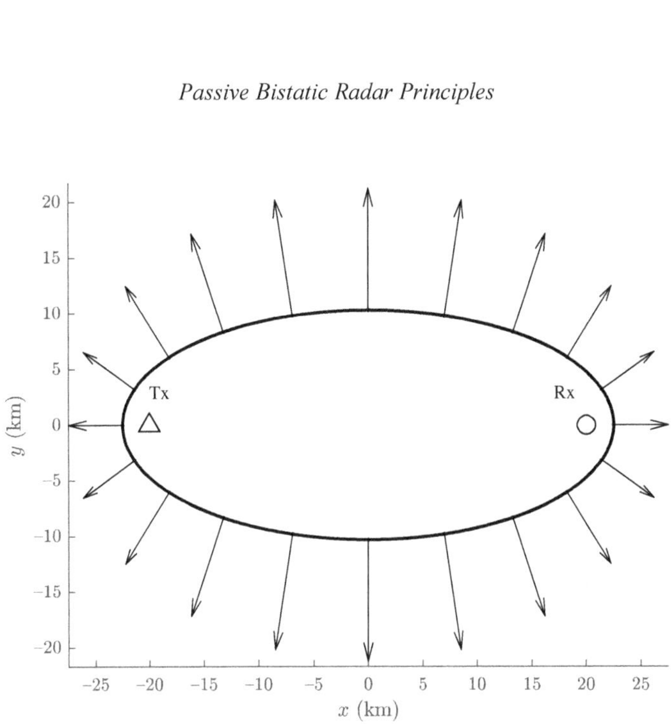

Figure 2.5 Example of Cartesian velocity vectors producing constant bistatic velocity.

As can be seen, the bistatic velocity resolution depends on the wavelength and applied integration time. The wavelength is again determined by the transmitter, similar to the signal bandwidth. The integration time can, however, be controlled by the radar operator, which provides, to some extent, freedom in determining radar performance. The limits of the extension of the integration time will be discussed in Chapter 4.

The bistatic range resolution (2.8) determines the ability to distinguish signal delay. However, it does not determine directly the ability to distinguish range in the Cartesian coordinates. This fact is illustrated in Figure 2.6. The plot at the top shows two bistatic ellipses, corresponding to a bistatic range of 3 km and 4 km, respectively. As can be clearly seen, a constant difference of bistatic range, equal to 1 km, results in varying range difference in Cartesian coordinates. The separation of the bistatic ellipses at the both sides is minimum, whereas the separation at the top and bottom is maximum. The bottom plots (Figure 2.6(b, c)) show the zoom of two areas. The Cartesian range separation for the two locations is 0.5 km and 1.28 km, respectively.

It turns out that the Cartesian rage resolution is related to the bistatic range resolution by the bistatic angle β. The plot of the bistatic angle β is shown in Figure 2.7. The bistatic angle approaches 180° as the target approaches the baseline.

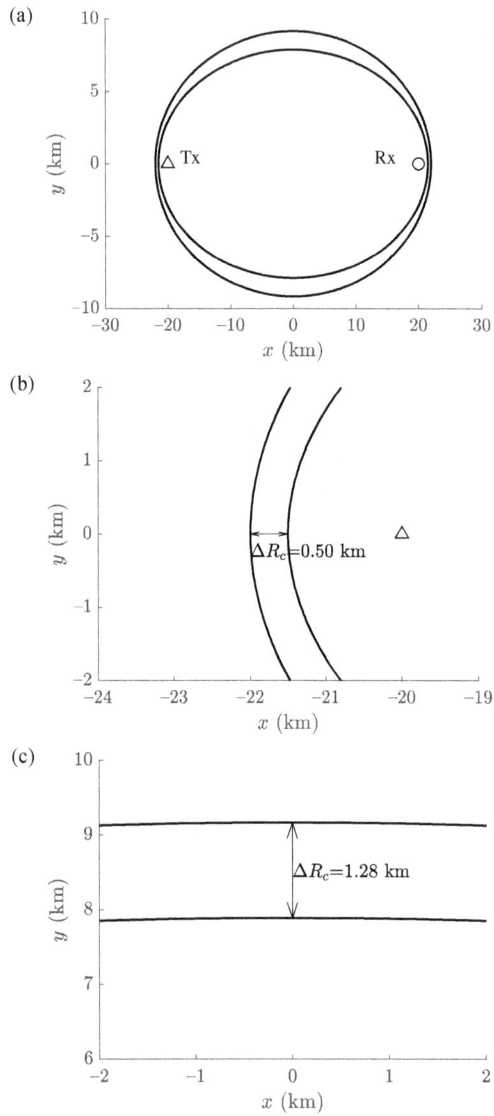

Figure 2.6 Two bistatic ellipses with bistatic range difference equal to 1 km: (a) whole ellipses, (b) zoom for $\beta = 0°$, and (c) zoom for $\beta \approx 130°$.

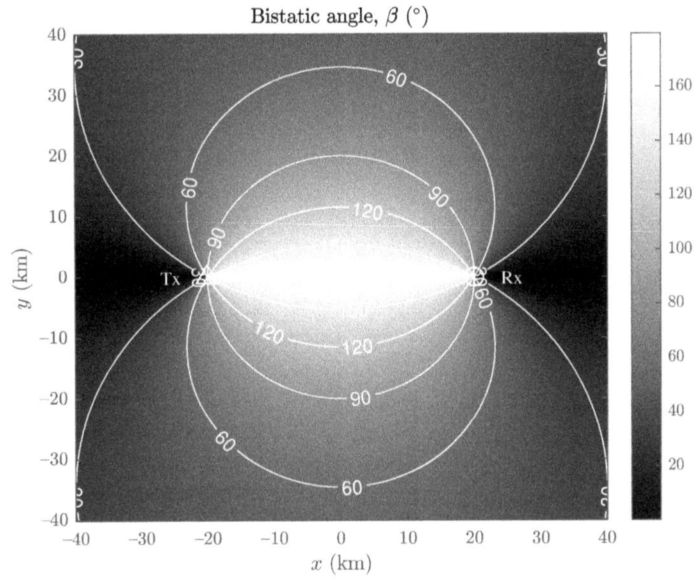

Figure 2.7 Values of bistatic angle in relation to transmitter and receiver.

However, when a target is located at a line including the transmitter and receiver, but outside the transmitter-receiver segment, the bistatic angle is $0°$.

The Cartesian range resolution can be calculated as:

$$\Delta R_c = \frac{c}{2\cos(\beta/2)B} = \frac{\Delta R}{2\cos(\beta/2)} \tag{2.10}$$

In a similar fashion, the Cartesian velocity resolution is calculated as:

$$\Delta V_c = \frac{\lambda}{2\cos(\beta/2)T} = \frac{\Delta V}{2\cos(\beta/2)} \tag{2.11}$$

In Figure 2.8 plots with Cartesian range resolution and Cartesian velocity resolution are shown. The bistatic range resolution ΔR was set to 300m. The minimum obtained Cartesian range resolution ΔR_c is 150m, half of the bistatic range resolution, which corresponds to the bistatic angle $\beta = 0°$. When a target is approaching the baseline, $\cos(\beta/2)$ in (2.10) approaches 0, and ΔR_c tends to infinity. The same behavior is observed for the Cartesian velocity resolution. The minimum Cartesian

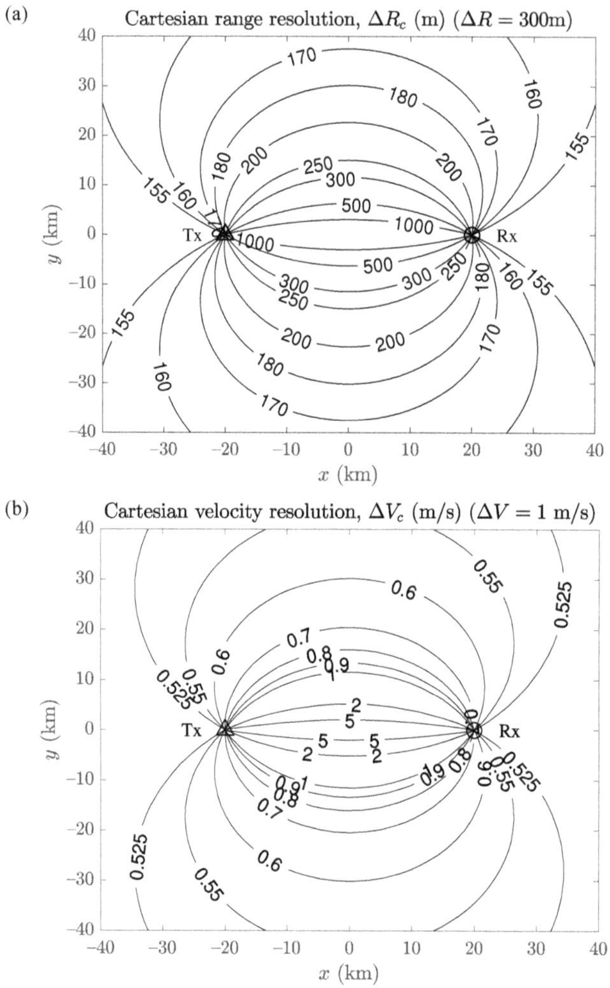

Figure 2.8 (a) Cartesian range resolution for bistatic range resolution $\Delta R = 300$m. (b) Cartesian velocity resolution for bistatic range resolution $\Delta V = 1$ m/s.

velocity resolution ΔV_c is 0.5 m/s, which is half of the assumed value of the bistatic velocity resolution ΔV equal to 1 m/s. Close to the baseline, the Cartesian resolution tends to infinity.

Presented results show that resolution in bistatic geometry depends not only on the signal (bandwidth, wavelength) and processing (integration time) parameters, but also on target position in relation to transmitter and receiver, more specifically, on the bistatic angle β.

2.3 BISTATIC RANGE EQUATION

Let us consider the sketch shown in Figure 2.9. The transmitter-target range is R_1 and the target-receiver range is R_2. The transmitter is emitting power P_t. If the power was radiated isotropically, that is with the same strength in each direction, the power density at range R_1 from the transmitter would be equal to the total emitted power P_t divided by the area of the sphere of radius R_1 (i.e., $4\pi R_1^2$). The directivity of the antenna G_t increases the power density in the direction of maximum radiation G_t times. Therefore, assuming that the target is at range R_1 in the direction of maximum radiation, the power density at the target can be calculated as:

$$S_1 = \frac{P_t G_t}{4\pi R_1^2} \tag{2.12}$$

The amount of power reflected by the target is determined by its *radar cross-section (RCS)*, σ. More specifically, the RCS, expressed in units of area, such as

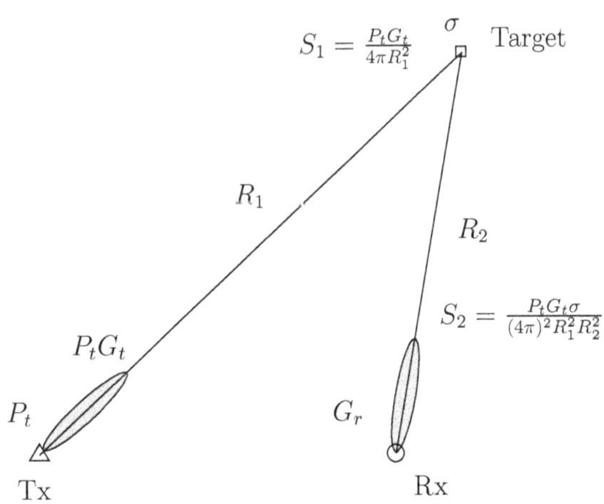

Figure 2.9 Sketch of scenario used for analysis of bistatic range equation.

square meter, is defined as:

$$\sigma = \lim_{R\to\infty} 4\pi R^2 \frac{S_s}{S_i} \tag{2.13}$$

where S_s is the scattered power density, and S_i is the incident power density. The scattered power is assumed to be radiated isotropically. This is a theoretical assumption used for the purpose of idealistic model of RCS, but in practice it is often not satisfied. In the above equation, the term $4\pi R^2$ is an area of a sphere with the radius of R. This area multiplied by the scattered power density S_s yields the total power scattered by the target. The RCS is therefore defined at a ratio of the scattered power to the incident power density. In passive radar the RCS that has to be considered is the *bistatic RCS*. Bistatic in this context means that the incident wave and scattered wave propagate in different directions, which correspond to the direction of the transmitter and receiver, respectively.

Assuming that the incident power density is S_1, the reflected power density at range R_2 from the target with the bistatic RCS equal to σ is:

$$S_2 = \frac{P_t G_t \sigma}{(4\pi)^2 R_1^2 R_2^2} \tag{2.14}$$

The power received by the radar is determined by the power density S_2 and the effective area of the receiving antenna of the radar A_{ef}:

$$P_r = \frac{P_t G_t \sigma A_{\text{ef}}}{(4\pi)^2 R_1^2 R_2^2} \tag{2.15}$$

Usually the parameter that is used to characterize the receiving antenna is its gain. The gain G_r is related to the effective area A_{ef} in the following way:

$$A_{\text{ef}} = \frac{G_r \lambda^2}{4\pi} \tag{2.16}$$

After substituting the effective area (2.16) into (2.15), the echo power P_r received by the radar can be calculated as [4]:

$$P_r = \frac{P_t G_t \sigma G_r \lambda^2}{(4\pi)^3 R_1^2 R_2^2} \tag{2.17}$$

The target echo power P_r competes with the noise of power P_n of various nature. Some of the most common noise sources include:

- **Thermal noise.** Thermal noise exists in all electronic devices which operate at temperatures above absolute zero. The power of the thermal noise is usually calculated as:

$$P_n = k_B T_0 B_r, \tag{2.18}$$

where k_B is the Boltzmann constant equal to 1.38×10^{-23} J/K, T_0 is the effective noise temperature of the receiver, and B_r is the bandwidth of the receiver.

The effective noise temperature depends on the *noise figure* of the receiver. The noise figure determines how much the signal-to-noise ratio is reduced when comparing input and output. The effective noise temperature can be calculated as:

$$T_0 = T_{\text{ref}} \cdot \left(10^{\left(\frac{N_f}{10}\right)} - 1 \right), \tag{2.19}$$

where N_f is the noise figure expressed in dB, and T_{ref} is the reference temperature, usually assumed to be 290K.

- **Man-made noise.** At higher frequency bands at which active radars usually operate, such as L, S, C, and X, the received noise is usually dominated by the thermal noise of the receiver. At lower frequency bands, where passive radars usually operate, such as VHF or UHF, man-made noise can dominate the receiver noise [5–9]. Man-made noise usually consists of unintentional emissions from devices such as electric motors, computers, power suppliers, and generators. The level of this noise varies with frequency, location, time of the day, and numerous other factors. As it turns out, the level of man-made noise can exceed the thermal noise level in the VHF and UHF bands by a few or even tens of decibels [10].

- **Background noise.** Natural noise resulting from mostly from lightning strikes.

- **Cosmic noise.** Cosmic noise originates outside of the Earth's atmosphere. The source of cosmic noise can be stars, including the Sun, quasars, or cosmic microwave background radiation.

Apart from the noise component, the detection capabilities of the radar are deteriorated by various losses. Some of the most important types of losses include:

- Propagation losses [11–14];
- Cable and transmission line losses;

- Antenna radiation pattern losses;
- Processing losses;
- Straddle (scalloping) losses;
- Detection losses.

Taking into account the noise power P_n and all losses L, the signal-to-noise ratio (SNR) can be calculated as:

$$\text{SNR} = \frac{P_t G_t \sigma G_r \lambda^2}{(4\pi)^3 R_1^2 R_2^2} \frac{BT}{P_n L}, \tag{2.20}$$

In the above equation, the processing gain, equal to BT, was taken into account. As it turns out, correlating the signal with bandwidth B over time T results in an increase in SNR equal to BT (this property will be analyzed in detail in Chapter 4).

Often the relationship between the target bistatic RCS and SNR is rearranged, so that minimum detectable RCS (or detectable RCS for short) is calculated assuming a certain level of SNR, which is enough to exceed the detection threshold. The detectable RCS σ_{det} can be calculated as:

$$\sigma_{\text{det}} = \frac{\text{SNR}_{\text{min}} (4\pi)^3 R_1^2 R_2^2 P_n L}{P_t G_t G_r \lambda^2 BT} \tag{2.21}$$

where SNR_{min} is the minimum SNR required for target detection.

Due to large variations of the detectable RCS, its values are expressed in logarithmic scale. Usually dBsm (decibels with respect to a square meter) are used, defined as:

$$\sigma_{\text{det}}\,(\text{dBsm}) = 10 \log_{10} \left(\frac{\sigma_{\text{det}}}{1\ \text{m}^2} \right) \tag{2.22}$$

Often in practice the transmitter power P_t and transmitter antenna gain G_t are not known separately. Instead, transmitter operators reveal information on the *equivalent isotropic radiated power* (EIRP) value. This is the power that would have to be fed to an isotropic antenna in order to provide the same power density at a certain range as in the case of an actual directional transmitter antenna in the direction of maximum radiation. From the point of view of the above analysis, we can write that $\text{EIRP} = P_t G_t$, under the assumption that the target is at the direction of maximum radiation of the transmitter.

Let us examine an example of detectable RCS calculated for a realistic passive radar scenario. The calculations were carried out for: $\text{EIRP} = 10$ kW, $G_r = 1$,

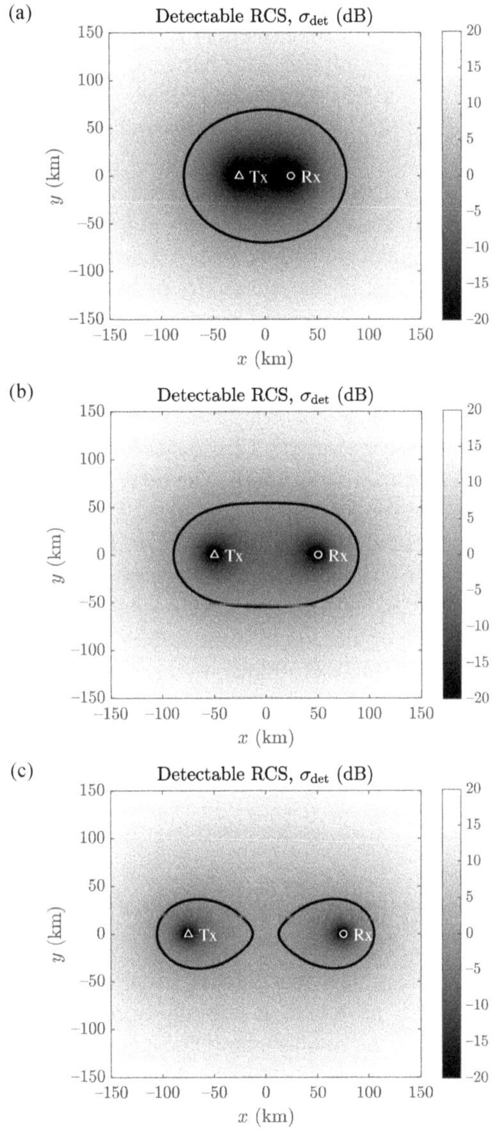

Figure 2.10 Detectable RCS for baseline length (a) 50 km, (b) 100 km, and (c) 150 km; black contour indicates RCS of 0 dBsm.

$\lambda = 3$m, $\text{SNR}_{\text{min}} = 12$ dB, $B = 200$ kHz, $B_r = 200$ kHz, $T_{\text{ref}} = 290$K, $L = 10$ dB, $T = 1$s, and $N_f = 5$ dB. The results are shown in Figure 2.10. Three plots correspond to baseline length (distance between the transmitter and receiver) equal to 50, 100, and 150 km. The RCS values are shown with the shades of gray, and saturated at -20 dBsm (0.01 m^2) and $+20$ dBsm (100 m^2). The black contour shows the RCS equal to 0 dBsm (1 m^2).

These results show an important feature of bistatic radar, which is quite different than in the case of monostatic radar. The received power in passive radar is proportional to $1/(R_1^2 R_2^2)$. Figures created from such a dependency are called *Cassini ovals*. It means that even if one of the ranges, R_1 or R_2, is large, the other one can be small, resulting in high received power. In practice this can mean that even if the target is far from the passive radar (R_2 is large), but close to the transmitter (R_1 is small), the received power is high enough to provide target detection. This is different from a typical monostatic scenario, where the received power is proportional to $1/R^4$ (where R is the range to the target). This means that if the target is far from the radar, the received power is inevitably small. The three cases shown in Figure 2.10 reveal another important feature of bistatic geometry. As the baseline increases, the total detection coverage is smaller. For example, let us define radar coverage by the minimum required RCS of 0 dBsm (1m^2). This corresponds to the black contour in the figure. For the shortest baseline (50 km), a single figure surrounded by the contour is formed. The area of this figure is the largest, which indicates the largest coverage. For the longest baseline (150 km), two separate areas around the transmitter and receiver are formed. The total area of those figures is smaller than in the short baseline case. The practical tip that follows from this analysis is that in order to provide long detection range, transmitters that are closer to the radar should be chosen. However, short distance to the transmitter may become the limiting factor due to the receiver dynamic range, as discussed later in this chapter.

2.3.1 Influence of Transmitter Azimuth Radiation Patterns

The parameters of the transmitters that passive radar uses, such as FM radio or digital television, can often be found in publicly accessible databases available on the Internet [15]. One of the main parameters specified for a transmitter is its power. As mentioned earlier, usually the EIRP value is specified. This value is often used for the prediction of passive radar coverage and omnidirectional radiation is assumed, as in the last example. In practice the situation is more complicated. The transmitters do not emit power isotropically. They are designed to illuminate the area of interest, which is usually located somewhere on the ground. For this reason, the transmitters are equipped with antennas with radiation patterns matched to the requirements. Both azimuth and elevation radiation patterns play important roles in the prediction of passive radar coverage. Let us examine the azimuth radiation patterns first. Elevation patterns will be investigated in the next section.

Figure 2.11 shows two examples of the azimuth radiation patterns of a DVB-T transmitter's antennas. The patterns were obtained from a public database [16]. The patterns are shown in polar coordinates, with 0 dB denoting the maximum radiation level, corresponding to the EIRP level. The pattern at the top is referred to in the transmitter description as "omnidirectional" in the azimuth plane. As can be seen, the actual pattern deviates from the omnidirectional one by as much as 5 dB. The pattern at the bottom is described as "sector" illumination. The focusing of transmitted power in the upper sector can be clearly seen.

In general, the dependency of the transmitter and receiver radiation pattern should be taken into account in the range equation. Therefore, if the dependency of the radiation pattern on the angle is considered, the equation for detectable RCS can be reformulated into:

$$\sigma_{\mathrm{det}} = \frac{\mathrm{SNR}_{\min}(4\pi)^3 R_1^2 R_2^2 P_n L}{P_t G_t(\phi,\theta) G_r(\phi,\theta) \lambda^2 BT} \tag{2.23}$$

where $G_t(\phi,\theta)$ is the transmitter radiation pattern depending on the azimuth angle ϕ and elevation angle θ (towards the target from the transmitter position), whereas $G_r(\phi,\theta)$ is the receiver radiation pattern depending on the azimuth and elevation angles (towards the target from the receiver position).

The result of the application of the azimuth transmitter radiation pattern to the range equation is shown in Figure 2.12 (the receiver radiation pattern was assumed to be omnidirectional). The top plot (Figure 2.12(a)) shows the reference scenario, where the azimuth pattern is omnidirectional. The middle plot (Figure 2.12(b)) represent the detectable RCS with the application of the realistic omnidirectional radiation pattern from Figure 2.11(a). The irregular shape of the 0 dBsm detection region marked with a black curve results directly from the dents in the radiation pattern. The bottom plot (Figure 2.12(c)) shows the RCS plot corresponding to the sector radiation pattern from Figure 2.11(b). In this case the influence of the azimuth radiation pattern is clearly visible as well; the 0 dBsm detection region is shrunk from the bottom.

The results presented in this section show that the azimuth radiation pattern of the transmitter's antenna influences the performance of passive radar substantially. A simplifying assumption on the omnidirectional azimuth radiation, often applied when no detailed information on the transmitter is available, can lead to overly

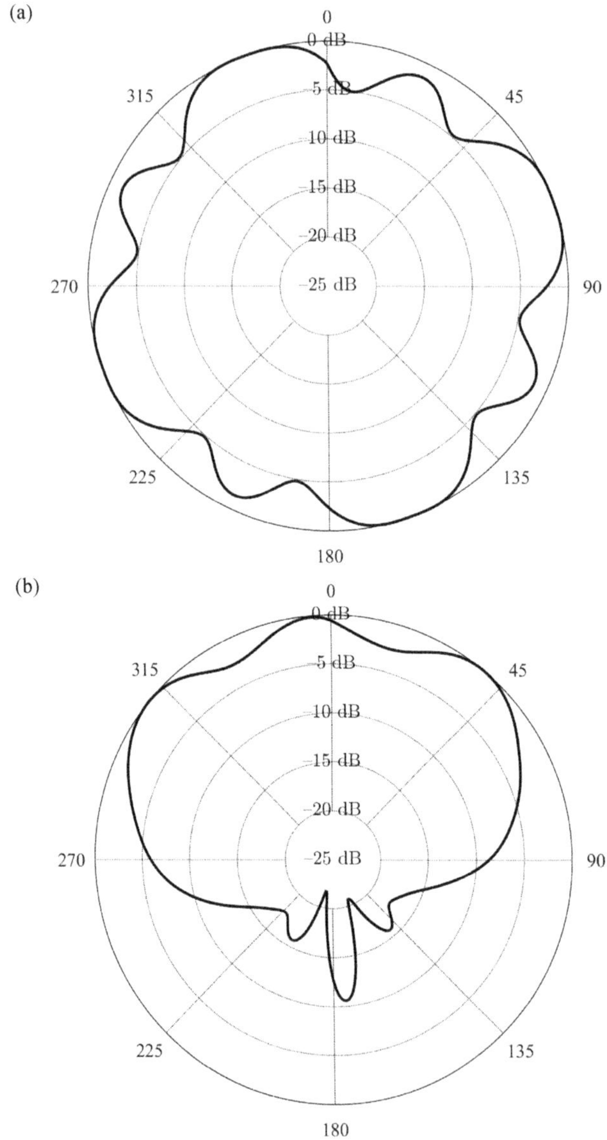

Figure 2.11 Two examples of antenna radiation patterns used in the DVB-T transmitter. (a) Azimuth radiation pattern that is referred to as "omnidirectional". (b) A "sector" radiation pattern.

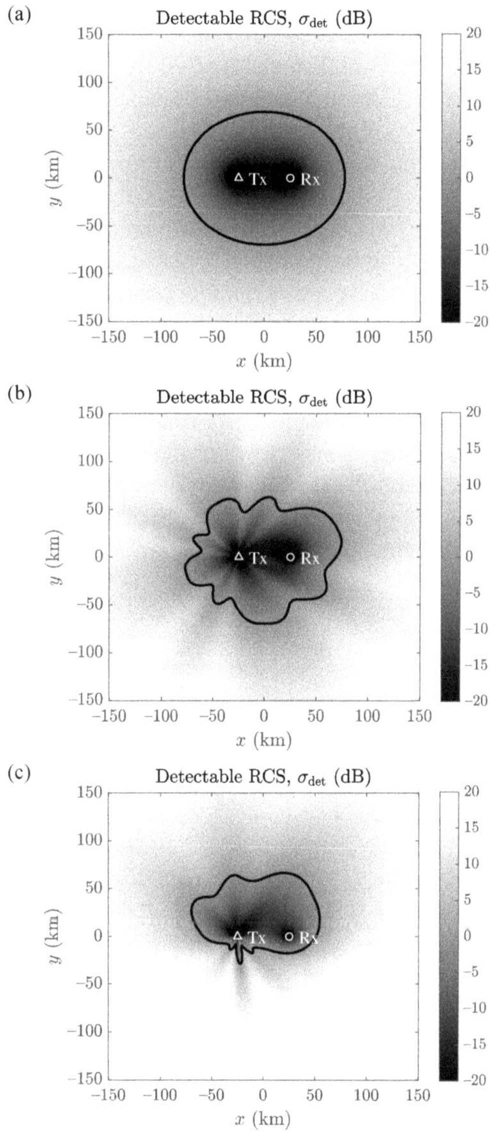

Figure 2.12 Azimuth cross-section of the detectable RCS for: (a) theoretical omnidirectional antenna, (b) realistic omnidirectional antenna, and (c) sector antenna; black contour indicates RCS of 0 dBsm.

optimistic performance prediction. If radiation patterns are available, they can be easily included in the calculation of the detectable RCS (or SNR). This, however, substantially complicates the overall analysis of passive radars from the operational point of view, as it implies additional geometrical dependencies.

2.3.2 Influence of Transmitter Elevation Radiation Patterns

Let us, in turn, analyze the influence of the elevation pattern on the detection capabilities of passive radar. The most popular types of transmitters used by passive radar, such as radio or television, are preferably placed on high masts, buildings, or mountains to provide better coverage. The receivers are typically placed on the Earth's surface; therefore, the usual practice is to focus the transmitted power in the elevation plane in a narrow sector. Because the transmitters are often placed relatively high, the elevation radiation pattern are usually tilted downwards, typically by 0.5–1°, so that a transmitter "looks" towards the ground [17–19].

The antennas used with the radio or television transmitters are usually constructed by stacking several radiating elements vertically. Individual segments containing the radiating elements are referred to as *bays*. This forms a uniform lin-

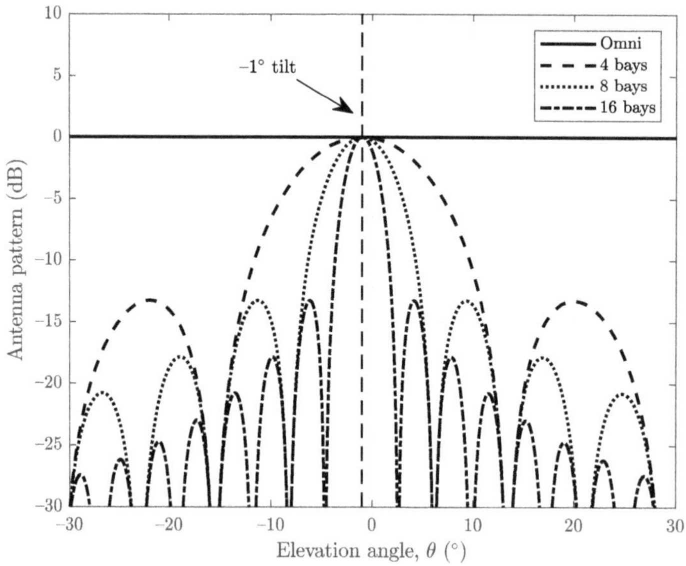

Figure 2.13 Examples of antenna radiation patterns for different number of radiating elements separated by $\lambda/2$. The patterns are shifted by 1° downwards.

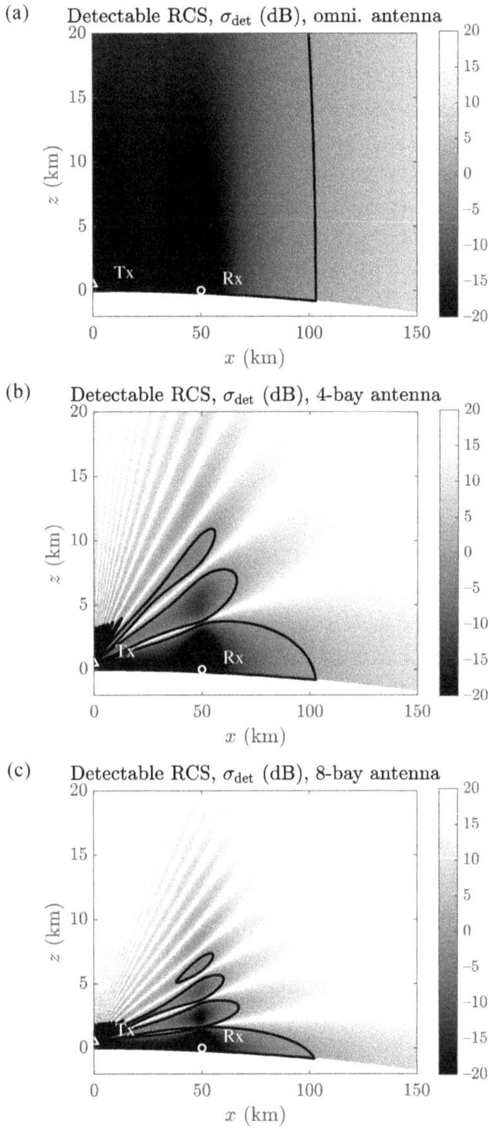

Figure 2.14 Elevation cross-section of the detectable RCS for: (a) omnidirectional elevation pattern, (b) 4-bay elevation pattern, and (c) 8-bay elevation pattern. Black contour in the figures indicates detectable RCS of 0 dBsm. Note different scale of the x and z axes.

ear array in the elevation direction. An example of a theoretical elevation radiation pattern is shown in Figure 2.13.[1] It was assumed that the radiating elements are separated by $\lambda/2$ (half wavelength). The number of bays is 4, 8, and 16, numbers often encountered in real life. The patterns are tilted downwards by 1°. The maximum pattern value is at 0 dB. This value corresponds to the EIRP. This value is usually provided by the broadcasters describing the transmitted power. The EIRP includes the actual transmitted power P_t and the absolute antenna gain G_t. This gain is usually several dBd (decibels with respect to the dipole antenna) for FM radio antennas and up to 18 dBd for television antennas [18, 19]. As can be seen in the figure, the increasing number of radiating elements decreases the width of the main beam. The 3-dB beamwidths for 4, 8, and 16-element antenna are 12.8°, 6.4°, and 3.2°, respectively. This means that the transmitted energy is focused in a relatively narrow range of elevation angles. As passive radar is used primarily for the detection of aircraft, target illumination by transmitters such as radio or television is far from optimum.

Let us investigate the influence of the transmitter antenna elevation patterns on the performance of passive radar. Figure 2.14 shows the elevation cross-section plots of detectable RCS. In all cases the same value of EIRP was assumed; the only difference was the shape of the elevation pattern. The horizontal axis corresponds to the x coordinate (aligned with the transmitter-receiver line), whereas the vertical axis corresponds to the z coordinate (note different scale on both axes). The plot in Figure 2.14(a) shows the detectable RCS for an omnidirectional pattern in the elevation direction. The calculated values of the RCS result directly from the dependency on the $R_1^2 R_2^2$ factor in (2.23). The plot in Figure 2.14(b) shows the detectable RCS calculated when taking into consideration the elevation pattern of a 4-bay antenna (tilted 1° downwards). The black contour of 0 dBsm reflects the elevation pattern shape. The nulls of the elevation pattern result in areas where target detection capabilities are significantly reduced. The next plot shows the results for an 8-bay antenna. The power emitted by the transmitter in this case is focused more for low elevation angles. This decreases the possibilities of detecting high-altitude targets even more in comparison to the 4-bay antenna.

2.3.3 Influence of DPI Removal on Effective Receiver Radiation Pattern

One of the issues characteristic for passive radar is the fact that the echo signal is received continuously at the same time as the reference signal. For this reason, pas-

1 More details about calculating radiation patterns for antenna arrays can be found in Chapter 3.

sive radar receiver should have a very high dynamic range in order to accommodate both a very strong reference signal and relatively very weak target echoes. This is especially true for the surveillance channel, in which weak target echoes compete with the direct signal, which leaks in through the sidelobes of the radiation pattern. Let us investigate the requirements for the dynamic range of the receiver [10].

The power of the *direct path interference* (DPI) received in the surveillance channel can be calculated as:

$$P_{DPI} = \frac{P_t G_t G_r(\phi_{\text{tx}})\lambda^2}{(4\pi R_b)^2}, \tag{2.24}$$

where P_t is the transmitted power, G_t is the transmitter antenna gain, $G_r(\phi_{\text{tx}})$ is the gain of the receiving antenna in the direction of the transmitter, and R_b is the range to the transmitter. The thermal noise power can be calculated using (2.18).

In order to obtain sensitivity at the level of the thermal noise, the dynamic range of the receiver should be at least:

$$D = \frac{P_{DPI}}{P_n} \tag{2.25}$$

An example of the dynamic range calculated for parameters typical for FM radio and DVB-T television is shown in Figure 2.15. The difference between the two cases is the receiver bandwidth ($B_r = 200$ kHz for FM radio and $B_r = 10$ MHz for DVB-T television) and the carrier frequency ($f_c = 100$ MHz for FM radio and $f_c = 600$ MHz for DVB-T television). In the calculations, the value of EIRP $= P_t G_t$ equal to 1, 10, and 100 kW is assumed. The temperature T_0 is set to 627K, which corresponds to 5 dB of the noise figure of the receiver. The gain of the receiver antenna in the direction of the transmitter is $G_r(\phi_{\text{tx}}) = 0$ dBi, a realistic value in usual scenarios. The dynamic range is calculated versus the range to the transmitter, R_b. As can be seen, the required dynamic range is increasing when the passive radar receiver is closer to the transmitter. It can reach over 100 dB for FM radio and close to 100 dB for DVB-T television for very short ranges. If the dynamic range of the receiver is not sufficient, attenuation has to be introduced in order to avoid receiver saturation. This, however, reduces the sensitivity of the radar by the value close to the introduced attenuation.

The dynamic range defined above relates to the passive radar receiver itself. Let us now consider the required dynamic range of the whole system, including the processing gain, in relation to the target echo power. By calculating the ratio of the DPI power expressed by (2.24) to the power of the target echo calculated with (2.17)

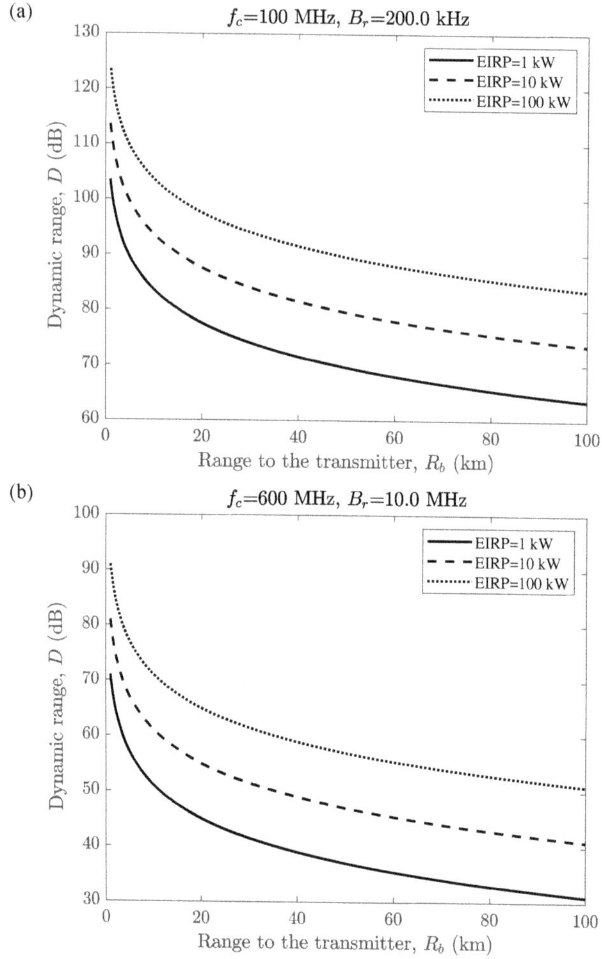

Figure 2.15 Required dynamic range of the receiver for: (a) FM radio and (b) DVB-T television.

(taking into account angle-dependent receiver antenna gain), the following relationship can be found:

$$\frac{P_{DPI}}{P_r} = \left(\frac{R_1 R_2}{R_b}\right)^2 \frac{4\pi}{\sigma} \frac{G_r(\phi_{\text{tx}})}{G_r(\phi_{\text{targ}})} \tag{2.26}$$

where $G_r(\phi_{\text{tx}})$ and $G_r(\phi_{\text{targ}})$ are the gains of the receiver antenna in the direction of the transmitter and target, respectively. An example of the DPI to target echo power ratio is plotted in Figure 2.16 under the assumption of 1 m^2 target RCS and the baseline of 50 km. It was also assumed that the $G_r(\phi_{\text{tx}})/G_r(\phi_{\text{targ}})$ ratio was constant and equal to -10 dB. In reality, that value changes depending on the target position. As can be seen, the ratio in the vicinity of the transmitter or the receiver is smaller as the target echo power increases. However, for large areas the ratio is very high, exceeding 100 dB. In order to detect a target echo in the presence of strong DPI, a sufficient dynamic range of the whole system has to be provided. This is facilitated by the coherent integration realized in passive radar, which provides BT gain, where B is the signal bandwidth and T is the integration time. Typical values of BT for passive radar reach 40–70 dB. When this value is compared with the ratio from Figure 2.16, it is clear that the DPI is still too strong, and it has to be removed from the echo channel. This process is called DPI removal. As a result of DPI removal, the effective radiation pattern is changed, which will be analyzed in this section [20].

Let us consider the problem of DPI removal by investigating the geometry shown in Figure 2.17. Two complex amplitude radiation patterns are presented: ref-

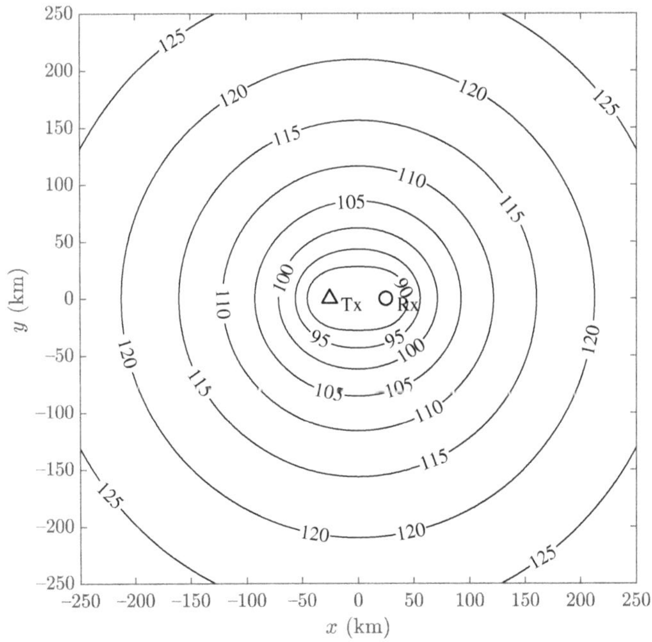

Figure 2.16 Ratio of the DPI to the target echo power for a 1 m^2 target.

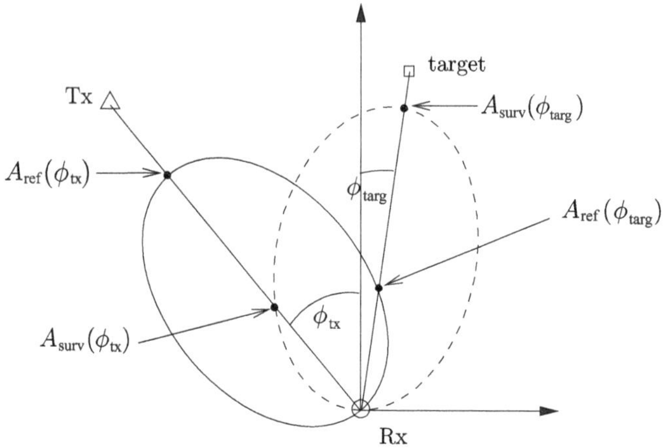

Figure 2.17 Reference (solid line) and surveillance (dashed line) beams with respect to the transmitter and target.

erence (solid line) $A_{\text{ref}}(\phi)$ and surveillance (dashed) $A_{\text{surv}}(\phi)$. Two marked angles, ϕ_{tx} and ϕ_{targ}, correspond to the direction of the transmitter and receiver.

The reference signal $x_r(t)$ can be modeled in the following way:

$$x_r(t) = A_{\text{ref}}(\phi_{\text{tx}})x_{\text{tx}}(t) + A_{\text{ref}}(\phi_{\text{targ}})x_{\text{targ}}(t) \tag{2.27}$$

It has two components. The first one, $x_{\text{tx}}(t)$, is the signal received directly from the transmitter. This signal is multiplied by the complex radiation pattern of the reference antenna $A_{\text{ref}}(\phi)$ for the angle corresponding to the transmitter direction, ϕ_{tx}. The reference signal also contains the target echo signal $x_{\text{targ}}(t)$. It is multiplied by the complex radiation pattern of the reference antenna for the angle corresponding to the target, $A_{\text{ref}}(\phi_{\text{targ}})$. Ideally, the radiation pattern of the reference antenna should be highly directional, with a narrow main beam and low sidelobes. In practice, however, the presence of the echo signal in the reference channel can be substantial (due to the limited width of the main beam and sidelobes), and can lead to the degradation of passive radar performance, as explained below.

The signal from the surveillance channel $x_e(t)$ also contains contributions from the direct signal $x_{\text{tx}}(t)$ and echo signal $x_{\text{targ}}(t)$:

$$x_e(t) = A_{\text{surv}}(\phi_{\text{tx}})x_{\text{tx}}(t) + A_{\text{surv}}(\phi_{\text{targ}})x_{\text{targ}}(t) \tag{2.28}$$

The direct signal $x_{\text{tx}}(t)$ is multiplied by the surveillance amplitude radiation pattern $A_{\text{surv}}(\phi)$ in the direction of the transmitter ϕ_{tx}. The target echo signal $x_{\text{targ}}(t)$

is multiplied by the surveillance radiation pattern $A_{\text{surv}}(\phi)$ in the direction of the target ϕ_{targ}. Usually the amplitude of the surveillance radiation pattern in the direction of the target is much larger than in the direction of the transmitter, that is, $|A_{\text{surv}}(\phi_{\text{tx}})| \ll |A_{\text{surv}}(\phi_{\text{targ}})|$. However, even a small presence of the reference signal in the surveillance channel is very disadvantageous (as explained in the Chapter 5) and has to be taken care of. This is done by subtracting the reference signal from the surveillance signal, usually using a kind of adaptive filter. The reference signal is, however, not a "clean" copy of the transmitted signal, but a combination of the transmitted and echo signals, as shown in (2.27). Because of this, by subtracting the reference signal from the surveillance channel signal, a part of the echo signal is also removed.

In order to remove the reference signal from the echo signal, the following subtraction has to be performed [20]:

$$
\begin{aligned}
x_e'(t) = x_e(t) - \frac{A_{\text{surv}}(\phi_{\text{tx}})}{A_{\text{ref}}(\phi_{\text{tx}})} x_r(t) = \\
A_{\text{surv}}(\phi_{\text{targ}}) x_{\text{targ}}(t) - A_{\text{ref}}(\phi_{\text{targ}}) \frac{A_{\text{surv}}(\phi_{\text{tx}})}{A_{\text{ref}}(\phi_{\text{tx}})} x_{\text{targ}}(t) = \\
\left(A_{\text{surv}}(\phi_{\text{targ}}) - A_{\text{ref}}(\phi_{\text{targ}}) \frac{A_{\text{surv}}(\phi_{\text{tx}})}{A_{\text{ref}}(\phi_{\text{tx}})} \right) x_{\text{targ}}(t)
\end{aligned}
\tag{2.29}
$$

In this way, the transmitted signal $x_{\text{tx}}(t)$ is completely removed from the surveillance echo signal $x_e(t)$. However, the target echo signal $x_{\text{targ}}(t)$ is also modified during this operation. This modification is reflected by the expression in the brackets in the last line of (2.29). As the target angle ϕ_{targ} changes, a different influence can be expected. The effective surveillance radiation pattern $A^{\text{ef}}_{\text{surv}}(\phi)$ can be calculated as:

$$
A^{\text{ef}}_{\text{surv}}(\phi) = A_{\text{surv}}(\phi) - A_{\text{ref}}(\phi) \frac{A_{\text{surv}}(\phi_{\text{tx}})}{A_{\text{ref}}(\phi_{\text{tx}})} \tag{2.30}
$$

The above equation means that the reference radiation pattern $A_{\text{ref}}(\phi)$ is subtracted from the surveillance radiation pattern $A_{\text{surv}}(\phi)$ with a complex coefficient equal to $A_{\text{surv}}(\phi_{\text{tx}})/A_{\text{ref}}(\phi_{\text{tx}})$.

An example of the influence of the DPI removal on the surveillance channel radiation pattern is shown in Figure 2.18. For simplicity it was assumed that the radiation pattern of the surveillance channel is omnidirectional in the azimuth direction. The reference antenna radiation pattern is directional, with the main beam pointing to the transmitter at 0°. The width of the main beam is approximately 44°. The effective surveillance radiation pattern is modified by the DPI removal as shown in

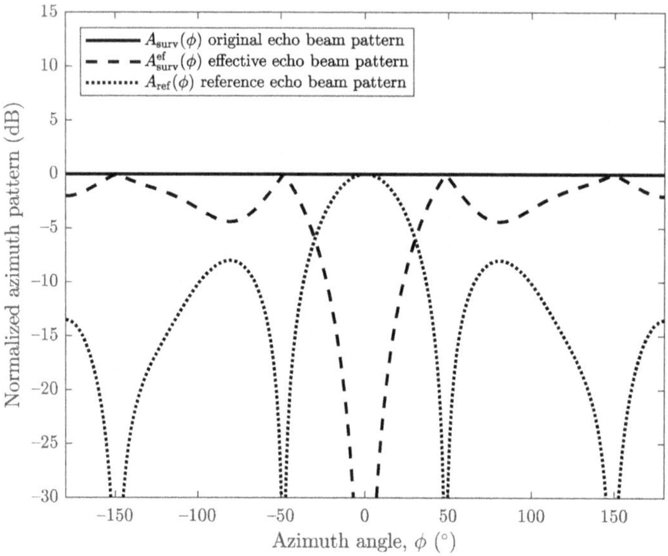

Figure 2.18 Influence of DPI removal on the effective receiver radiation pattern.

the figure. At the angle of 0°, which corresponds to the maximum of the reference pattern, a null is created in the surveillance beam. This means that the radar is blind in the direction of the transmitter. At angles where the reference pattern has nulls (i.e., $\pm 50°$ and $\pm 150°$), there is no influence on the surveillance pattern. At angles corresponding to substantial sidelobes of the reference pattern, noticeable losses in the surveillance pattern can be seen.

The application of the effective surveillance pattern to the bistatic range equation is shown in Figure 2.19. In the calculations, equation (2.23) was used with the receiver antenna gain defined as $G_r(\phi, \theta) = |A^{ef}_{surv}(\phi)|^2$. The contours of constant RCS are no longer the Cassini ovals. The null created in the direction of the transmitter in $A^{ef}_{surv}(\phi)$ resulted in the aforementioned blindness of the radar. In practice it means that a certain sector of angles surrounding the direction of the transmitter is excluded from observation.

A practical question is: how can this blindness be eliminated, or at least alleviated? One of the solutions is a reconstruction of the reference signal [21–25]. In the case of digital signals, such as DVB-T or DAB, the received signal can be decoded to the bit level. If the signal-to-noise ratio is high enough, the bit error rate can be very low, which means that reconstruction can be very precise. In addition, error correction built in the signal standard can be used to further improve the qual-

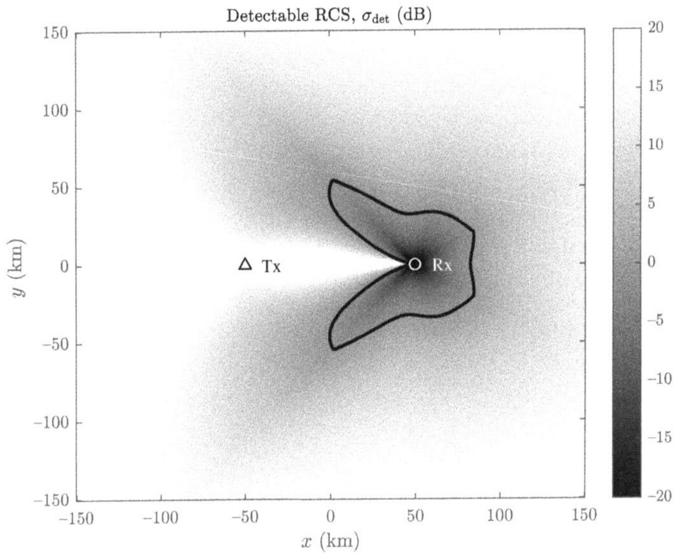

Figure 2.19 Detectable RCS with the influence of effective radiation pattern after DPI removal.

ity of the decoded information. Next, the digital bit stream can be reconstructed to a time-domain signal. In this way, a clean copy of the transmitted signal is obtained, without the presence of the echoes in the reference signal. In such a case, the blindness of the radar does not occur.

2.4 EVALUATION OF THE ILLUMINATING SIGNAL USING THE AMBIGUITY FUNCTION

The standard way of evaluating a signal for the purposes of radar is by the *ambiguity function*. It is defined in the following way[2] [26]:

$$\chi(\tau, f_d) = \int_{-\infty}^{+\infty} x(t)x^*(t-\tau)\mathrm{e}^{\mathrm{j}2\pi f_d t}\mathrm{d}t \tag{2.31}$$

The ambiguity function can be interpreted as the correlation of the signal with its copy that is shifted in time and frequency. The ambiguity function is closely

2 Various, slightly different, definitions of the ambiguity function exist in literature. One of them is used here. The main idea, however, is the same.

related to the cross-ambiguity function, which is calculated between the reference and echo signals during normal passive radar operation. For this reason, the analysis of ambiguity function allows the performance indicators of passive radar to be assessed, such as sidelobe level, resolution and potential periodicities.

It is convenient to express time delay τ and Doppler shift f_d in (2.31) in terms of bistatic range R and bistatic velocity V, by using substitutions $\tau = R/c$ and $f_d = -V/\lambda$, respectively, where c is the speed of light and λ is the wavelength. Moreover, in practice, the ambiguity function is calculated for finite time, called *integration time* T. After these modifications, the ambiguity function can be expressed in the following form:

$$\chi(R, V) = \int_{-T/2}^{+T/2} x(t)x^*\left(t - \frac{R}{c}\right) \mathrm{e}^{-\mathrm{j}2\pi \frac{V}{\lambda} t} \mathrm{d}t \tag{2.32}$$

2.4.1 The Ambiguity Function of a Noise Signal

Before evaluating a signal corresponding to individual transmitter types, let us analyze a noise signal[3] [27]. Noise represents an idealistic signal, having somewhat perfect properties, which are desirable of all signals. By identifying the properties of a noise signal, it is possible to distinguish them from features corresponding to specific characteristics depending on the modulation, pilots, and cyclic prefixes.

Consider a white noise signal. It is a signal whose power spectral density (PSD) is constant over all frequencies. The name comes from the analogy to white light in the visible spectrum, where all wavelength (and thus frequencies) contribute to what is referred to as white light. A signal with flat PSD is actually a purely theoretical concept, as it has infinite power (which results from integrating PSD over all frequencies); therefore, it is not possible in practice. Despite its lack of practical representation, white noise is widely used in theoretical analyses. Now let us consider the autocorrelation function, which is the cross-section of the ambiguity function for the zero Doppler shift (i.e., $\chi(\tau, 0)$). Under the assumption of signal ergodicity, the autocorrelation of white noise with PSD at a constant level of 1, which is the inverse Fourier transform of PSD, has the form of the Dirac delta:

$$\mathcal{F}^{-1}\{1\} = \delta(\tau) \tag{2.33}$$

where $\mathcal{F}^{-1}\{\cdot\}$ denotes the inverse Fourier transform operator. A sketch of white noise PSD and corresponding autocorrelation is shown in Figure 2.20(a).

3 The analysis in this section has engineering, illustrative character, rather than a strict theoretical one.

In radar, the cross-correlation of the transmitted and received signals is calculated in order to perform target detection. It is closely related to the autocorrelation. In fact, the cross-correlation can be viewed as the convolution of the autocorrelation with the impulse response of the radio channel. Therefore, investigation of autocorrelation allows the expected performance of the radar to be evaluated. In the case of a white noise signal with autocorrelation in the form of the Dirac delta, the convolution of its autocorrelation with radio channel's impulse response would result in a perfect representation of the radio channel, as convolving a signal with the Dirac delta does not change the signal. For this reason, theoretical white noise could provide perfect knowledge of the radio channel with infinite resolution. Unfortunately, as mentioned earlier, white noise is only a theoretical concept, which cannot be realized in practice.

A more realistic signal is a band-limited noise. In the simplest case, the shape of the band-limited noise PSD is rectangular. In this case the autocorrelation has a shape of sinc$(\cdot)$ function (cardinal sine function, or $\text{sinc}(x) = \sin(x)/x$):

$$\mathcal{F}^{-1}\left\{\Pi\left(\frac{f}{B}\right)\right\} = B \cdot \text{sinc}\left(\pi B t\right) \tag{2.34}$$

The pair of PSD and corresponding autocorrelation function for band-limited noise is shown in Figure 2.20(b). As the autocorrelation is no longer perfect Dirac delta, the measured shape of the radio channel impulse response will be distorted by convolution with the sinc$(\cdot)$ function.

Until now is has been assumed that the theoretical autocorrelation for analyzed signals is calculated, which involves analysis for infinite time. In practice, the time of analysis is always finite. The main consequence of this fact is the presence of the *correlation sidelobes*, *correlation noise floor*, or *residual fluctuations* [27]. These are sidelobes resulting from the fact that even when two independent noise signals are compared by means of correlation calculated for finite time, the result is nonzero. It turns out that the mean level of residual fluctuations in relation to the main peak depends on the integration time T and signal bandwidth B. The ratio of the main peak to the sidelobes can be calculated as BT. The situation with band-limited noise and finite integration time is depicted in Figure 2.20(c). As can be seen the sidelobes in the autocorrelation are of dual nature. The sinc-shaped sidelobes result from the rectangular PSD, and they are deterministic. In contrast, the second type of sidelobes (i.e., the residual fluctuations), have a random nature and are not dependent on the PSD shape, but on the time-bandwidth product.

The fourth example in Figure 2.20(d) shows a signal with Gaussian-shaped PSD. In this case, the autocorrelation, which is the inverse Fourier transform of the

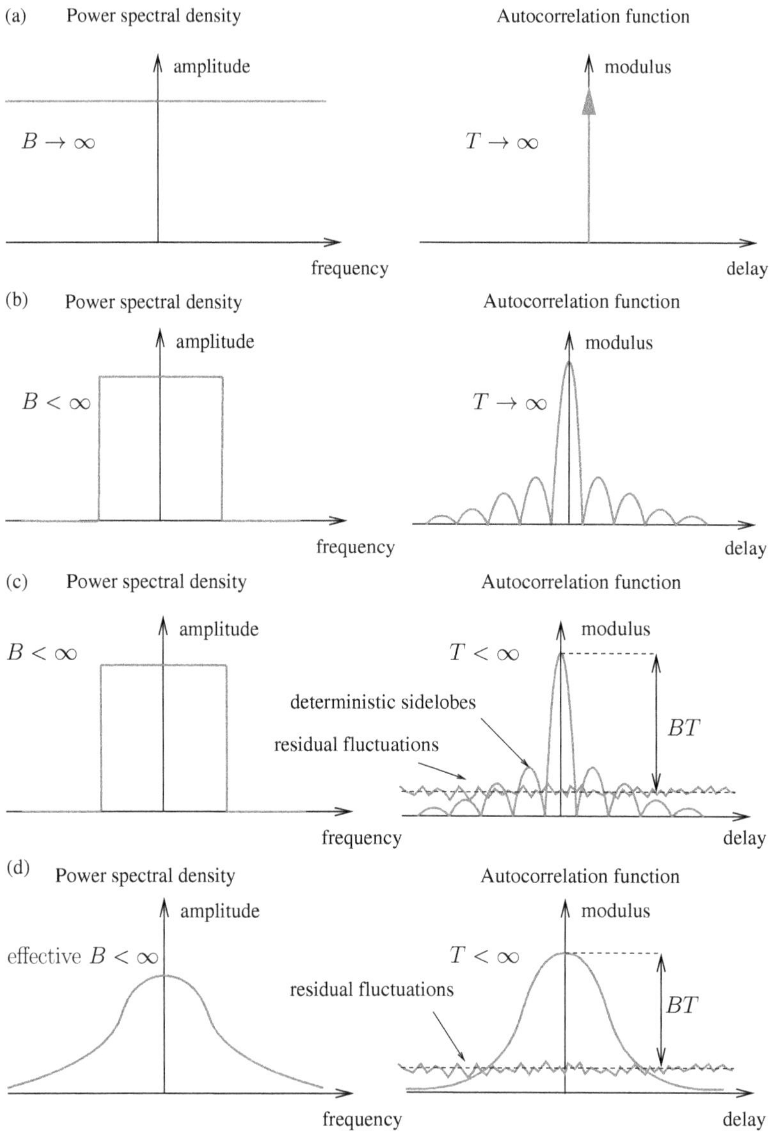

Figure 2.20 PSD (left) and corresponding modulus of the autocorrelation function (right). (a) Infinite band and infinite integration time. (b) Rectangular spectrum and infinite integration time. (c) Rectangular spectrum and finite integration time. (d) Gaussian spectrum and finite integration time.

PSD, is also Gaussian shaped:

$$\mathcal{F}^{-1}\left\{\exp\left(-2(\pi\alpha f)^2\right)\right\} = \frac{1}{\sqrt{2\pi\alpha^2}}\exp\left(-\frac{t^2}{2\alpha^2}\right) \tag{2.35}$$

where α is the shape parameter. Gaussian-shaped autocorrelation is smooth, therefore has no deterministic sidelobes, although the main lobe has a width inversely proportional to the width of the PSD. Theoretically the bandwidth of the signal is infinite. The effective bandwidth (see discussion in Section 2.5.1), however, is finite. The residual fluctuations, at the level of BT below the main peak, are present due to a finite integration time.

Let us analyze a practical example of a noise signal and its ambiguity function. The baseband representation of a discrete-time signal under investigation is:

$$x[n] = x_I[n] + \mathrm{j}x_Q[n] \tag{2.36}$$

where $x_I[n]$ and $x_Q[n]$ are real-valued Gaussian-distributed noise signals with white spectral distribution. The spectrum (calculated using the fast Fourier transform) of a realization of a noise signal is shown in Figure 2.21, revealing flat PSD. As the

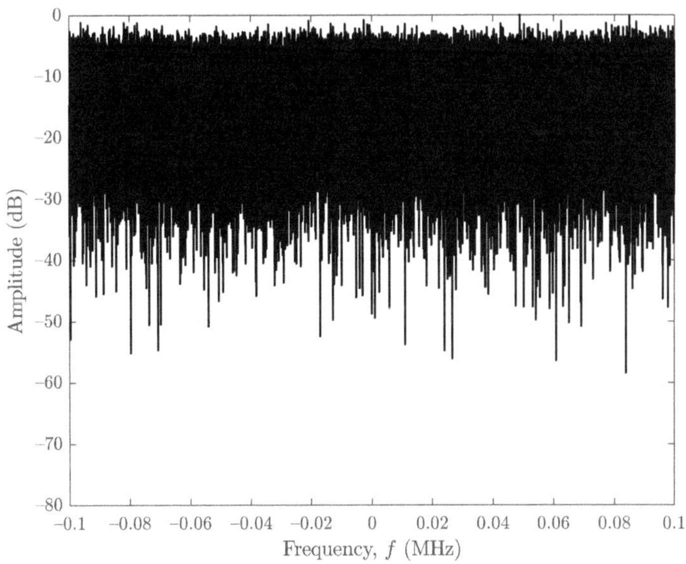

Figure 2.21 Spectrum of a realization of a noise signal (calculated using the fast Fourier transform).

signal is discrete in time, the spectrum is periodic, with unambiguous representation in $(-f_s/2, +f_s/2)$ range, where f_s is the sampling frequency (here $f_s = 200$ kHz).

Figure 2.22 shows the ambiguity function modulus of the noise signal calculated for finite integration time $T = 1$s. The deterministic sidelobes are not visible due to the discretization of the ambiguity function calculation; because the signal

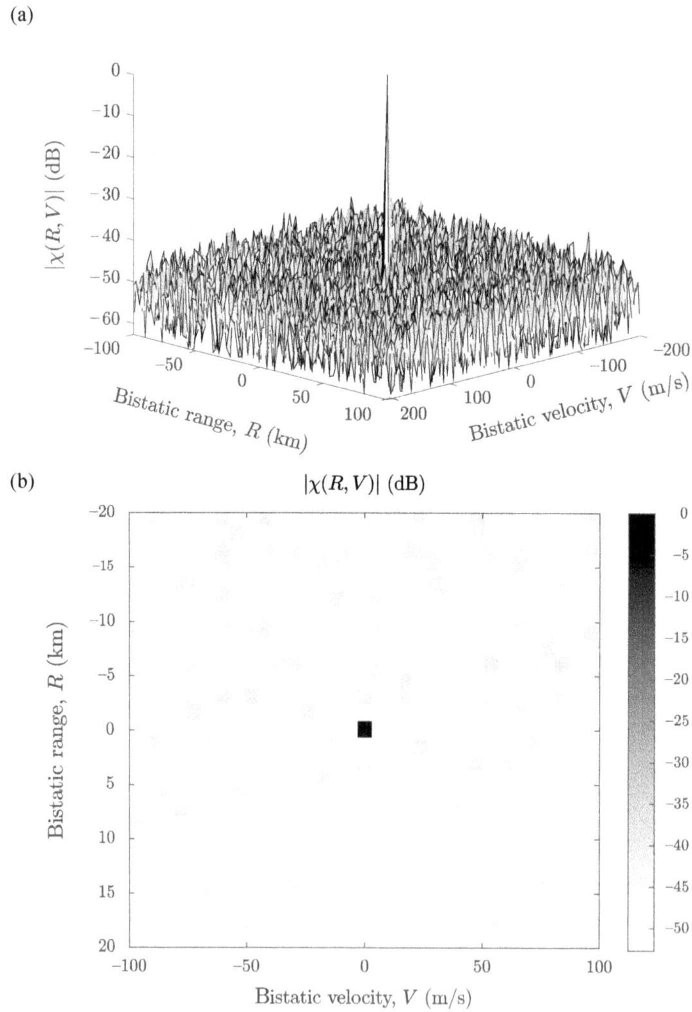

Figure 2.22 Ambiguity function modulus of a noise signal. (a) 3D view. (b) Top view.

spectrum occupies the whole bandwidth, the points where the ambiguity function is calculated correspond to zeros of the sinc(·) function. The residual fluctuations span over the whole range-velocity space. They are at the level of $BT = 200\,\text{kHz} \cdot 1\,\text{s} = 2 \cdot 10^5$, which corresponds to 53 dB.

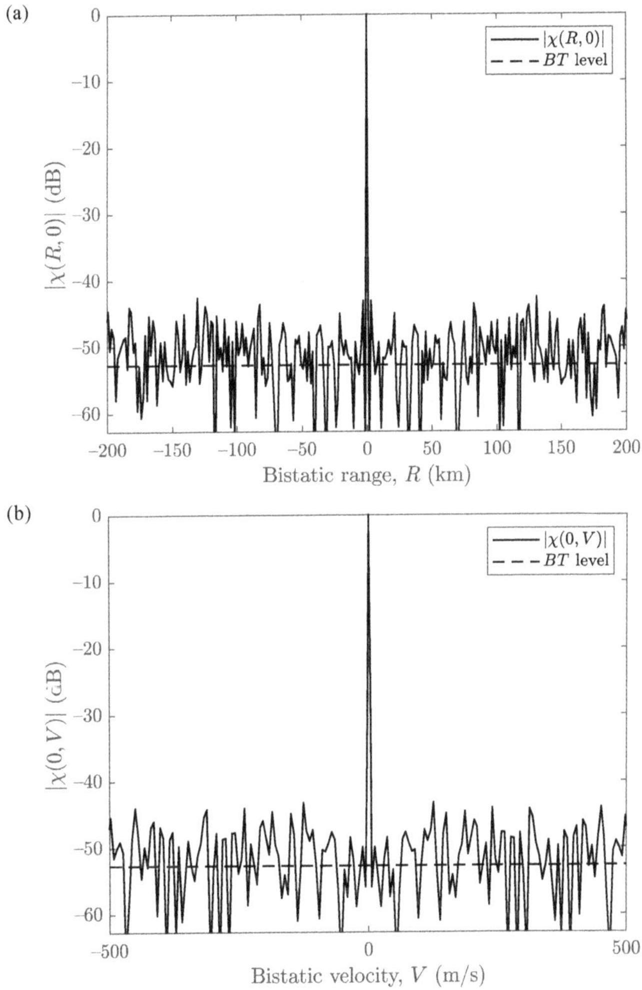

Figure 2.23 (a) Range and (b) velocity cross-section of the ambiguity function modulus of a noise signal.

The slices of the ambiguity function for zero velocity and zero range are shown in Figure 2.23. A single peak at zero range and zero velocity is visible. Random sidelobes, at the constant level of -53 dB are visible.

Figure 2.24 shows the influence of integration time on the level of random sidelobes. The integration time for the four plots is 5, 50, 500, and 5000 ms, which corresponds to 10^3, 10^4, 10^5, and 10^6 samples, respectively, for the sampling frequency of 200 kHz. The BT product for the four considered cases is 30, 40, 50, and 60 dB. The level of the random sidelobes visible in the figures matches very well to the theoretical level of BT plotted with dashed line.

Now consider a band-limited noise signal, whose spectrum is shown in Figure 2.25. The bandwidth of the signal is approximately 0.2 of the sampling fre-

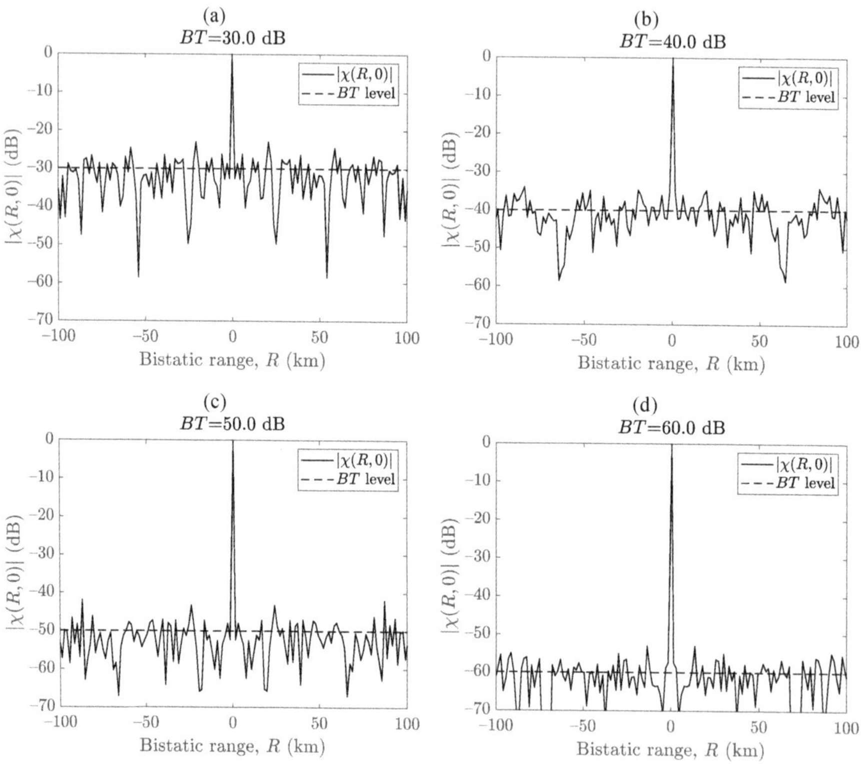

Figure 2.24 Range cross-section of the ambiguity function modulus of a noise signal for: (a) 10^3 samples, (b) 10^4 samples, (c) 10^5 samples, and (d) 10^6 samples.

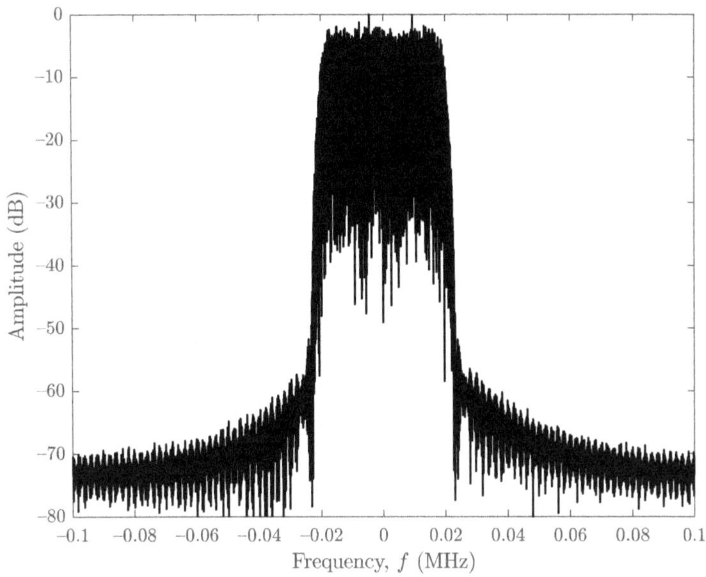

Figure 2.25 Spectrum of a band-limited noise signal.

quency. This has been obtained by filtering the white noise signal with a finite impulse response (FIR) filter.

As indicated in the analysis above, and shown in Figure 2.20(c), a band-limited noise signal has a sinc-like autocorrelation function with additional random sidelobes. This is confirmed by the ambiguity function shown in Figure 2.26. As in the case of white noise, a uniform level of random sidelobes is present over all ranges and velocities. However, additional deterministic sidelobes are visible in the range dimension. These are directly connected with the quasi-rectangular shape of the signal spectrum.

The two types of sidelobes (i.e., deterministic and random) are visible in the range slice of the ambiguity function shown in Figure 2.27(a). The main peak is surrounded by the sinc-like sidelobes. The maximum level of the sidelobes is -13 dB, a typical value for a rectangular-shaped spectrum. The sidelobe level decreases further away from the main peak. At a certain distance, the deterministic sinc-like sidelobes are dominated by random sidelobes, which are at the level of approximately -46 dB. The velocity cross-section of the ambiguity function is very similar as in the case of white noise; there is a single main peak for $V = 0$ m/s surrounded by random sidelobes.

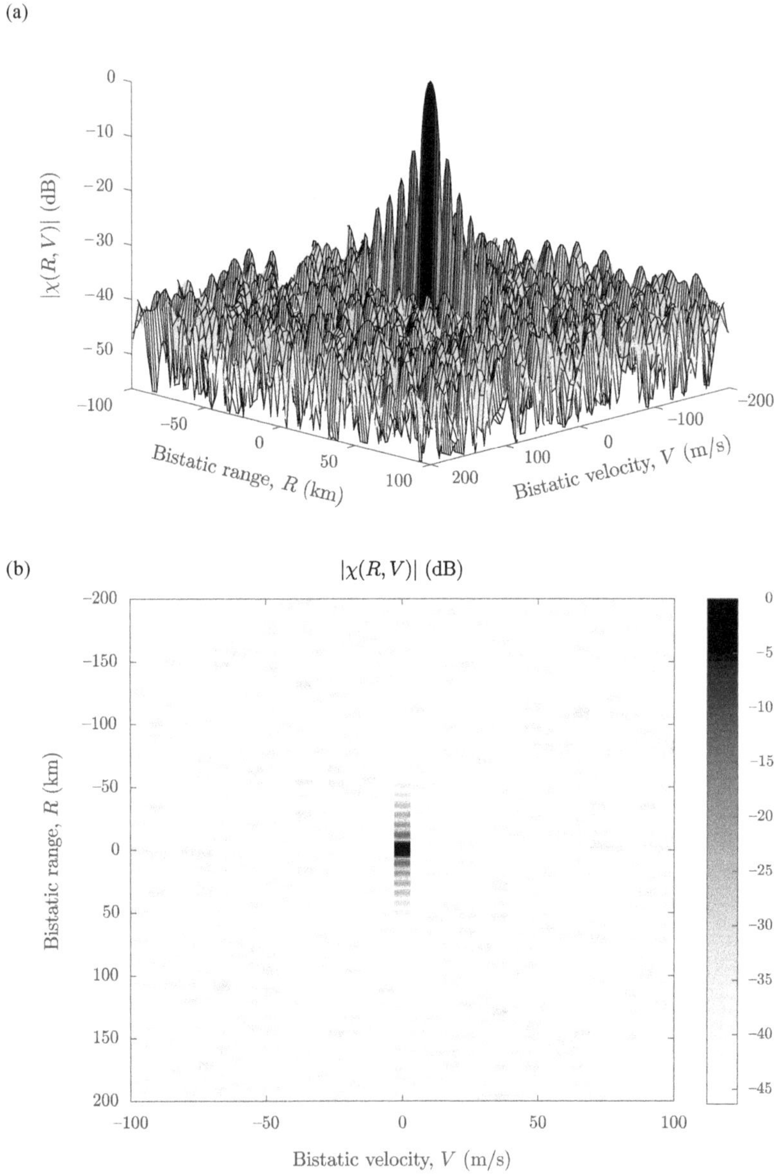

Figure 2.26 Ambiguity function modulus of a band-limited noise signal. (a) 3D view. (b) Top view.

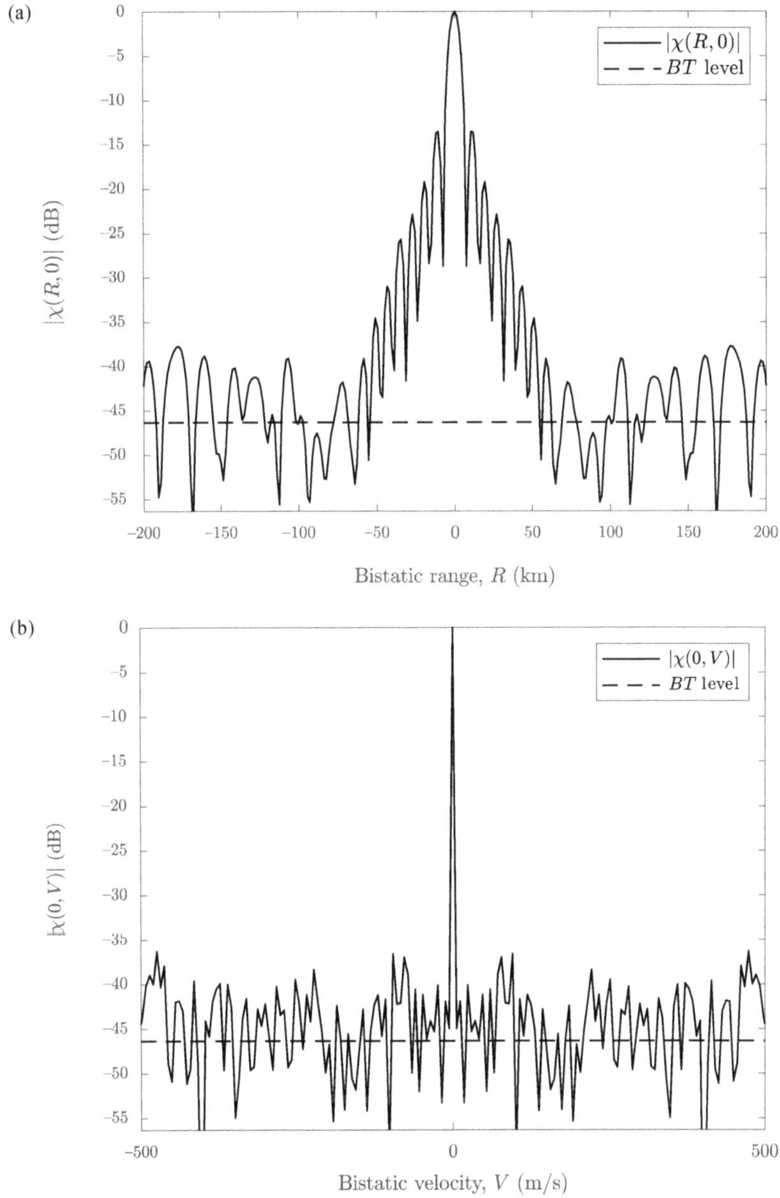

Figure 2.27 (a) Range and (b) velocity cross-section the ambiguity function modulus of a band-limited noise signal.

2.5 ILLUMINATORS OF OPPORTUNITY

There are numerous types of transmitters that can be used for the purpose of passive radar. When choosing the appropriate illuminator type, various properties have to be considered, such as transmitted power, frequency band, signal bandwidth, and modulation. In this book, the main focus is on the detection of moving targets (mainly aerial) at ranges of tens or hundreds of kilometers. From this point of view, several types of transmitters are potentially suitable. The most common illuminators will be characterized.

2.5.1 Analog Radio (FM)

FM radio [28] is one of the most frequently used illuminators of opportunity for passive radar [1, 29–34]. It operates in the 88–108-MHz frequency range. The transmitter powers reach tens or hundreds of kilowatts, providing long detection range. One of the greatest shortcomings of FM radio as a source of illumination for radar, however, is the low bandwidth of the signal. In addition, the bandwidth of the signal changes in time due to the analog modulation applied. The nominal bandwidth of the signal is 150 kHz, with 200 kHz interchannel spacing. However, the instantaneous bandwidth of the signal strongly depends on the content of the transmitted program. For speech, the bandwidth can be very narrow, which results in degraded range resolution [35, 36]. Despite this disadvantage, FM radio is often used for passive radar, mainly because of its large coverage area. A detailed analysis of the FM signal from the point of view of passive radar can be found in [37].

Figure 2.28 shows an example of the full spectrum of the FM radio signal, ranging from 88 MHz to 108 MHz. Each FM station is visible in the spectrum as a single peak, as the width of a single channel is 150 kHz, which is very small compared to the whole span of 20 MHz. In this particular example, tens of stations are visible, each with varying amplitudes. Some of the signals are transmitted from the same locations, which limits the spatial diversity, helpful in target localization.

An example of a single channel of the FM radio signal is shown in Figure 2.29. The signal has been downconverted to the baseband. The plot at the top shows the absolute value of the time representation of the signal. The amplitudes are saturated at a constant value. This is the result of the modulation scheme applied: FM radio uses frequency modulation. The information is carried in the changes of the frequency of the signal, while the amplitude is kept at a constant level. The spectrum of the signal is shown at the bottom of the figure. The bandwidth of the signal is relatively wide, as the analyzed signal was fast music. However, as mentioned, the

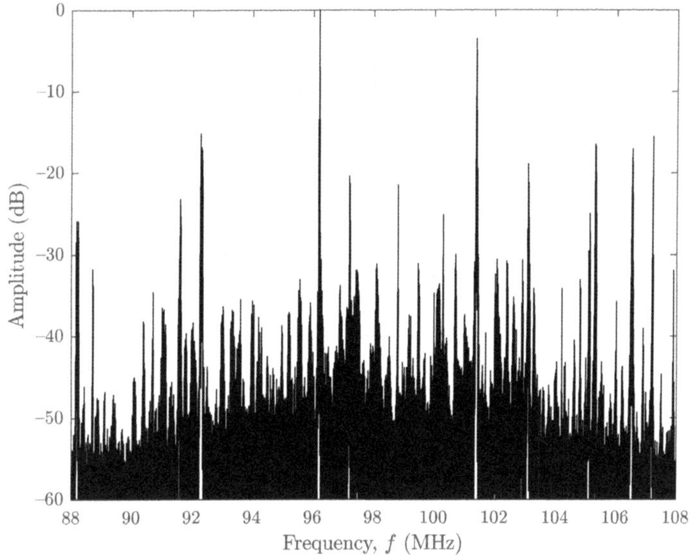

Figure 2.28 Example of the full spectrum of the FM signal.

signal bandwidth of the FM radio signal can vary substantially depending on the transmission content.

Let us examine the FM radio signal in detail. The FM radio signal in the baseband representation can be expressed as:

$$x(t) = \mathrm{e}^{\mathrm{j}\phi(t)} \tag{2.37}$$

where instantaneous phase $\phi(t)$ is defined as:

$$\phi(t) = 2\pi \int_{-\infty}^{t} f(\tau)\mathrm{d}\tau \tag{2.38}$$

The instantaneous frequency can be calculated as derivative of the instantaneous phase divided by 2π:

$$f(t) = \frac{1}{2\pi}\frac{\mathrm{d}\phi(t)}{\mathrm{d}t} \tag{2.39}$$

The instantaneous frequency of the FM radio signal is defined as:

$$f(t) = \left(0.9\left(\frac{x_L(t) + x_R(t)}{2} + \frac{x_L(t) - x_R(t)}{2}\sin(4\pi f_p t)\right) + 0.1\sin(2\pi f_p t) + \sin(6\pi f_p t)x_{rds}(t)\right)\Delta f \tag{2.40}$$

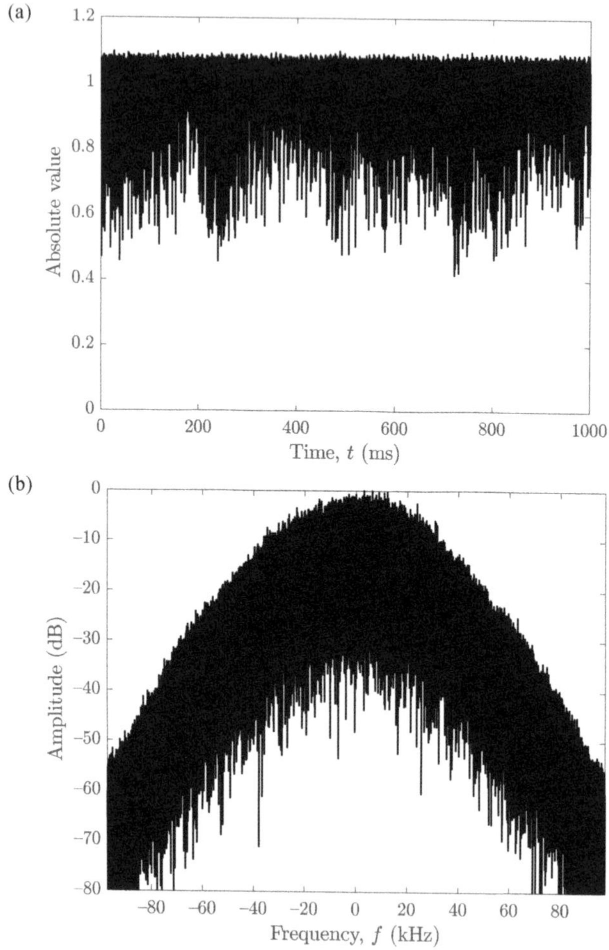

Figure 2.29 FM radio signal. (a) Time domain (absolute value). (b) Spectrum.

where $x_L(t)$ and $x_R(t)$ are the left and right channel audio signals, f_p is the pilot frequency equal to 19 kHz, and Δf is the frequency deviation equal to 75 kHz. The signal consists of four main components. The $(x_L(t) + x_R(t))/2$ part is the average of the left and right channel signals, or the mono signal. The $(x_L(t) - x_R(t))/2 \sin(4\pi f_p t)$ component is the differential signal of the left and right channels, which is amplitude-modulated onto the 38-kHz carrier frequency. The modulation used here is double sideband, suppressed carrier. For the purpose of demodulation, the pilot frequency of 19 kHz, which is half of the 38-kHz carrier frequency, is also included in the signal. This is the third main component of the signal, represented in (2.40) as $0.1 \sin(2\pi f_p t)$. The last component is the signal of Radio Data Stream (RDS), a standard for sending digital information connected with the current transmitter program [38]. The RDS signal component is modulated onto the third harmonic of the 19-kHz pilot frequency, $3f_p = 57$ kHz.

An example of the demodulated FM radio signal spectrum is shown in Figure 2.30. The four components described above can be identified in the spectrum. In the center, between -15 kHz and $+15$ kHz, the mono signal spectrum is visible, marked as L+R. At ± 19 kHz the pilot signal peak can be identified. The differential signal, marked as L-R, is centered around ± 38 kHz. The RDS signal is located around third harmonic of the pilot frequency of ± 57 kHz. The peaks at ± 76 kHz and ± 95 kHz are most probably the fourth and fifth harmonics of the pilot frequency, respectively.

Let us consider the ambiguity function of the FM radio signal. An example of the ambiguity function is shown in Figure 2.31. As can be seen, there is a clear peak at zero range and zero velocity. Apart from this, there are sidelobes uniformly spanning across all ranges and velocities, exactly in the same way as in the case of

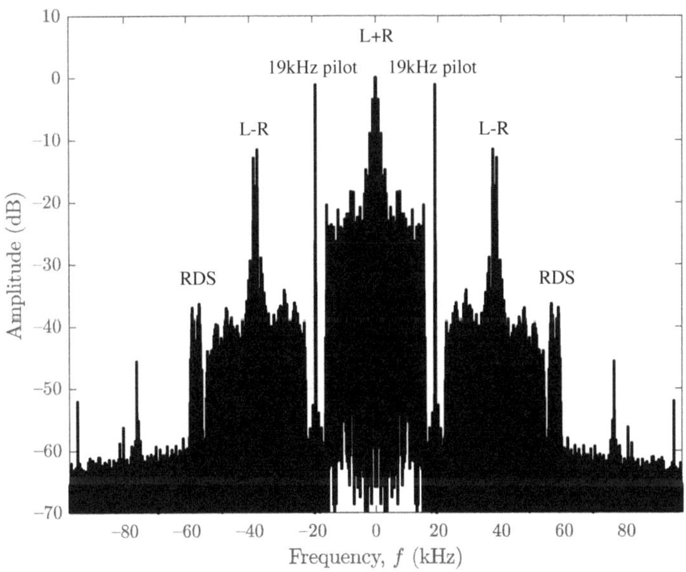

Figure 2.30 Spectrum of the demodulated FM radio signal.

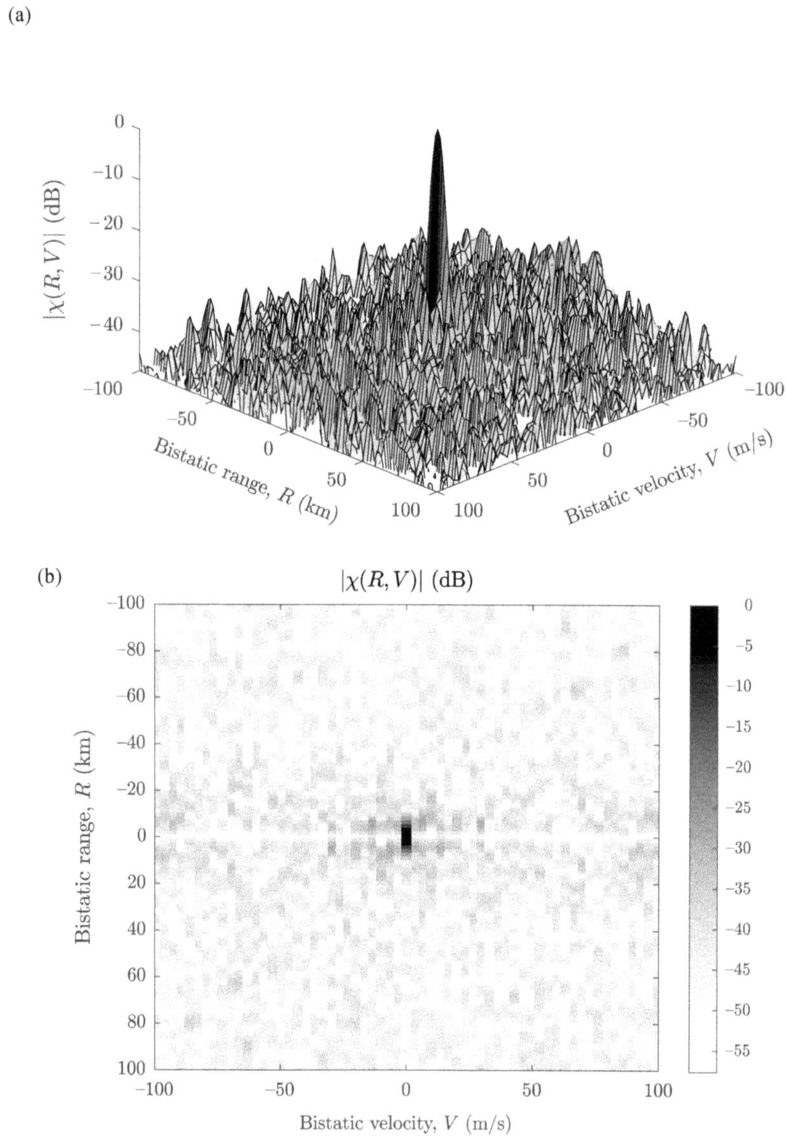

Figure 2.31 Ambiguity function modulus of the FM radio signal. (a) 3D view. (b) Top view.

the noise signal analyzed previously. The level of the sidelobes is BT below the main peak. In this particular case, the bandwidth of the signal was approximately 100 kHz, and the integration time was 1s. This corresponds to the BT of nearly 50 dB. These sidelobes are the cause of the masking effect, which is one of the main problems in passive radar. This problem will be analyzed in Chapter 5.

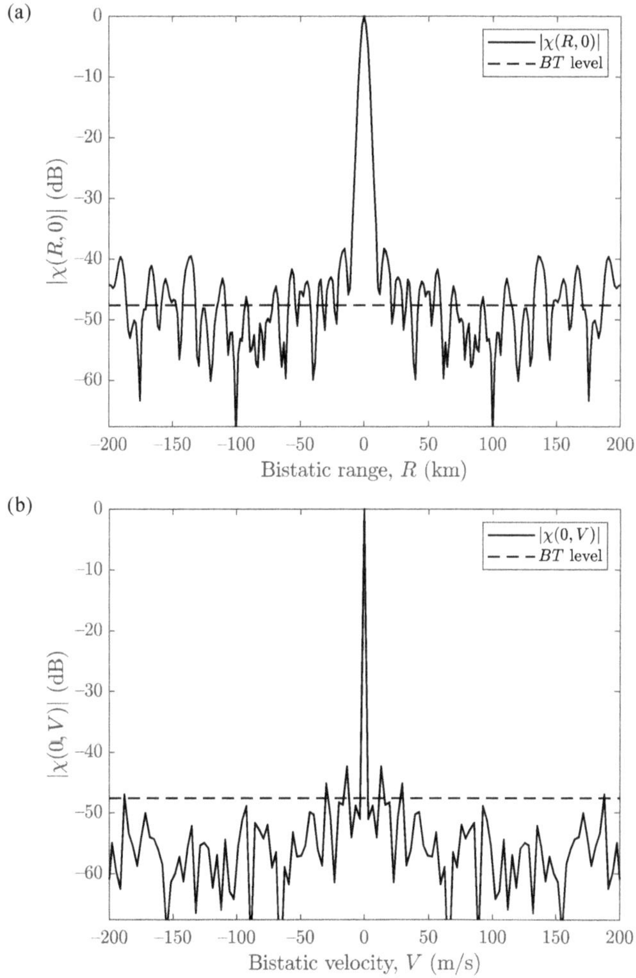

Figure 2.32 (a) Range and (b) velocity cross-section of the ambiguity function modulus of the FM radio signal.

The cross-sections of the ambiguity function for zero velocity ($\chi(R, 0)$) and for zero range ($\chi(0, V)$) are shown in Figure 2.32. Again, the sidelobes at the level of BT below the main peak are visible.

The signal bandwidth directly affects the range resolution of passive radar. This is a particularly important issue for FM radio, as the signal bandwidth depends on the transmitted content. For this reason, a detailed analysis will be carried out.

Consider the problem of measuring effective signal bandwidth. There are many ways that a signal bandwidth can be defined. One of the approaches is to measure effective bandwidth as a frequency range corresponding to a 3-dB reduction of the signal spectrum with respect to the maximum value. Another approach is to define a certain power percentage (e.g., 99%), that the effective bandwidth should contain. Yet another approach is to define the effective bandwidth in the following way: let us consider the power spectral density (PSD) $S(f)$. The effective bandwidth can be defined as a width of rectangular spectrum, whose height is equal to the maximum value of the analyzed power spectral density $S_{\max} = \max S(f)$. The area of the rectangular shape should be equal to the total power of the signal, that is integrated power spectral density $S(f)$.

$$S_{\max} \cdot B_{\mathrm{ef}} = \int_{-\infty}^{+\infty} S(f)\mathrm{d}f \tag{2.41}$$

The concept of effective bandwidth measurement corresponding to (2.41) is shown in Figure 2.33.

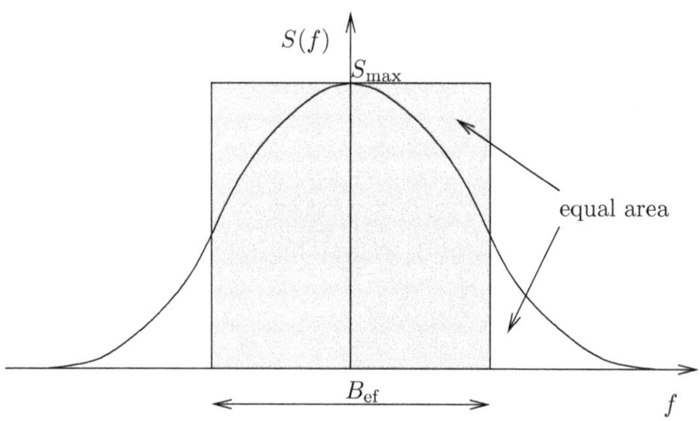

Figure 2.33 Concept of effective bandwidth measurement.

An example showing the varying nature of the FM signal is demonstrated in Figure 2.34. The plot at the top shows the spectrum of a signal corresponding to fast pop music. The effective bandwidth of this signal calculated using (2.41) is approximately 50 kHz. This signal will provide relatively fine range resolution (for FM radio-based radar). The plot at the bottom shows an example of the spectrum of speech. As can be seen, the spectrum is spiky and does not have a continuous support.

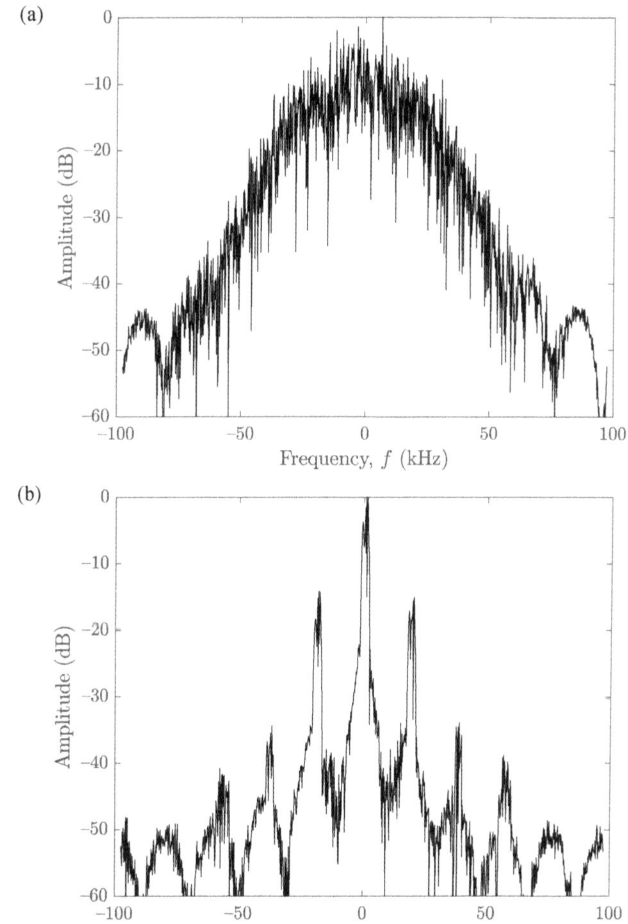

Figure 2.34 Comparison of spectra of FM signals: (a) fast pop music and (b) speech.

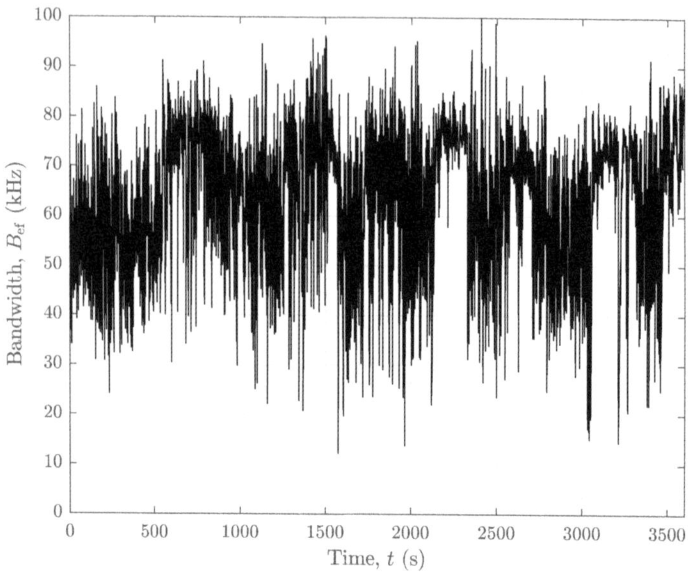

Figure 2.35 Effective bandwidth of an FM radio signal over period of 1h.

The effective bandwidth of this signal is approximately 10 kHz, much narrower than in the previous example.

Figure 2.35 shows the FM radio signal bandwidth over time. The bandwidth was calculated for an interval of 1 second by using (2.41). Over the 1-hour period, the program content ranged from music to commercials and news. The measured bandwidth varied from 12 kHz to 100 kHz. These results clearly show how the performance of FM-based passive radar can vary over time.

Figure 2.36 shows bistatic range resolution calculated for the same FM signal over a period of 100 seconds. It was calculated in two ways. One of them was to substitute calculated effective bandwidth (2.41) into (2.8) (i.e., c/B_{ef}). The other way was to calculate the bistatic range resolution from the ambiguity function (2.31). The bistatic range resolution was calculated as the width of the ambiguity function peak corresponding to a 3-dB reduction with respect to the maximum. As can be seen, both approaches, that is, through measurement of signal bandwidth according to (2.41) and directly measuring the width of the ambiguity function peak, yield similar results. In the presented example, the effective bandwidth was approximately 50 kHz. This corresponds to a bistatic range resolution in the order of 5–6 km. This result is much poorer than in the case of most modern active radars, where a range resolution of tens of meters is not uncommon. This, however, does not disqualify

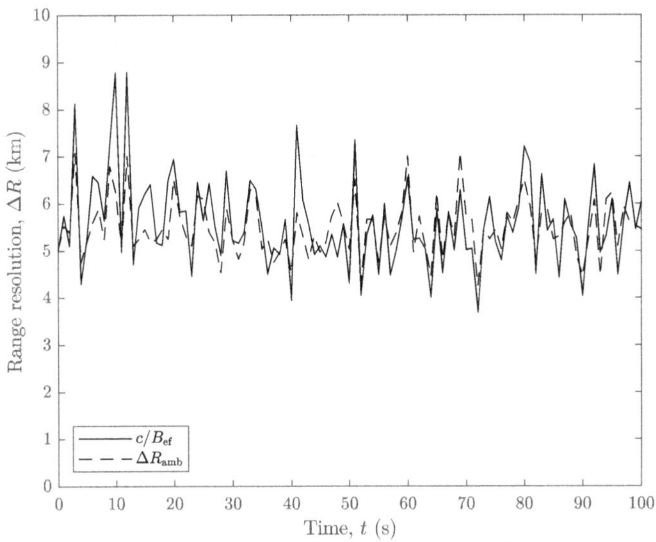

Figure 2.36 Bistatic range resolution corresponding to an FM radio signal over time measured using effective bandwidth and the ambiguity function.

FM-based radar. One of the advantages of FM-based radar is its fine velocity resolution. Typically, an integration time in the order of 1 second is applied, which results in a bistatic velocity resolution of approximately 3 m/s. Moreover, if a radar is used for long-range detection (e.g., > 100 km), this kind of range resolution may turn out to be sufficient.

2.5.2 Digital Television (DVB-T)

One of the most popular standards of digital television signals worldwide is DVB-T (Digital Video Broadcasting-Terrestrial). This standard is used in Europe, Africa, Asia, Australia, and South America. DVB-T is broadcast in the UHF band, typically in the 470–860-MHz frequency range. The power of the transmitters is in the order of tens of kilowatts, occasionally reaching a hundred kilowatts. The signal bandwidth is up to 7.61 MHz, which provides very good range resolution, comparable to active surveillance radars. The modulation standard used in DVB-T is OFDM (orthogonal frequency-division multiplexing). The modulation itself is relatively advantageous from the point of view of passive radar, as the OFDM signal has properties similar to a band-limited noise. The signal, however, contains certain elements, such as pilots or cyclic prefix, which have negative effect on the ambiguity function. These effects can be relatively easily eliminated by appropriate signal processing [39, 40]. Another difficulty which can be encountered when using DVB-T is the use of the single frequency network (SFN) transmission scheme [41, 42]. It consists of transmitting the same signal on the same carrier frequency by different transmitters. This gives rise to ambiguity in the association of target detection to transmitter. Nevertheless, due to its advantages, the DVB-T standard is often used for the purposes of passive radar [43–45].

The signal structure of the DVB-T signal is defined in [46]. The data is sent in packets, which correspond to individual OFDM symbols. The main idea of OFDM is to divide a high-rate data stream into multiple low-rate streams. These streams are then used for modulation of multiple subcarriers. The length of one symbol is matched to the spacing between the subcarriers, so that signals corresponding to different subcarriers are orthogonal, hence the name of the OFDM modulation.

Because the high-rate data stream is converted to numerous low-rate data streams, it is possible to use a guard interval, which reduces the effect of intersymbol interference. The guard interval is filled with a part of the next OFDM symbol, which forms a cyclic prefix. The idea of a cyclic prefix is shown in Figure 2.37.

The standard defines 5, 6, 7, and 8-MHz channels; however, the 8-MHz channel is used most frequently. In each case 2k and 8k modes can be used, which differ in the number of subcarriers. The number of subcarriers is 1,705 and 6,817, respectively. The symbol length is 224 μs for 2k and 896 μs for 8k, which correspond to a carrier spacing of 4,464 Hz and 1,116 Hz, respectively. Possible guard interval lengths are 1/4, 1/8, 1/16, and 1/32 of the OFDM symbol length.

An important part of DVB-T signal demodulation is channel equalization. In order to facilitate the equalization process, the signal contains pilots, known values of the signal at certain subcarriers. There are two types of pilots: scattered and continual. The position (subcarrier number) of scattered pilots changes in each symbol. The position of continual pilots is fixed for all symbols. The pilots are transmitted with a boosted power level.

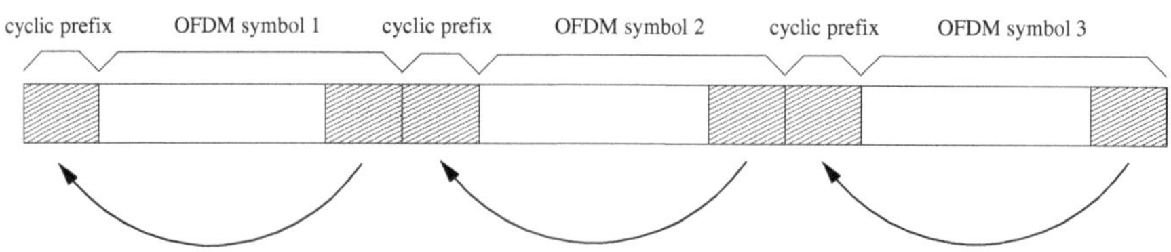

Figure 2.37 The concept of cyclic prefix used in the DVB-T television standard. A part of the OFDM symbol is used as the guard interval.

The data on each subcarrier is modulated using QPSK (quadrature phase shift keying) or 16- or 64-QAM (quadrature amplitude modulation). Each QPSK symbol corresponds to 2 bits, which gives 4 possibilities. In the case of 16-QAM, each symbol corresponds to 4 bits and consequently 16 possibilities. For 64-QAM the symbol corresponds to 6 bits, which gives 64 possibilities. The modulation schemes are represented with constellations, which show the amplitude combinations of the I (in-phase) and Q (quadrature) components. The constellations for QPSK, 16-QAM, and 64-QAM are shown in Figure 2.38.

Figure 2.39 shows an example of a real DVB-T signal spectrum. It has a characteristic rectangular shape. It represents an 8-MHz, 8k mode signal with a 1/4 guard interval. The increased amplitudes of the pilots are visible in the spectrum. The constellation of the demodulated signal is shown in Figure 2.40. In this case QAM-64 modulation has been applied. The OFDM data are transmitted using one of 64 symbols, which are placed on clearly visible 8×8 grid. Apart from the main data stream symbols, the pilot symbols are visible in $\pm 4/3 + 0\mathrm{j}$ points. Pilots are modulated using Differential Binary Phase Shift Keying (DBPSK) with the amplitude of 4/3. In addition, transmission parameter signaling (TPS) symbols are visible in $\pm 1 + 0\mathrm{j}$ points. TPS symbols are modulated using DBPSK with the amplitude equal to 1.

The ambiguity function of the DVB-T signal is shown in Figure 2.41. As in the case of noise and FM signals, random sidelobes at the level of BT are also visible. In this case with $B = 7.6$ MHz and $T = 0.1$s, the BT level is approximately 59 dB. The peaks corresponding to the pilot frequencies can be seen on the ambiguity function.

Range and velocity slices of the ambiguity function are shown in Figure 2.42. In the case of DVB-T, the most predominant sidelobes are the deterministic sidelobes in the range directions due to the rectangular shape of the signal spectrum. The peaks in the range slice at ± 3.5 km correspond to two SFN transmitters.

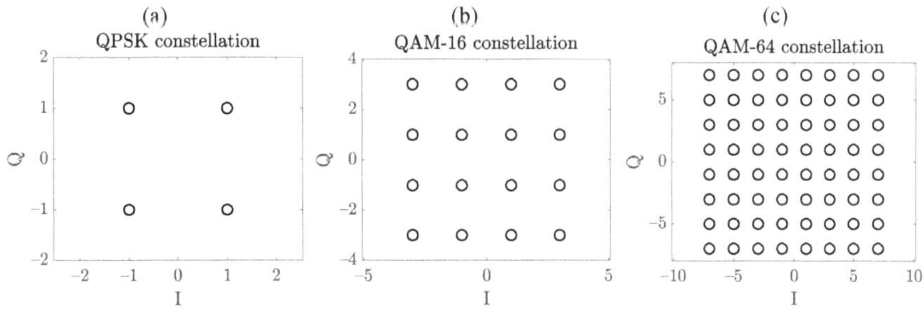

Figure 2.38 Constellations used in DVB-T standard. (a) QPSK, (b) 16-QAM, and (c) 64-QAM.

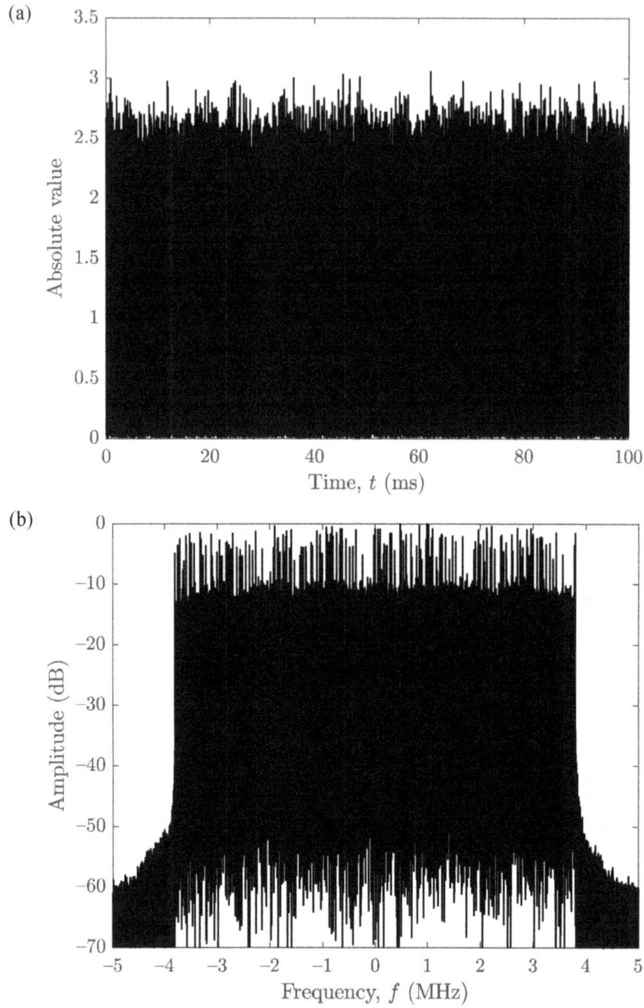

Figure 2.39 DVB-T television signal. (a) Time domain (absolute value). (b) Spectrum.

In some countries, the DVB-T standard is replaced by its second version, Digital Video Broadcasting – Second Generation Terrestrial (DVB-T2). The standard is an extension of its first version, which allows for higher throughput to be obtained. This can be used to realize high-definition television (HDTV) transmission. In the standard, QPSK, 16-QAM, 64-QAM, or 256-QAM modulations are

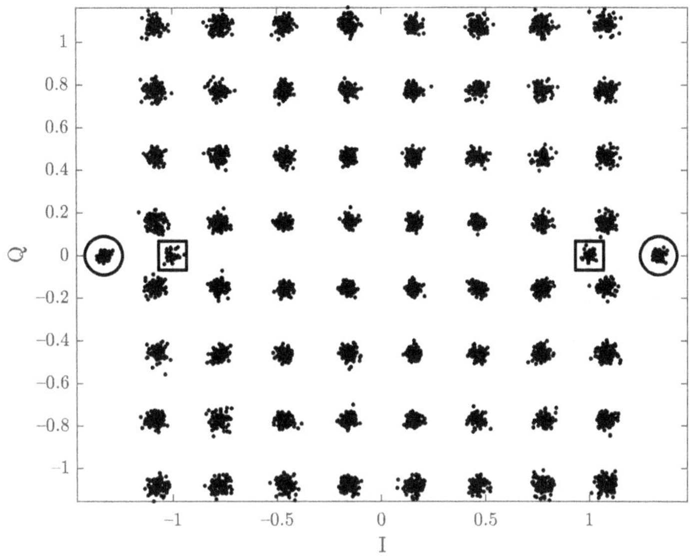

Figure 2.40 Constellation of decoded DVB-T signal with QAM-64 modulation.

used. The OFDM mode can be: 1k, 2k, 4k, 8k, 16k, and 32k. The guard interval can be 1/128, 1/32, 1/16, 19/256, 1/8, 19/128, and 1/4. In general, the standard is much more complex than the DVB-T, providing greater flexibility. This, in turn, means that the demodulation and remodulation for the passive radar purposes are much more difficult. The DVB-T2 has been used for several years now and a substantial body of work on this topic in the context of passive radar has been developed [23, 24, 47–57]. However, further work is required due to the more complicated nature of the standard.

The standard corresponding to DVB-T that is used in North America is called Advanced Television Systems Committee (ATSC) [58]. The DVB-T and ATSC standards are not compatible with each other, but some of the signal characteristics from the point of view of passive radar are similar. The ATSC signal is 6 MHz wide and uses 8VSB (vestigial sideband modulation) modulation, which is based on 8-level amplitude-shift keying scheme. The bit data stream is coded on 3 bits, which results in 8 levels, which are then used to modulate the signal with the symbol rate of 10.76 MHz. The root-raised-cosine filter is then used to filter out the lower band, leaving the upper band. As the filter is not perfect, the vestige of the lower band remains in the signal, hence the name of the VSB modulation. The randomization of the data results in the signal having characteristics similar to a noise signal. There

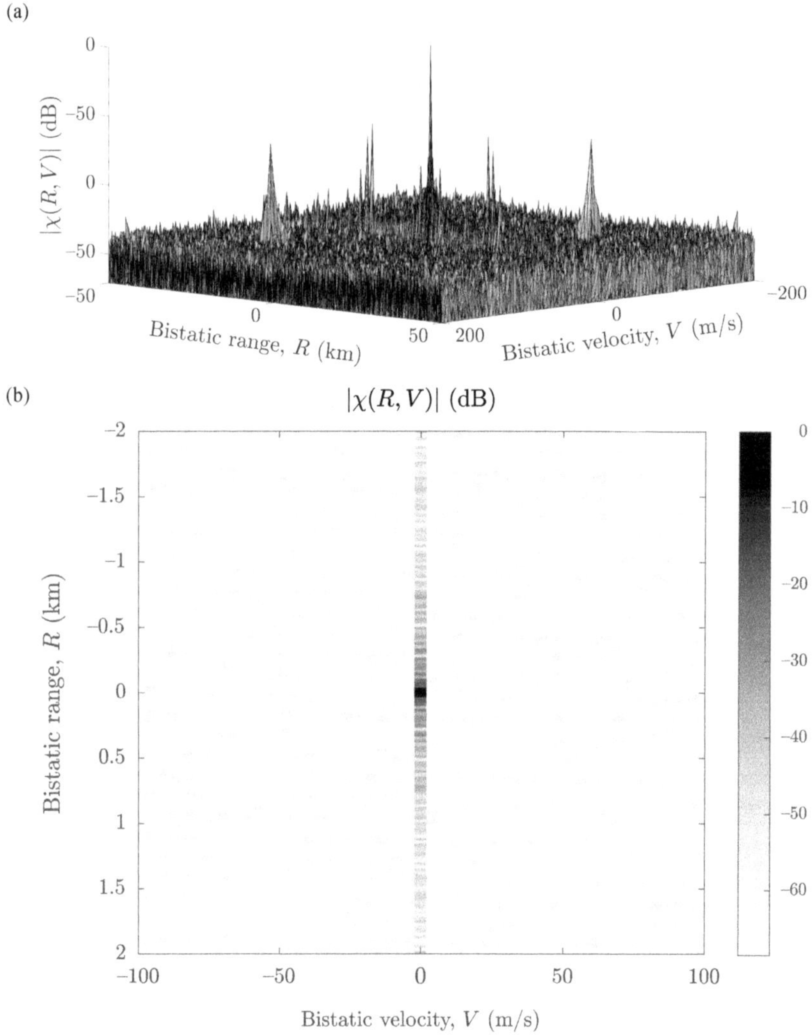

Figure 2.41 Ambiguity function modulus of the DVB-T signal. (a) 3D view. (b) Top view.

are, however, repetitive elements of the signal, such as synchronization sequences occurring every 77.3 μs and 24.2 ms, or the carrier frequency peak, which deteriorate the ambiguity function from the point of view of passive radar processing. These elements can be relatively easily removed from the signal, similar as in the case of

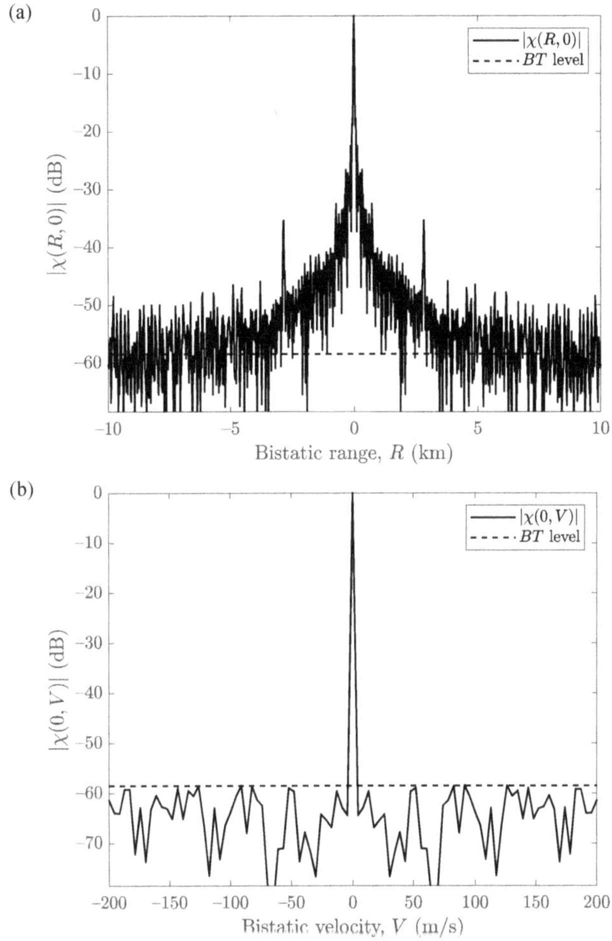

Figure 2.42 (a) Range and (b) velocity cross-section of the ambiguity function modulus of the DVB-T television signal.

DVB-T. The ATSC signal is transmitted in the VHF or UHF band, similar as the DVB-T signal. Examples of ATSC-based passive radar and related analyses can be found in [59–61].

Integrated Services Digital Broadcasting – Terrestrial (ISDB-T) is a Japanese digital television standard [62] that was later extended to ISDB-T International, which is used mainly in South America. The ISDB-T signal has a bandwidth of

5.6 MHz (however, different variants are also possible) and uses Band Segmented Transmission-OFDM (BST-OFDM), which consists of dividing the signal band into several subbands and transmitting separate OFDM-modulated streams. The ISDB-T signal is usually transmitted in the UHF band, but different frequencies can be also used. Some examples of using ISDB-T signal for passive radar purposes can be found in [63–66].

Another digital television standard is the Digital Terrestrial Multimedia Broadcasting (DTMB), which is used in China and Cuba [67]. The DTMB signal utilizes time-domain synchronous orthogonal frequency division multiplexing (TDS-OFDM). One of the main differences between the DVB-T and DTMB is that the latter one uses a pseudo-noise sequence as the guard interval. The subcarriers are modulated using 4, 16, 32, 64-QAM and 4-QAM-NR (4-QAM with Nordstrom-Robinson coding). The 4k mode uses 3,780 carriers. The signal bandwidth is 7.56 MHz. The transmitters operate in the UHF band. Similar to other digital television standards, the ambiguity function of the DTMB signal is relatively good from the point of view of passive radar processing; however, some artifacts are present due to the repetitive pseudo-noise sequence or power variation between different segments of the signal. In [25, 68–70], some examples of DTMB-based passive radars can be found.

2.5.3 Digital Radio (DAB)

The digital radio standard, DAB (Digital Audio Broadcasting), uses the upper part of the VHF band (typically around 200 MHz). The format of the signal is defined in [71]. The transmitted power is lower than in FM radio, usually in the order of a few kilowatts. The modulation used in the DAB standard is OFDM, as in DVB-T, which is relatively advantageous from the point of view of passive radar. This is because of the (band-limited) noise-like nature of the OFDM-modulated signal. The main disadvantage of the DAB standard is a smaller area coverage due to few medium- or low-powered transmitters. In contrast to the successful introduction of digital TV, there has been no widespread switch over from FM to DAB, nor is there likely to be in the foreseeable future. Nonetheless, systems based on DAB transmitters are being constructed, especially in countries where their popularity is relatively high [43, 45, 72–74].

An example of a real DAB signal is shown in Figure 2.43. The signal in the time domain is shown at the top. The signal resembles noise, a feature characteristic for OFDM-modulated signals. However, the time structure can also be recognized. The data stream is organized in frames 96 ms in length. The consecutive frames

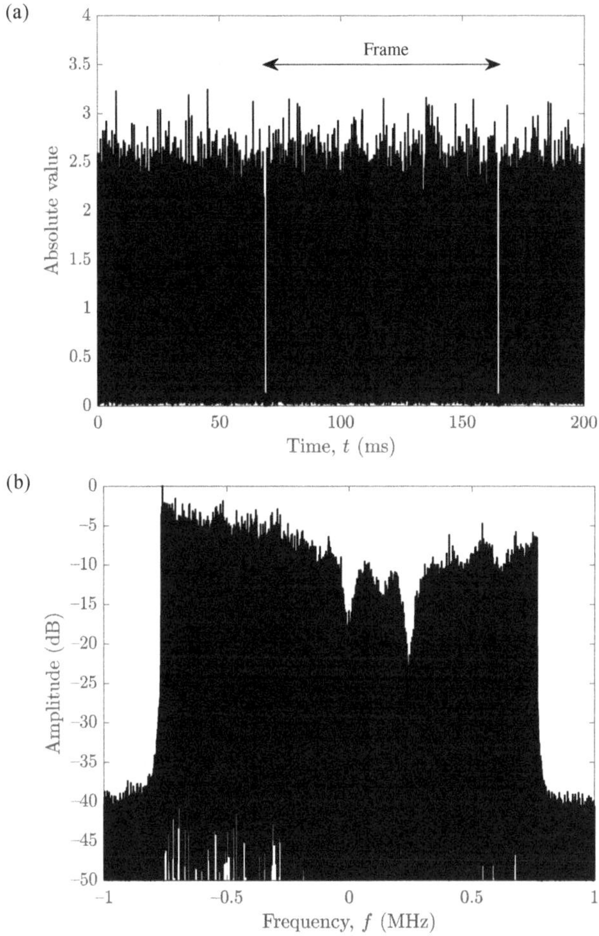

Figure 2.43 DAB radio signal. (a) Time domain (absolute value). (b) Spectrum.

can be recognized by a gap between them, as marked in the figure. In the frame, there are 76 OFDM symbols. The plot at the bottom shows the DAB signal in the frequency domain. A rectangular envelope of the signal, typical for OFDM signals, can be distinguished. In the DAB standard, the data streams are transmitted on 1,536 subcarriers, which form a rectangular-shaped spectrum. The modulation on each of the subcarriers is D-QPSK (differential quadrature phase shift keying). The bandwidth of the DAB signal is 1.5 MHz. The rectangular shape of the spectrum

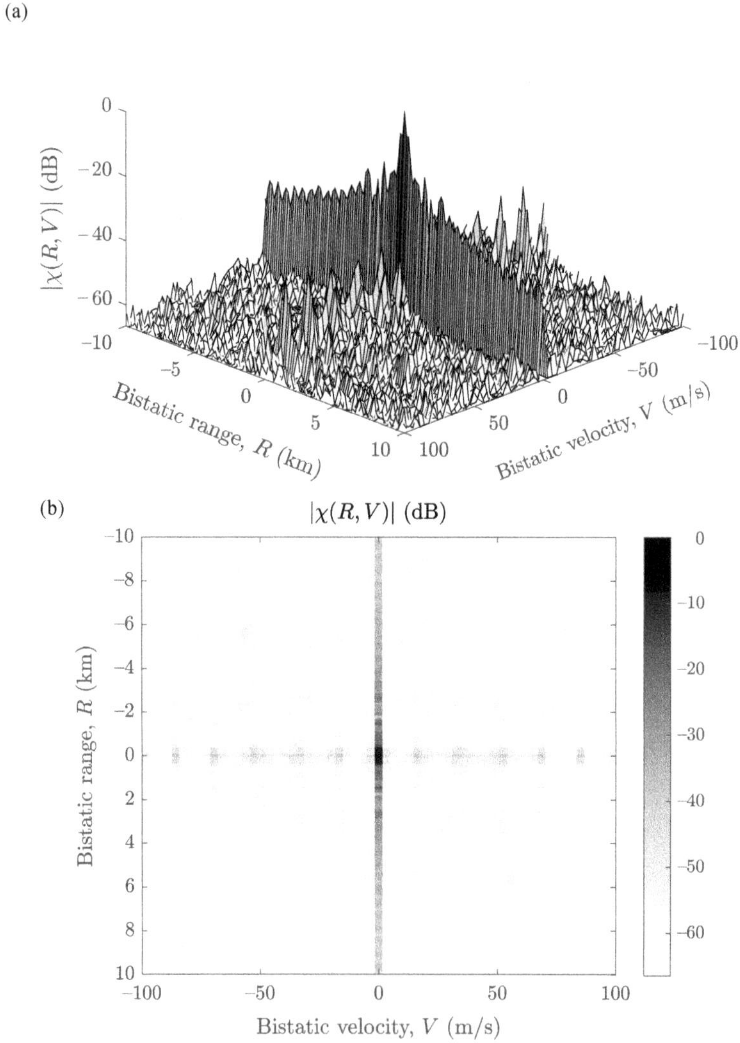

Figure 2.44 Ambiguity function modulus of the DAB radio signal. (a) 3D view. (b) Top view.

is highly distorted by the multipath effect. As the signal was recorded in an urban environment, apart from the direct path signal, reflected signal components were also present. This results in a frequency response of the radio channel that is not flat, which is the effect visible in the figure.

The ambiguity function of the DAB signal is presented in Figure 2.44. The integration time was 0.5s, which results in a BT level of -59 dB. The random sidelobes spread over all ranges and velocities are visible at BT below the main peak. High sidelobes in the range direction are present. Moreover, some sidelobes in the velocity direction are also visible. The nature of those sidelobes is better investigated on the cross-sections of the ambiguity function, shown in Figure 2.45.

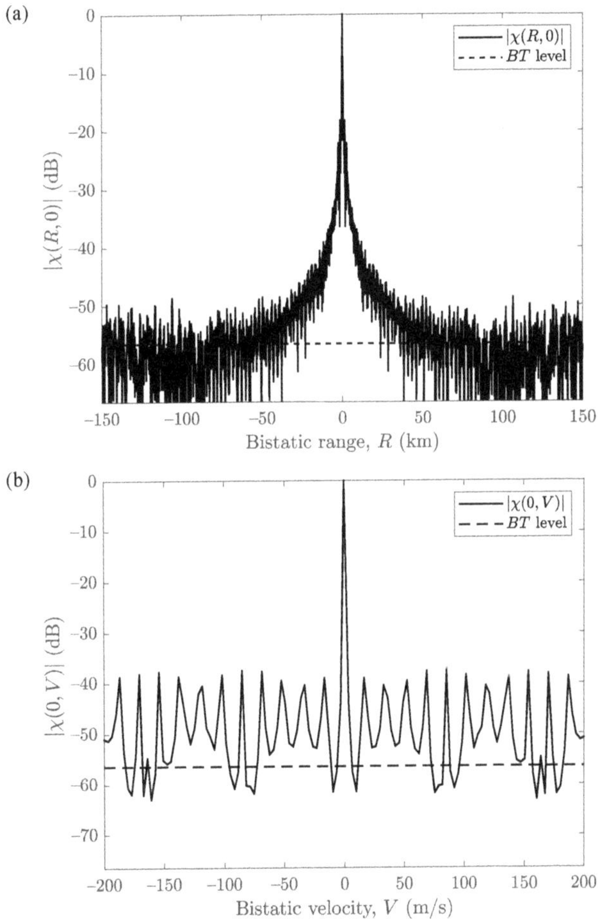

Figure 2.45 (a) Range and (b) velocity cross-section of the ambiguity function modulus of the DAB radio signal.

The range cross-section of the ambiguity function (Figure 2.45(a)) reveals deterministic sidelobes in the vicinity of 0 km. The sidelobes result from the rectangular shape of the signal spectrum. At further ranges the deterministic sidelobes are dominated by the random sidelobes. In the velocity cross-section, periodic sidelobes are visible at a level higher than the BT level. These sidelobes are associated with the periodic nature of the signal: 96-ms-long frames, which correspond to 10.42 Hz in the frequency domain. The presented signal example was recorded at the carrier frequency of 183 MHz; thus, the wavelength is 1.64m. The periodicity of 10.42 Hz can be expressed in bistatic velocity as $\lambda f_d \approx 17$ m/s (compare (2.7)), which is observed in the figure.

2.5.4 Cellular Telephony (GSM)

GSM (Global System for Mobile Communications) is one of the most popular standards of cellular networks. It is used in Africa, Europe, the Middle East, Asia, and Oceania. The system can operate in different frequency bands, but the most popular are GSM-900 operating around 900 MHz and GSM-1800 operating around 1,800 MHz. More specifically, in the case of GSM-900 the uplink frequency range is 890–915 MHz, whereas the downlink frequency range is 935–960 MHz. The frequency bands are divided into 124 channels, each 200 kHz wide. Each frequency channel is further divided into 8 time slots. The physical channel is created by combining the frequency band with a time slot number. The corresponding uplink and downlink channels are separated by 45 MHz, and the time slot number is shifted by 3. In the case of GSM-1800, the uplink corresponds to 1,710–1,785 MHz and downlink to 1,805–1,880 MHz. There are 374 frequency channels. The uplink and downlink channels are separated by 95 MHz.

In general, the GSM-900 is used at longer distances (e.g., between cities). GSM-1800 is devoted to shorter distances, and is focused on higher capacity. In relation to passive radar use, GSM-900 is more suitable for target detection as the transmitter power is higher (several hundred watts) than in GSM-1800 (tens of watts).

In the time domain, the smallest part of the signal is one symbol, corresponding to 1 bit, which is 3.69 μs. The standard uses GMSK (Gaussian minimum shift keying) modulation for encoding bits. The symbols are combined into packets, which correspond to 148 bits (in some cases 88 bits). The packet, together with a guard interval, forms a time slot, which is 0.58 ms long. Eight time slots constitute a GSM frame, which is 4.6 ms long. The frames are combined into multiframes, multiframes into superframes, and superframes into hyperframes.

Figure 2.46 shows the absolute value of the time sequence of a recorded GSM signal, as well as the spectrum of the GSM signal. The time plot shows gaps in the signal, which correspond to time slots (0.58 ms). The shape of the spectrum is similar to that of FM radio (with fast music content). Not only is the shape similar to FM radio, but also the effective bandwidth is comparable.

The ambiguity function and its range and velocity slices are shown in Figures 2.47 and 2.48. As can be seen, apart from a uniform noise floor and the main

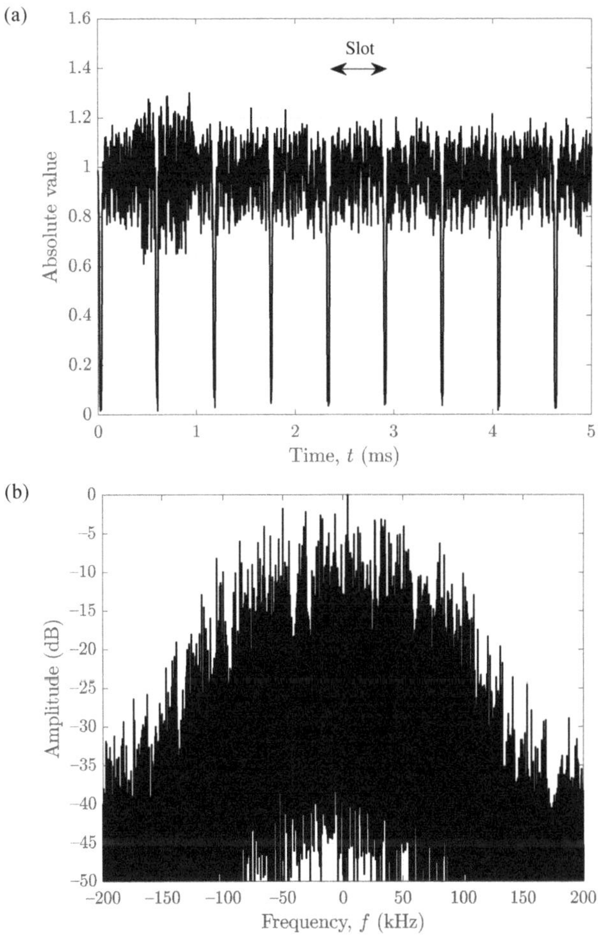

Figure 2.46 GSM signal. (a) Time domain (absolute value). (b) Spectrum.

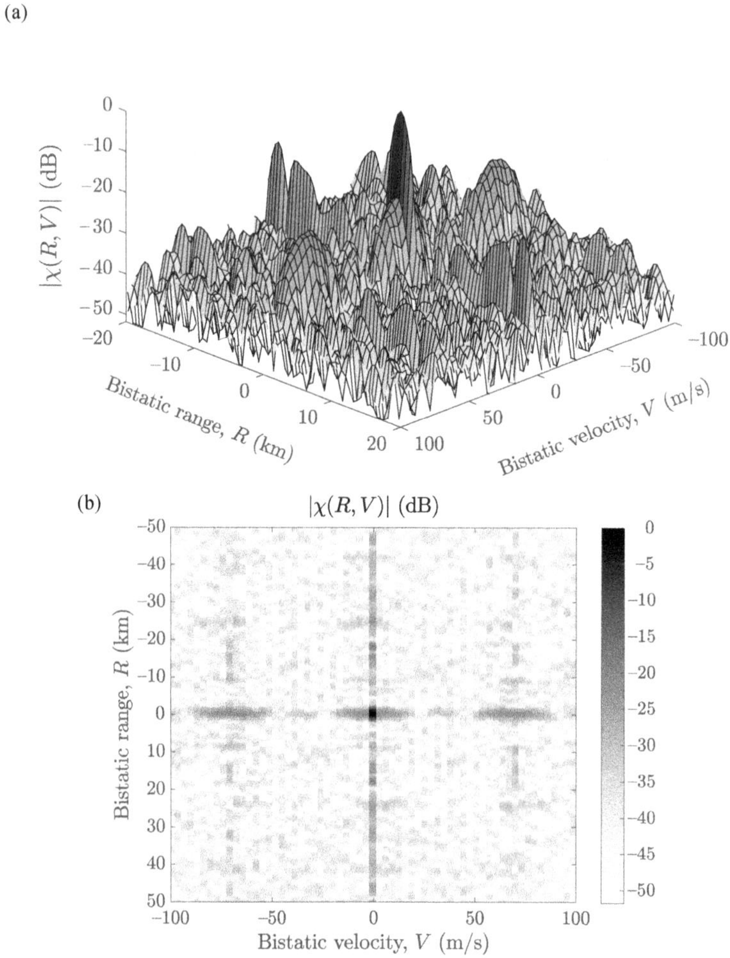

Figure 2.47 Ambiguity function modulus of the GSM signal. (a) 3D view. (b) Top view.

correlation peak at (0, 0), there are multiple peaks. They result from the complicated structure of the signal. They will negatively influence passive radar operation, as those additional peaks can be erroneously treated as other target echoes.

As has been demonstrated the bandwidth of the GSM signal is comparable to the one used in FM radio; however, due to the digital modulation, the signal characteristics are much more constant. The power of transmitters used in GSM is much lower than in, for example FM radio and DVB-T. For this reason, the detection

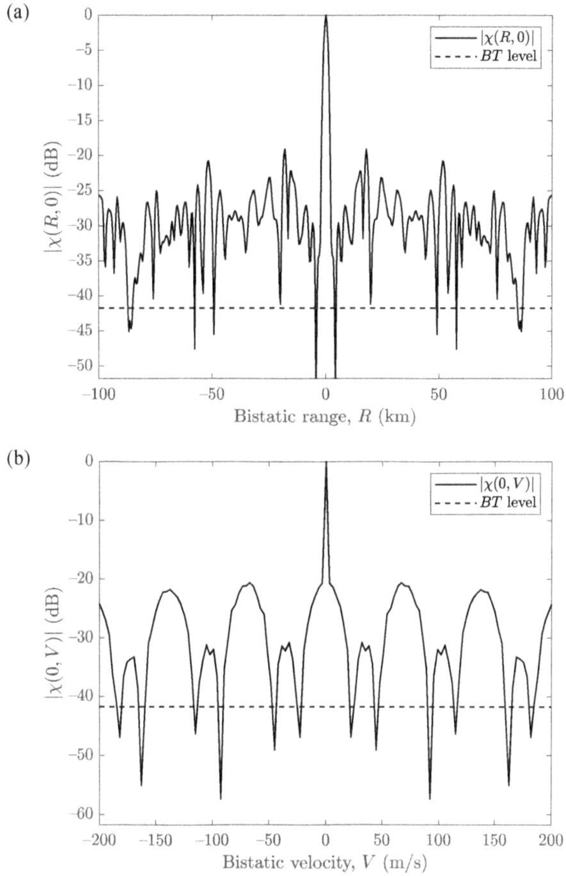

Figure 2.48 (a) Range and (b) velocity cross-section of the ambiguity function modulus of the GSM telephony signal.

range in comparison with the size of the resolution cell is much shorter in GSM than in FM. As a result, the GSM-based passive radar can be used for short-range surveillance with a detection range in the order of a few kilometers [21, 75–78].

2.5.5 Wireless Networks (WiFi)

WiFi (wireless fidelity) is a family of standards used for wireless local area networks. There are numerous versions of the standard, under the common name of IEEE

802.11. Different versions are denoted with different letter(s) (e.g., IEEE 802.11a, IEEE 802.11g, or IEEE 802.11n). Most of the versions operate in 2.4-GHz or 5-GHz frequency bands, however, other options are also possible, depending on the standard's version.

Let us consider the IEEE 802.11g standard version as an example. This is not the newest version of the standard, but it is widespread. The time- and frequency-domain signals are shown in Figure 2.49. The signal has been recorded in the 2.4-

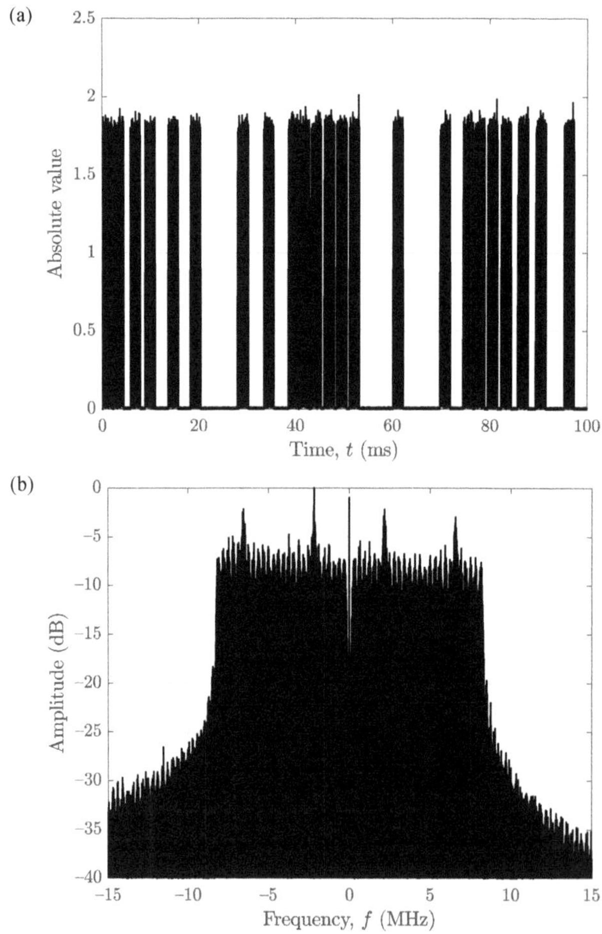

Figure 2.49 WiFi signal. (a) Time domain (absolute value). (b) Spectrum.

GHz frequency band. Unlike most transmission types used in passive radar, WiFi has a packet structure, with possible long gaps between the packets. This feature makes this signal more similar to classical active pulse radar waveform. This, in turn, implies signal processing that resembles the classical matched filter approach. The density of packets depends on the network traffic, which is a feature similar to FM radio with respect to the dependency of passive radar performance on the transmission content. The IEEE 802.11g standard uses OFDM modulation, which

(a)

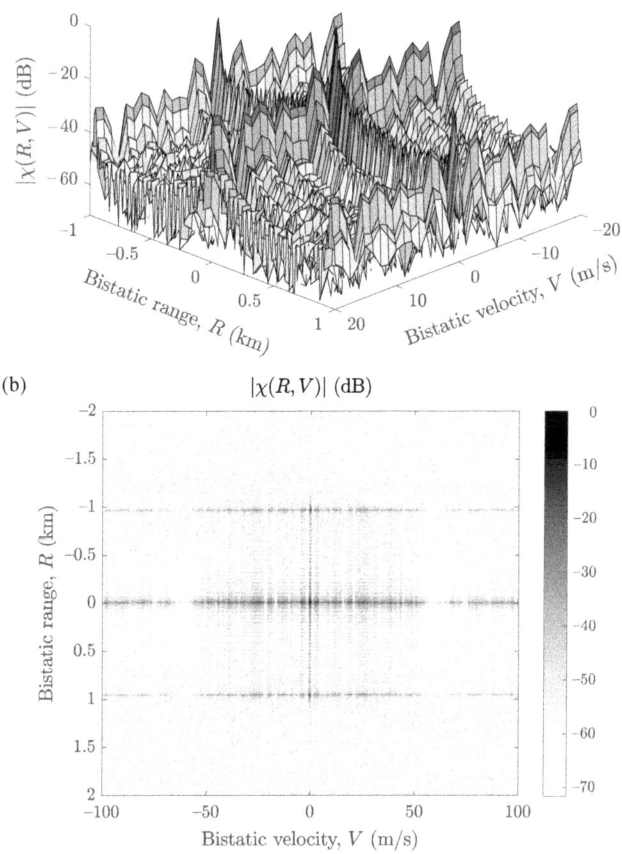

Figure 2.50 Ambiguity function modulus of the WiFi signal. (a) 3D view. (b) Top view.

is reflected by the rectangular shape of the spectrum. The signal spectrum shown in Figure 2.49(b) has a bandwidth of approximately 16.6 MHz.

The cross-ambiguity function calculated for the WiFi signal is shown in Figure 2.50, and the range and velocity slices are shown in Figure 2.51. The wide bandwidth of the signal provides a fine bistatic range resolution cell, in the order of 18m, a value twice as good as in the case of a DVB-T signal. Values of the cross-

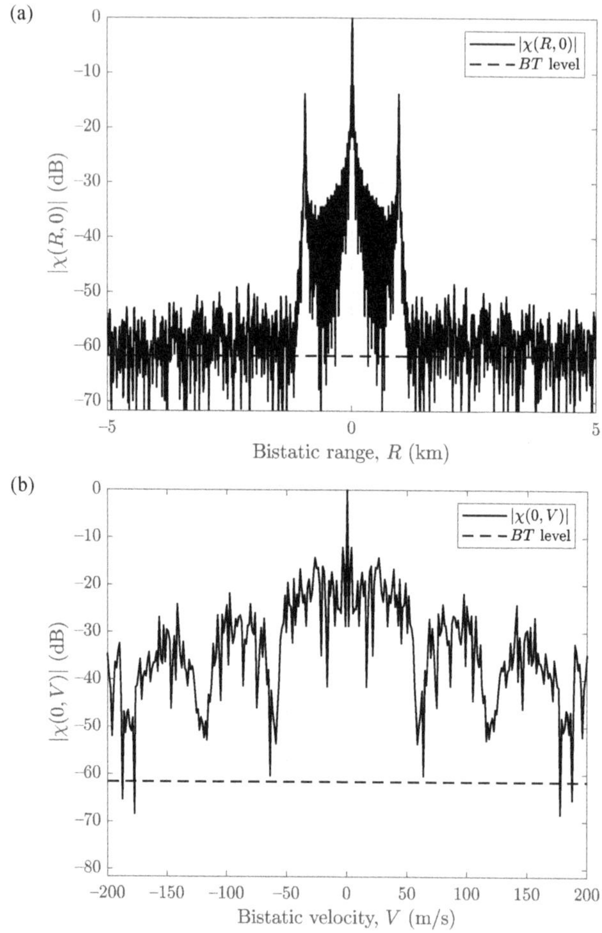

Figure 2.51 (a) Range and (b) velocity cross-section of the ambiguity function modulus of the WiFi signal.

ambiguity function above the BT (with respect to the peak) are in the range of ± 1 km. This corresponds to the length of the OFDM symbol, equal to 3.2 μs, which translates to 960m. The peaks at the end of this 960-m interval are caused by the cyclic prefix. For larger delays (or ranges), a constant level of sidelobes BT below the peak can be observed. The velocity slice of the cross-ambiguity function, shown in Figure 2.51(b), reveals slowly decreasing sidelobes. This is disadvantageous from the point of view of target detection, as the sidelobes of strong target echoes can mask weaker echoes.

Due to its high range resolution and low transmission power, WiFi illumination is useful for short-range surveillance, such as small airport monitoring, pedestrians or vehicle detection [79–83].

2.5.6 Comparison of Different Transmitter Types

The transmitter types analyzed above are not the only ones used for passive radar. However, the aim of the presented analysis was not a comprehensive review, but a representative sample. Table 2.1 lists the main parameters of different signal types. The parameters can vary with region. For example, DAB radio is usually used in the VHF band (174–230 MHz); however, the L-band (1,452–1,480 MHz) version is also possible. The parameters listed in the table can also be different for different versions of the standard, for example, the DVB-T standard defines 5, 6, 7, and 8-MHz channel bandwidths, but the 8-MHz version is most frequently used.

A comparison of the parameters from the table leads to the conclusion that, depending on the illuminator of opportunity, the performance of passive radar can

Table 2.1
Parameters of Different Transmitter Types

Transmitter Type	*Frequency Band*	*Signal Bandwidth*	*Transmitter Power*	*Comments*
FM	88–108 MHz	Varying, typ. 10–100 kHz	Up to 100 kW	Low accuracy, long det. range
DAB	174–230 MHz, 1,452–1,480 MHz	1.5 MHz	Up to tens of kW	Medium accuracy, medium det. range
DVB-T	470–860 MHz	7.6 MHz	Up to 100 kW	High accuracy, medium det. range
GSM	935–960 MHz 1,805–1,889 MHz	200 kHz	Up to 320W	Low accuracy, short det. range
WiFi	2,400–2,500 MHz 5,000–6,000 MHz	16 MHz	Up to 200 mW	High accuracy, short det. range

vary substantially. FM radio provides poor range resolution, which in addition varies in time with the content of the transmitted program. However, FM-based passive radar can provide long detection range thanks to the high transmitter power, which can reach up to and over 100 kW. DVB-T signal has a wide bandwidth, comparable to that of a typical active surveillance radar, which gives very fine range resolution. The transmitted power reaches tens of kilowatts and occasionally over 100 kW. Although the transmitted power in FM radio and DVB-T television is comparable, higher frequencies used in DVB-T cause the detection range to be shorter than for FM radio. DAB-based passive radar, with a bandwidth of 1.5 MHz, can obtain a reasonable range resolution. The power of the DAB transmitter is typically a few or tens of kilowatts, which provides medium detection range. GSM signal bandwidth is approximately 200 kHz, which is very similar to FM radio. The difference is that GSM uses digital modulation, which results in a constant bandwidth, in contrast to the analog frequency modulation of FM radio. In any case, the range resolution obtainable with GSM is rather poor. This, in combination with relatively low transmitted power, in the order of hundreds of watts, makes GSM-based passive radar suitable only for short-range surveillance with low range accuracy. WiFi illumination provides very fine range resolu-

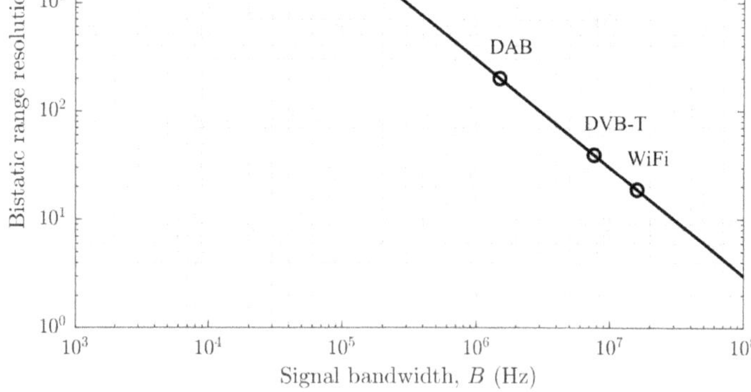

Figure 2.52 Bistatic range resolution for different types of transmitters.

tion thanks to its wide signal bandwidth. The transmitted power level is, however, very low compared to the transmitters already mentioned. This renders WiFi-based passive radar suitable only for short-range detection. In this application, however, high range accuracy can be expected, which allows for even a single person to be detected.

A visual representation of the range resolution for different types of transmitters is shown in Figure 2.52. The horizontal axis shows the bandwidth of the signal. The vertical axis represents the bistatic range resolution, calculated according to (2.8). The markers on the plot correspond to different transmitter types. In the case of FM radio, a span of bandwidth and corresponding range resolution is shown in order to stress the variability of FM signal bandwidth. As can be seen, the bistatic range resolution of passive radar for different illuminators of opportunity spans across several orders of magnitude. This is one of the important aspects of designing passive radar, but it has to be considered together with another fundamental parameter: detection range. In order to compare the detection range of different types of transmitters, a common measurement, referred to here as *equivalent monostatic range*, R_{me}, will be introduced. As analyzed earlier, the bistatic range equation depends on $R_1^2 R_2^2$. If it is assumed that a quasi-monostatic scenario is dealt with (i.e.,

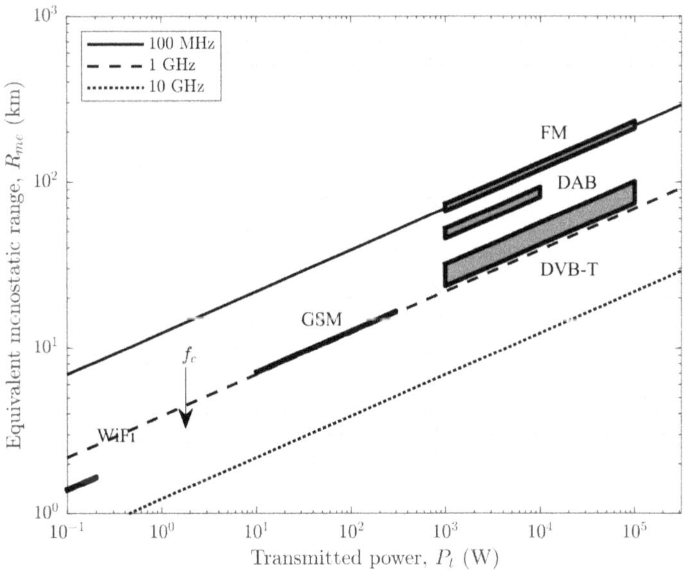

Figure 2.53 Equivalent monostatic detection range for different types of transmitters.

the transmitter is in the same place as the receiver), then $R_1 = R_2$. In such a case the monostatic equivalent range can be calculated as $\sqrt{R_1 R_2}$. By rearranging (2.21), monostatic equivalent detection range R_{me} can be obtained as:

$$R_{me} = \sqrt{R_1 R_2} = \sqrt[4]{\frac{\sigma P_t G_t G_r \lambda^2 T}{\text{SNR}_{\text{min}} (4\pi)^3 k_B T_0 L}} \tag{2.42}$$

Here it was assumed that the signal bandwidth B is equal to receiver bandwidth B_R; thus, the bandwidth in the processing gain BT is reduced with the receiver bandwidth in the noise power $k_B T_0 B_R$. In order to compare different transmitter types, a set of common parameters was assumed: $\sigma = 1$ m^2, $G_t = 0$ dB, $G_r = 5$ dB, $T = 1$s, $\text{SNR}_{\text{min}} = 12$ dB, and $L = 10$ dB. For most of the illuminator types, these are reasonable parameters. The results of calculations of equivalent monostatic range R_{me} computed with (2.42) are shown in Figure 2.53. With the parameters listed above fixed, the detection range still depends on two variables: the transmitted powerP_t and wavelength λ. This dependency is represented in the figure the following way. Each of the three inclined lines correspond to a different frequency, equal to 100 MHz, 1 GHz, and 10 GHz, respectively. These lines can serve as a reference. The detection range for individual transmitters is represented with gray areas. The horizontal span corresponds to a typical range of values of the transmitted power. The span in the vertical direction represents variability of frequency. As seen in the figure, FM radio provides the highest detection range, in the order of 200–300 km. Slightly lower detection range, ≤ 100 km, is obtained with DAB and DVB-T. Typically, DVB-T transmitters use higher power than DAB, however, this is compensated by the lower frequency of DAB. GSM-based passive radar is able to detect targets at the ranges of 10–20 km. The shortest detection range, in the order of a few kilometers, is provided by WiFi illumination.

2.6 SUMMARY

In the chapter, basic geometric relationships for passive radar were introduced. As the transmitter and the receiver in passive radar are separated, bistatic geometry applies, which leads to certain relationships different than in classical monostatic radar (e.g., concerning the range resolution). The range equation, one of the fundamental relationships in radar, has also its own form in passive radar. In addition, factors such as the azimuth and elevation patterns of the transmitter, or the direct path interference removal, modify the detection range of the passive radar, which was also analyzed.

The ambiguity function, which is used for the analysis of the radar waveform, was introduced. It was applied to an idealistic noise signal and then to signals from different illuminators of opportunity used by passive radar. The various types of transmitters were also compared with respect to the transmitted power, as well as the signal bandwidth. The analysis shows that certain types of transmitters can be used for long-range detection (e.g., FM or DVB-T), whereas others (e.g., WiFi or GSM) are suitable only for short-range applications.

The following chapters will focus on individual steps of signal processing applied in passive radar, from beamforming to Cartesian tracking.

References

[1] P. E. Howland, D. Maksimiuk, and G. Reitsma, "FM Radio Based Bistatic Radar," *IEE Proc. Radar, Sonar and Navigation*, vol. 152, pp. 107–115, June 2005.

[2] H. Griffiths and C. Baker, *An Introduction to Passive Radar*. Norwood, MA: Artech House, 2017.

[3] M. Malanowski and K. Kulpa, "Digital Beamforming for Passive Coherent Location Radar," in *Proc. RadarCon 2008*, (Rome, Italy), p. CD, May 2008.

[4] H. D. Griffiths and C. J. Baker, "Passive Coherent Location Radar Systems. Part 1: Performance Prediction," *IEE Proc. Radar, Sonar and Navigation*, vol. 152, pp. 153–159, June 2005.

[5] M. J. Paradie and J. Pernic, "Airborne VHF Environment Noise Measurements," in *Proceedings of the 1996 Tactical Communications Conference. Ensuring Joint Force Superiority in the Information Age*, pp. 389–395, April 1996.

[6] R. Dalke, R. Achatz, Y. Lo, P. Papazian, and G. Hufford, "Measurement and Analysis of Man-Made Noise in VHF and UHF Bands," in *Proceedings of 1997 Wireless Communications Conference*, pp. 229–233, August 1997.

[7] N. G. Riley and K. Docherty, "Modelling and Measurement of Man-Made Radio Noise in the VHF-UHF Bands," in *1995 Ninth International Conference on Antennas and Propagation, ICAP '95 (Conf. Publ. No. 407)*, vol. 2, pp. 313–316, April 1995.

[8] J. Zoellner, J. Robert, M. Slimani, P. Schlegel, and M. Pulsmeier, "Analysis of the Impact of Man-Made Noise on DVB-T and DVB-T2," in *IEEE International Symposium on Broadband Multimedia Systems and Broadcasting*, pp. 1–6, June 2012.

[9] E. N. Skomal, "Analysis of Airborne VHF/UHF Incidental Noise over Metropolitan Areas," *IEEE Trans. Electromagnetic Compatibility*, vol. EMC-11, pp. 76–83, May 1969.

[10] M. Malanowski, K. Kulpa, J. Kulpa, P. Samczyński, and J. Misiurewicz, "Analysis of Detection Range of FM-Based Passive Radar," *IET Radar, Sonar Navigation*, vol. 8, pp. 153–159, February 2014.

[11] T. Dabrowski, W. C. Barott, and B. Himed, "Effect of Propagation Model Fidelity on Passive Radar Performance Predictions," in *2015 IEEE Radar Conference (RadarCon)*, pp. 1503–1508, May 2015.

[12] W. C. Barott, K. M. Scott, and B. Himed, "Effects of Atmospheric Refractivity and Variability on Passive Radar Performance Prediction," in *2018 IEEE Radar Conference (RadarConf18)*, pp. 0987–0992, April 2018.

[13] M. Inggs, Y. Paichard, and G. Lange, "Passive Coherent Location System Planning Tool," in *2009 International Radar Conference "Surveillance for a Safer World" (RADAR 2009)*, pp. 1–5, October 2009.

[14] M. Edrich, "FM/DAB/DVB-T Multiband Multistatic Passive Radar System – Design Considerations and Lessons Learnt," in *Proc. EuRAD 2017*, October 2016.

[15] https://fmscan.org/, "fmscan.org," November 2018.

[16] https://www.uke.gov.pl/, "Urzad Komunikacji Elektronicznej (Office of Electronic Communications)," November 2018.

[17] D. W. O'Hagan, H. D. Griffiths, S. M. Ummenhofer, and S. T. Paine, "Elevation Pattern Analysis of Common Passive Bistatic Radar Illuminators of Opportunity," *IEEE Trans. Aerospace and Electronic Systems*, vol. 53, pp. 3008–3019, December 2017.

[18] https://www.kathrein.com, "Broadcast-Solutions, Professional Antennas and Combiners," April 2019.

[19] https://www.tesla.cz, "TESLA Hloubětín a.s., ANTENNA SYSTEMS," April 2019.

[20] K. Kulpa and M. Malanowski, "The Concept of Simple MIMO PCL Radar," in *Proc. European Radar Conference – EURAD*, (Amsterdam, The Netherlands), pp. 240–243, October 2008.

[21] P. Krysik, K. Kulpa, M. Malanowski, and P. Samczyński, "Reconstruction of GSM Reference Signals for PCL Processing," in *Proc. 3rd FHR Focus Days on PCL*, (Wachtberg, Germany), pp. 1–19, May 2011.

[22] M. Bączyk and M. Malanowski, "Reconstruction of the Reference Signal in DVB-T-Based Passive Radar," *International Journal of Electronics and Telecommunications, PAN*, vol. 57, no. 1, pp. 43–48, 2011.

[23] D. W. O'Hagan, M. Setsubi, and S. Paine, "Signal Reconstruction of DVB-T2 Signals in Passive Radar," in *2018 IEEE Radar Conference (RadarConf18)*, pp. 1111–1116, April 2018.

[24] D. Bok, "Reconstruction and Reciprocal Filter of OFDM Waveforms for DVB-T2 Based Passive Radar," in *2018 International Conference on Radar (RADAR)*, pp. 1–6, August 2018.

[25] W. Xianrong, W. Junfang, H. Sheng, and T. Hui, "Reconstruction of Reference Signal for DTMB-Based Passive Radar Systems," in *Proceedings of 2011 IEEE CIE International Conference on Radar*, vol. 1, pp. 165–168, October 2011.

[26] P. M. Woodward, *Probability and Information Theory with Applications to Radar*. Dedham, MA: Artech House, second ed., 1980.

[27] K. Kulpa, *Signal Processing in Noise Waveform Radar*. Norwood, MA: Artech House, 2013.

[28] "Transmission Standards for FM Sound Broadcasting at VHF," Tech. Rep. RECOMMENDATION ITU-R BS.450-3, International Telecommunication Union, 2001.

[29] A. D. Lallo, A. Farina, R. Fulcoli, P. Genovesi, R. Lalli, and R. Mancinelli, "Design, Development and Test on Real Data of an FM Based Prototypical Passive Radar," in *Proc. RadarCon 2008*, (Rome, Italy), p. CD, May 2008.

[30] M. Malanowski, G. Mazurek, K. Kulpa, and J. Misiurewicz, "FM Based PCL Radar Demonstrator," in *Proc. International Radar Symposium 2007*, (Cologne, Germany), pp. 431–435, September 2007.

[31] B. D. Nordwall, " 'Silent Sentry' A New Type of Radar," *Aviation Week and Space Technology*, vol. 149, pp. 70–71, November 1998.

[32] S. Li, H. Wan, H. Yin, and Y. Huang, "Detection of Moving Target in FM Broadcast-Based Passive Radar," in *Proc. Int. Conf. on Wireless Communications, Networking and Mobile Computing – WiCom 2007*, (Shanghai, China), pp. 649–652, September 2007.

[33] M. Meyer and J. D. Sahr, "Passive Coherent Scatter Radar Interferometer: Implementation, Observations, and Analysis," *Radio Science*, vol. 39, pp. 1–10, May 2004.

[34] F. D. Lind, J. D. Sahr, and D. M. Gidner, "First Passive Radar Observations of Auroral E-region Irregularities," *Geophysical Research Letters*, vol. 26 (14), pp. 2155–2158, July 1999.

[35] C. J. Baker, H. D. Griffiths, and I. Papoutsis, "Passive Coherent Location Radar Systems. Part 2: Waveform Properties," *IEE Proc. Radar, Sonar and Navigation*, vol. 152, pp. 160–168, June 2005.

[36] H. D. Griffiths, C. J. Baker, H. Ghaleb, R. Ramakrishnan, and E. Willman, "Measurement and Analysis of Ambiguity Functions of Off-Air Signals for Passive Coherent Location," *Electronics Letters*, vol. 39, pp. 1005–1007, June 2003.

[37] A. Lauri, F. Colone, R. Cardinali, C. Bongioanni, and P. Lombardo, "Analysis and Emulation of FM Radio Signals for Passive Radar," in *2007 IEEE Aerospace Conference*, pp. 1–10, March 2007.

[38] "Specification of the Radio Data System (RDS) for VHF/FM Sound Broadcasting in the Frequency Range from 87,5 to 108,0 MHz," Tech. Rep. EN 50067, European Committee for Electrotechnical Standardization, 1998.

[39] R. Saini and M. Cherniakov, "DTV Signal Ambiguity Function Analysis for Radar Application," *IEE Proc. Radar, Sonar and Navigation*, vol. 152, pp. 133–142, June 2005.

[40] G. Zhiwen, T. Ran, M. Yongfeng, and S. Tao, "DVB-T Signal Cross-Ambiguity Functions Improvement for Passive Radar," in *Proc. IEEE CIE International Conference on Radar*, (Shanghai, China), pp. 1–4, October 2006.

[41] M. Daun and W. Koch, "Multistatic Target Tracking for Non-Cooperative Illumination by DAB/DVB-T," in *Proc. IEEE Radar Conference*, (Rome, Italy), pp. 1–6, May 2008.

[42] D. Poullin, M. Flécheux, and M. Klein, "3D Location of Opportunistic Targets Using DVB-SFN Network: Experimental Results," in *3rd FHR Focus Days on PCL*, (Wachtberg, Germany), p. CD, May 2011.

[43] H. Kuschel, M. Glende, J. Heckenbach, S. Müller, J. Schell, and C. Schumacher, "Experimental Passive Radar Systems Using Digital Illuminators (DAB/DVB)," in *Proc. International Radar Symposium 2007*, (Cologne, Germany), pp. 411–417, September 2007.

[44] D. Poullin and M. Flécheux, "Recent Progress in Passive Coherent Location (PCL) Concepts and Techniques in France," in *Proc. RadarCon 2008*, (Rome, Italy), p. CD, May 2008.

[45] A. Schroeder and M. Edrich, "CASSIDIAN Multiband Mobile Passive Radar System," in *Proc. International Radar Symposium 2010*, (Vilnius, Lithuania), pp. 286–291, June 2010.

[46] "Digital Video Broadcasting (DVB); Framing Structure, Channel Coding and Modulation for Digital Terrestrial Television," tech. rep., European Telecommunications Standards Institute (ETSI), 2009.

[47] K. Pölönen and V. Koivunen, "Detection of DVB-T2 Control Symbols in Passive Radar Systems," in *2012 IEEE 7th Sensor Array and Multichannel Signal Processing Workshop (SAM)*, pp. 309–312, June 2012.

[48] D. A. Kovalev and V. I. Veremyev, "Correction of DVB-T2 Signal Cross-Ambiguity Function for Passive Radar," in *2014 International Radar Conference*, pp. 1–4, October 2014.

[49] G. Cui, H. Li, and B. Himed, "A Correlation-Based Signal Detection Algorithm in Passive Radar with DVB-T2 Emitter," in *2014 48th Asilomar Conference on Signals, Systems and Computers*, pp. 1418–1422, November 2014.

[50] E. Vorobev, A. Barkhatov, and V. Kutuzov, "DVB-T2 Passive Coherent Location Radar," in *2016 IEEE NW Russia Young Researchers in Electrical and Electronic Engineering Conference (EIConRusNW)*, pp. 470–474, February 2016.

[51] A. Barkhatov, E. Vorobev, and A. Konovalov, "Experimental Results of DVB-T2 Passive Coherent Location Radar," in *2017 IEEE Conference of Russian Young Researchers in Electrical and Electronic Engineering (EIConRus)*, pp. 1229–1232, February 2017.

[52] J. Pidanic and K. Juryca, "Analysis of DVB-T2 Signal for Exploitation by Passive Coherent Location System," in *2017 27th International Conference Radioelektronika (RADIOELEKTRONIKA)*, pp. 1–4, April 2017.

[53] V. Winkler, C. Klöck, and M. Edrich, "Migration to the DVB-T2-Standard for Passive Radar," in *2017 18th International Radar Symposium (IRS)*, pp. 1–10, June 2017.

[54] E. Vorobev, A. Barkhatov, V. Veremyev, and V. Kutuzov, "DVB-T2 Passive Radar Developed at Saint Petersburg Electrotechnical University," in *2018 22nd International Microwave and Radar Conference (MIKON)*, pp. 204–207, May 2018.

[55] E. N. Vorobev, "Helicopter Recognition in DVB-T2 Passive Bistatic Radar," in *2018 IEEE Conference of Russian Young Researchers in Electrical and Electronic Engineering (EIConRus)*, pp. 1171–1173, January 2018.

[56] E. Vorobev, V. Veremyev, and V. Kutuzov, "Recognition of Propeller-Driven Aerial Targets in DVB-T2 Passive Bistatic Radar," in *2018 19th International Radar Symposium (IRS)*, pp. 1–8, June 2018.

[57] S. Paine, F. Schonken, M. Malape, D. W. O'Hagan, J. Swart, F. Louw, and M. Setsubi, "Multi Band FM and DVB-T2 Passive Radar Demonstrator," in *2018 19th International Radar Symposium (IRS)*, pp. 1–10, June 2018.

[58] "A/53: ATSC Digital Television Standard, Parts 1–6," tech. rep., Advanced Television Systems Committee, Inc., 2007.

[59] W. C. Barott and J. Engle, "Single-Antenna ATSC Passive Radar Observations with Remodulation and Keystone Formatting," in *2014 IEEE Radar Conference*, pp. 0159–0163, May 2014.

[60] W. C. Barott and B. Himed, "Cochannel Interference in ATSC Passive Radar," in *2015 IEEE Radar Conference (RadarCon)*, pp. 1270–1275, May 2015.

[61] K. M. Scott, W. C. Barott, and B. Himed, "Illuminator Selection Statistics Using ATSC Passive Radar with a Mobile Receiver," in *2018 IEEE Radar Conference (RadarConf18)*, pp. 0981–0986, April 2018.

[62] "Transmission System for Digital Terrestrial Television Broadcasting, ARIB STD-B31 Version 1.6," tech. rep., Association of Radio Industries and Businesses, 2005.

[63] K. Suwa, S. Nakamura, S. Morita, T. Wakayama, H. Maniwa, T. Oshima, R. Maekawa, S. Matsuda, and T. Tachihara, "ISAR Imaging of an Aircraft Target Using ISDB-T Digital TV Based Passive Bistatic Radar," in *2010 IEEE International Geoscience and Remote Sensing Symposium*, pp. 4103–4105, July 2010.

[64] J. Honda and T. Otsuyama, "Feasibility Study on Aircraft Positioning by Using ISDB-T Signal Delay," *IEEE Antennas and Wireless Propagation Letters*, vol. 15, pp. 1787–1790, 2016.

[65] W. Feng, J. Friedt, G. Cherniak, and M. Sato, "Novel Algorithm for High Resolution Passive Radar Imaging with ISDB-T Digital TV Signal," in *IGARSS 2018 – 2018 IEEE International Geoscience and Remote Sensing Symposium*, pp. 6687–6690, July 2018.

[66] W. Feng, J. Friedt, G. Cherniak, and M. Sato, "Passive Radar Imaging by Filling Gaps Between ISDB Digital TV Channels," *IEEE Journal of Selected Topics in Applied Earth Observations and Remote Sensing*, pp. 1–14, 2019.

[67] "Framing Structure, Channel Coding and Modulation for Digital Television Terrestrial Broadcasting System, Chinese National Standard GB 20600–2006," tech. rep., The Standardization Administration of the People's Republic of China, 2006.

[68] X. Wan, Y. Cheng, H. Tang, and Y. Liu, "The Effects of Non-ideal Factors on Clutter Cancellation in DTMB-Based Passive Radar," in *2017 IEEE Radar Conference (RadarConf)*, pp. 1440–1445, May 2017.

[69] H. Tang, X. Wan, L. Hong, W. Chen, and J. Yi, "Detection Improvement by Modified Modulation Error Rate of Reference Signal in Passive Radar," in *2014 XXXIth URSI General Assembly and Scientific Symposium (URSI GASS)*, pp. 1–4, August 2014.

[70] M. Lu, J. Yi, X. Wan, and W. Zhan, "Co-Channel Interference in DTMB-Based Passive Radar," *IEEE Trans. Aerospace and Electronic Systems*, pp. 1–11, 2018 (Early Access).

[71] "Radio Broadcasting Systems; Digital Audio Broadcasting (DAB) to Mobile, Portable and Fixed Receivers," tech. rep., European Telecommunications Standards Institute (ETSI), 2016.

[72] C. Coleman and H. Yardley, "Passive Bistatic Radar Based on Target Illuminations by Digital Audio Broadcasting," *IEE Proc. Radar, Sonar and Navigation*, vol. 2, pp. 366–375, October 2008.

[73] D. W. O'Hagan, H. Kuschel, J. Heckenbach, M. Ummenhofer, and J. Schell, "Characteristics and Properties of DAB Based Passive Radar," in *Proc. 2nd FHR Focus Days on PCL*, (Wachtberg, Germany), p. CD, November 2009.

[74] D. Gould, R. Pollard, C. Sarno, and P. Tittenso, "A Multiband Passive Radar Demonstrator," in *Proc. International Radar Symposium 2006*, (Cracow, Poland), pp. 657–660, May 2006.

[75] D. K. P. Tan, H. Sun, Y. Lu, M. Lesturgie, and H. L. Chan, "Passive Radar Using Global System for Mobile Communication Signal: Theory, Implementation and Measurements," *IEE Proc. Radar, Sonar and Navigation*, vol. 152, pp. 116–123, June 2005.

[76] P. Samczyński, K. Kulpa, M. Malanowski, P. Krysik, and L. Maślikowski, "A Concept of GSM-Based Passive Radar for Vehicle Traffic Monitoring," in *Proc. Microwaves, Radar and Remote Sensing Symposium – MRRS*, (Kiev, Ukraine), pp. 271–274, August 2011.

[77] U. Nickel, "System Considerations for Passive Radar with GSM Illuminators," in *IEEE International Symposium on Phased Array Systems and Technology – ARRAY*, (Waltham, Massachusetts, USA), pp. 189–195, October 2010.

[78] D. Tan, S. Hongbo, and L. Yilong, "Sea and Air Moving Target Measurements Using a GSM Based Passive Radar," in *Proc. IEEE Radar Conference*, (Arlington, Virginia, USA), pp. 783–786, May 2005.

[79] F. Colone, P. Falcone, and P. Lombardo, "Passive Bistatic Radar Based on Mixed DSSS and OFDM WiFi Transmissions," in *Proc. European Radar Conference – EURAD*, (Manchester, UK), pp. 154–157, October 2011.

[80] T. Martelli, F. Colone, and P. Lombardo, "First Experimental Results for a WiFi-Based Passive Forward Scatter Radar," in *2016 IEEE Radar Conference (RadarConf)*, pp. 1–6, May 2016.

[81] M. Stentella, A. Losito, T. Martelli, and F. Colone, "Stand-alone WiFi-Based Passive Forward Scatter Radar Sensor for Vehicles Classification," in *International Conference on Radar Systems (Radar 2017)*, pp. 1–6, October 2017.

[82] T. Martelli, F. Murgia, F. Colone, C. Bongioanni, and P. Lombardo, "Detection and 3D Localization of Ultralight Aircrafts and Drones with a WiFi-Based Passive Radar," in *International Conference on Radar Systems (Radar 2017)*, pp. 1–6, October 2017.

[83] S. Rzewuski, M. Wielgo, K. Kulpa, M. Malanowski, and J. Kulpa, "Multistatic Passive Radar Based on WIFI – Results of the Experiment," in *2013 International Conference on Radar*, pp. 230–234, September 2013.

Chapter 3

Beamforming

3.1 INTRODUCTION

In this chapter, beamforming for passive radar is considered [1]. The use of beamforming is one of the approaches to obtain antenna patterns providing the required spatial separation of the received signals. The two most popular types of antenna arrays, namely the uniform linear array and uniform circular array, are analyzed and compared.

An antenna in a passive radar, as in any radar or wireless communications system, is of fundamental importance. One of the main tasks of the antenna system in a passive radar is to provide spatial separation between the reference and echo signals. For this reason, a narrow main beam and low sidelobes are desired. However, the VHF and UHF bands, where passive radars usually operate, limit the possibilities of constructing antennas (or antenna arrays) with such properties due to long wavelength and thus large antenna size requirements. In some cases, such as when DVB-T signals are used, the array also has to operate in a wide band, which is a serious challenge.

There are two general approaches to constructing an antenna system for a passive radar. In the first approach, at least two directional, fixed antennas are used [2, 3], as shown in Figure 3.1(a). One of these antennas points towards the transmitter of opportunity, and it is used for the reception of the reference signal. The second antenna is directed towards the area of interest, in order to receive the target-echo signal. This approach is not very versatile; the antenna setup is matched to a specific scenario with a single transmitter and a particular area of surveillance. The use of multiple transmitters or omnidirectional coverage is very difficult to obtain is such a configuration.

(a)

(b)

Figure 3.1 Antenna systems for passive radar. (a) Two fixed directional antennas. (b) Antenna array with beamforming.

An alternative approach is to use an antenna array and the beamforming technique [4–9], as shown in Figure 3.1(b). In this case, multiple beams can be created and pointed towards different transmitters or surveillance areas. This approach is more complicated than using two directional antennas, as more coherent receiving channels are required, and a beamforming algorithm has to be applied. These difficulties, however, are compensated by the advantages provided by this solution (e.g., versatility and beam steering capability).

A comparison of two approaches to designing an antenna system for a passive radar is shown in Figure 3.2. The top plot shows a passive radar equipped with two fixed directional antennas, where one antenna is directed towards the transmitter

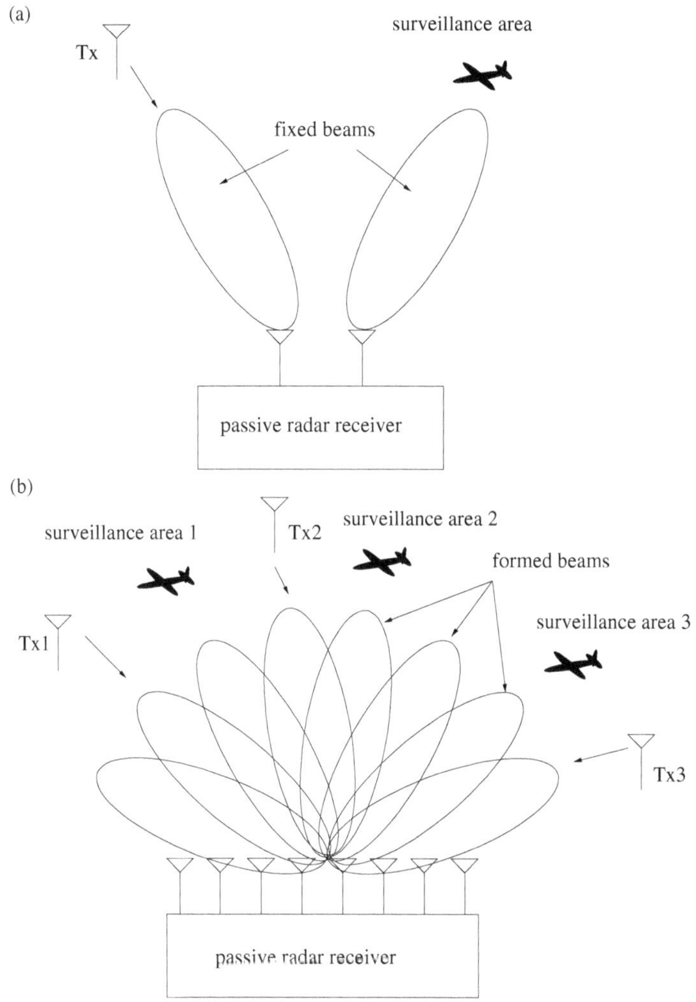

Figure 3.2 Concept of passive radar using (a) two fixed directional antennas and (b) multichannel receiver with beamforming.

and the other points to the surveillance area where the targets are expected. If using another transmitter from a different direction was required in such an approach, it would entail using an additional antenna. A multichannel receiver and beamforming are presented in Figure 3.2(b). The beams are created by appropriately summing the signals from all channels. In this case, different beams can be directed towards spatially distributed transmitters, as well as surveillance areas.

Beamforming can be realized either in the analog or digital domain. In the first case, the signals from individual array elements are weighted, phase-shifted, and summed using analog circuits. The signals corresponding to the formed beams are then sampled. Such an approach has been applied in [10, 11] for the suppression of the direct path interference in the echo channel. The limitations of the analog beamforming result from the typical limitations of analog circuits: inaccuracies of the parameters (differences in the gain or phase shift), lack of repeatability, and dependency of parameters on the temperature.

An alternative approach is digital beamforming (i.e., signals from individual elements of the antenna array are sampled, and beams are created in the digital domain by summing the received signals with appropriate weights). In this way, the beamforming coefficients, thus the beam steering, can be controlled in the software. Applying the beamforming coefficients in the digital domain provides much better flexibility, versatility, accuracy and repeatability than in analog beamforming. For this reason, virtually all modern passive radar systems use the digital version of beamforming. Therefore, the digital beamforming is also the focus of this chapter; however, some of the material is also applicable to analog beamforming.

The classical configuration of the antenna array is the uniform linear array (ULA) [12], whose elements are placed along a line with uniform spacing. Beam steering is obtained by choosing appropriate phase shifts for consecutive elements of the array. One of the properties of the ULA is that the beamwidth increases with the off-boresight angle. This may be a serious disadvantage in passive radar, where the ability to steer a beam with similar properties in all directions is often required. This ability is important because often multiple transmitters of opportunity are used simultaneously, which requires forming reference beams in many directions. Moreover, in order to enable omnidirectional target detection coverage, echo beams should be formed in all directions. For this reason, the ULA is used only in specific systems, where only a certain angle sector is of interest (e.g., DELIA system developed at Fraunhofer FHR [13]).

One of the alternative approaches to the ULA is the uniform circular array (UCA). In a UCA the array elements are placed in a circle, with uniform angular spacing. This configuration seems to be the most preferred in passive radar. This approach was used, for example, in CORA (Germany) [4], Silent Sentry (USA) [5], HA-100 (France) [6], PARADE (Germany) [7], AULOS (Italy) [14], system developed at TNO (The Netherlands) [8], *PaRaDe*[1] (Poland) [9], and PCL-PET system (Poland) [15]. The circular array provides omnidirectional coverage capabilities. Moreover, beams can be formed practically independently of their direction with respect to the array.

Structures other than the ULA and UCA can also be used. One of the approaches is to use an irregular structure to obtain good performance in a wide band. Another approach is to use a MIMO (multiple-input, multiple-output) approach, which can increase the angular resolution and system coverage [16].

In this chapter the ULA and UCA will be investigated. The beamforming will be analyzed from the signal processing point of view rather than the electromagnetic theory point of view.

3.2 UNIFORM LINEAR ARRAY (ULA)

Uniform linear array (ULA) is one of the simplest forms of antenna array and also very often used in many applications. It consists of a number of radiating[2] elements arranged in a straight line, spaced uniformly. An example of the geometry of a ULA for $N = 4$ elements is shown in Figure 3.3. The spacing between the elements is denoted with d. Usually this distance is expressed in wavelengths, as the relationship between d and wavelength λ influences the properties of the array rather than the absolute value of element spacing. The total length of the array is equal to $L = (N-1)d$. The angle between the wave direction and a normal to the array is denoted with ϕ.

In the analysis below, narrowband signals are assumed. This means that the signal bandwidth is small compared to the carrier frequency. In such a situation, the signal resembles a sine wave, and the delay of the signal can be represented by the phase shift. If the plane wave impinges on the array parallel ($\phi = 0°$), the phase of the signal in each element is the same, as the wave reaches each element in the same moment. If the wave incidence angle ϕ is different than zero, a certain phase shift between signals from different elements will be present. The phase shift can be calculated by considering the array geometry shown in Figure 3.3. The range difference for a wave between two elements separated by d can be calculated as

1 Systems named "Parade" have been developed at the Warsaw University of Technology and Cassidian. In this book the name "*PaRaDe*" is used for the system built at the Warsaw University of Technology and, "PARADE" for the system built at Cassidian.

2 In passive radar the elements do not radiate. However, as antennas behave in the same way when receiving and transmitting signals, often naming is used referring to transmission when describing reception.

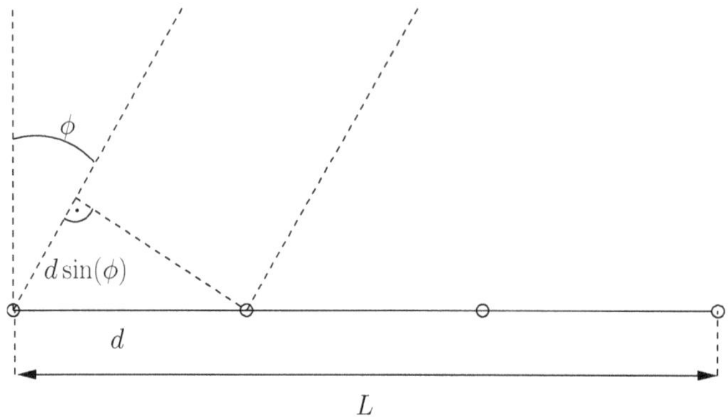

Figure 3.3 Geometry of a uniform linear array with 4 elements.

$d\sin(\phi)$. This distance results in a phase shift equal to $\frac{2\pi}{\lambda}d\sin(\phi)$. Using this result, the phase shift to the nth element with respect to the first element can be calculated as:

$$\alpha_n = \frac{2\pi}{\lambda}d(n-1)\sin(\phi) \tag{3.1}$$

The phase shifts (3.1) can be used to calculate the *steering vector* $\mathbf{a}(\phi) = [a_1(\phi), a_2(\phi), \ldots, a_N(\phi)]'$ ($'$ denotes matrix transpose):

$$a_n(\phi) = e^{j\alpha_n} = e^{j\frac{2\pi d}{\lambda}(n-1)\sin(\phi)} \tag{3.2}$$

The steering vector represents phase shifts for individual elements that a plane wave experiences for a given direction ϕ.

Initially it will be assumed that the individual elements of the array radiate isotropically. In such a case, the radiation pattern of the array depends only on the array geometry and applied beamforming coefficient. Beamforming is realized by summing the signals from individual elements with certain weights, w_n. The array pattern, or *array factor*, can be calculated as:

$$F(\phi, \mathbf{w}) = \sum_{n=1}^{N} w_n^* a_n(\phi) = \mathbf{w}^H \mathbf{a}(\phi), \tag{3.3}$$

where $\mathbf{w} = [w_1, w_2, \ldots, w_N]'$ is the vector of complex beamforming coefficients and ^{H} denotes Hermitian transpose (conjugate transpose).

In the simplest approach, the amplitudes of beamforming coefficients $|w_n|$ are the same and equal to 1. In order to steer the beam in a particular direction

ϕ_0, a phase shift is introduced to the consecutive beamforming coefficients. The shift is calculated in such a way as to compensate for the phase shifts of a wave coming from a certain direction ϕ_0. In this way, the amplitude of the wave from this particular direction is summed coherently, thus maximized. The beamforming coefficients vector calculated with this approach can be expressed as:

$$w_n = \mathrm{e}^{\mathrm{j}\frac{2\pi d}{\lambda}(n-1)\sin(\phi_0)} \tag{3.4}$$

Consider an example for $\phi_0 = 0°$, meaning beam pointing perpendicular to the array. In such a case, $w_n = 1$ for $n = 1, \ldots, N$, and the array radiation pattern can be calculated using the following sum:

$$F(\phi, \mathbf{w}) = \sum_{n=1}^{N} w_n^* a_n(\phi) = \sum_{n=1}^{N} \mathrm{e}^{\mathrm{j}\frac{2\pi d}{\lambda}(n-1)\sin(\phi)} \tag{3.5}$$

By calculating the sum of geometric series and using simple manipulations, the radiation pattern can be expressed as:

$$F(\phi, \mathbf{w}) = \frac{\sin\left(N\frac{\pi d}{\lambda}\sin(\phi)\right)}{\sin\left(\frac{\pi d}{\lambda}\sin(\phi)\right)} \mathrm{e}^{\mathrm{j}\frac{\pi d}{\lambda}(N-1)} \tag{3.6}$$

The absolute value of the radiation pattern can be calculated as:

$$|F(\phi, \mathbf{w})| = \left|\frac{\sin\left(N\frac{\pi d}{\lambda}\sin(\phi)\right)}{\sin\left(\frac{\pi d}{\lambda}\sin(\phi)\right)}\right| \tag{3.7}$$

For angles $\phi \rightarrow 0°$ the value of the radiation pattern $|F(\phi, \mathbf{w})| \rightarrow N$. This is a logical result, as signals from N elements are summed coherently.

An example of array radiation patterns for $N = 4$, 8, and 16 elements and $d = \lambda/2$ element spacing is shown in Figure 3.4. All patterns show a similar structure. There is a main beam pointing at $\phi_0 = 0°$ (as required) surrounded by sidelobes. There is also a symmetrical beam, located at 180°. This means that the array radiates in the same way to the front and back. Usually this property is undesirable, and is eliminated by applying radiating elements with a directional pattern. The plots show that a larger number of elements leads to a narrower beam. This is one of the fundamental properties of antennas; a larger antenna allows for the focusing of power in a narrower beam. The 3-dB beamwidth (in radians) can be approximated as $\lambda/L = \lambda/((N-1)d)$. The sidelobe level is -11.3 dB. -12.8 dB and -13.2 dB

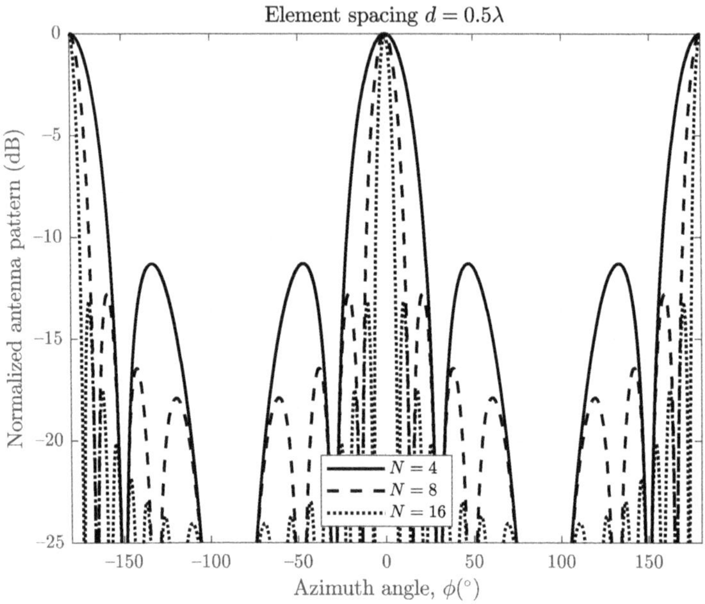

Figure 3.4 ULA radiation pattern for 4, 8, and 16 elements.

for $N = 4, 8, 16$, respectively. For a larger number of array elements N, the sidelobe level is saturated at approximately -13 dB. This value is often not satisfactory, as it does not allow out-of-the-main-beam signals to be attenuated enough. In order to reduce the sidelobe level, tapering of the beamforming coefficients is used, which is discussed in further detail later in this chapter.

The dependency of the radiation pattern on the element spacing is illustrated in Figure 3.5. The plots show patterns for $d = 0.25\lambda, 0.5\lambda, 1.0\lambda$. In all cases, the same number of elements $N = 8$ was used. Larger element spacing corresponds to a narrower pattern because the length of the array ($L = (N - 1)d$) increases. Too large element spacing can, however, lead to the undesirable effect of *grating lobes*. These are visible for $d = 1.00\lambda$ for angles $\pm 90°$. Their height is equal to the height of the main beams. The grating lobes are the result of too sparse sampling of space, which corresponds to the aliasing effect, known from the Nyquist sampling theory. In order to avoid the aliasing effect, the maximum element spacing should not exceed 0.5λ. This corresponds to the well-known condition for unambiguous sampling in the time domain ($0.5 f_s \geq f_{\max}$, where f_s is the sampling frequency, and $f_{\max}$ is the maximum frequency in the signal spectrum).

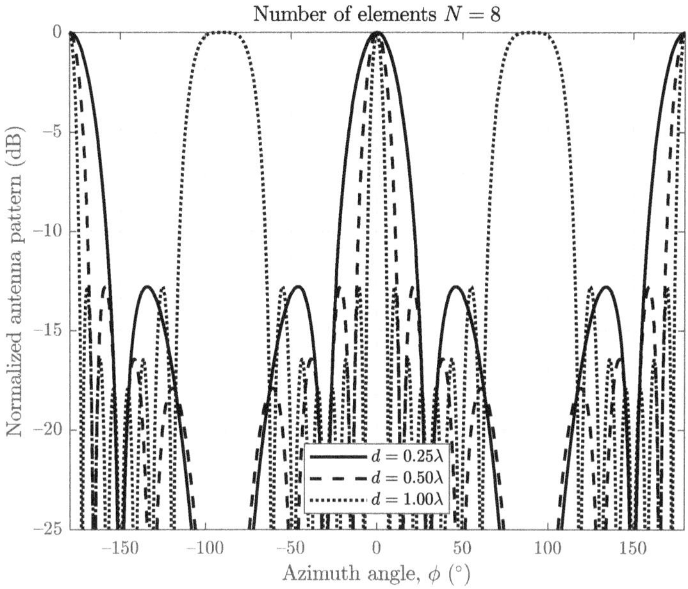

Figure 3.5 Radiation pattern for different element spacings of ULA.

From the results presented so far, it is clear that a longer array provides a narrower main beam, but the grating lobes can appear when the array is too sparse. These results are summarized in Figure 3.6. The solid curve shows a 3-dB beamwidth (the values are marked on the left vertical axis). The dashed curve indicates the sidelobe level (the values are marked on the right vertical axis). The horizontal axis shows the spacing between the elements. Four plots are created for 8, 12, 16, and 32 array elements. An increase in the element spacing results in a reduction in the beamwidth. The sidelobes are at a constant level, equal to approximately -13 dB for lower element spacing values. When the element spacing reaches λ, the grating lobes increase to the level of 0 dB, equal to the main beam level.

As mentioned before, the beam steering in the ULA is obtained by introducing a phase shift to the beamforming coefficients (see (3.4)). Let us consider an example of the beam steering by investigating radiation patterns shown in Figure 3.7. In the first case, the beam is directed at $\phi_0 = 0°$. As a result, peaks at 0° and 180° appear. When the beam is pointed to 30°, the beams at 30° and $180° - 30° = 150°$ are created. In the last case, the beam is pointed to 60°; thus, peaks appear at 60° and $180° - 60° = 120°$. For larger steering angles ϕ_0, the widening of the main beam can be observed. Moreover, when the beam angle approaches $\pm 90°$, the beam in the

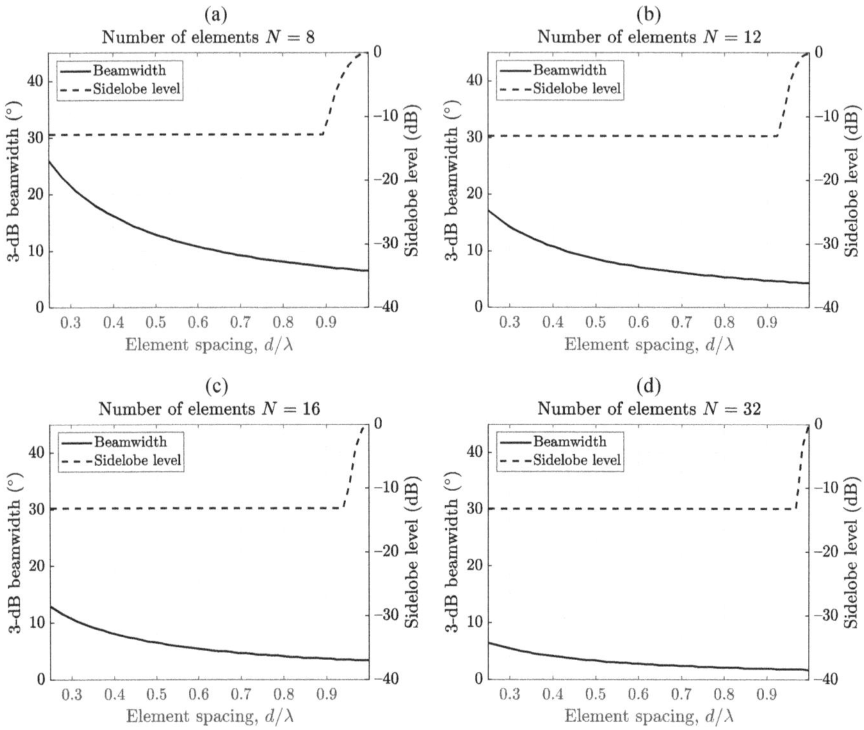

Figure 3.6 Beamwidth and sidelobe level for: (a) 8 elements, (b) 12 elements, (c) 16 elements, and (d) 32 elements.

desired direction and its mirror reflection merge together. These results lead to the conclusion that pattern properties for a ULA, especially the beamwidth, depend on the beam steering angle, which limits its versatility in terms of the observation angle. The dependency of the beamwidth on the steering angle is investigated in the next example. In practice the ULA is used to cover a certain sector of angles, typically 90–120°. The mirrored beam is attenuated by using directional elements that look in the desired direction, which will be discussed in the following part of this chapter. Alternatively, absorbing material can be used at the back of the array.

Figure 3.8 shows the beamwidth for different angles of the main beam. As observed in the previous example, the higher beam steering angle results in a wider beam. The abrupt increase in beamwidth, observed for example for 70° for $N = 16$, results from the fact that two symmetrical beams are merged and treated as one wide beam.

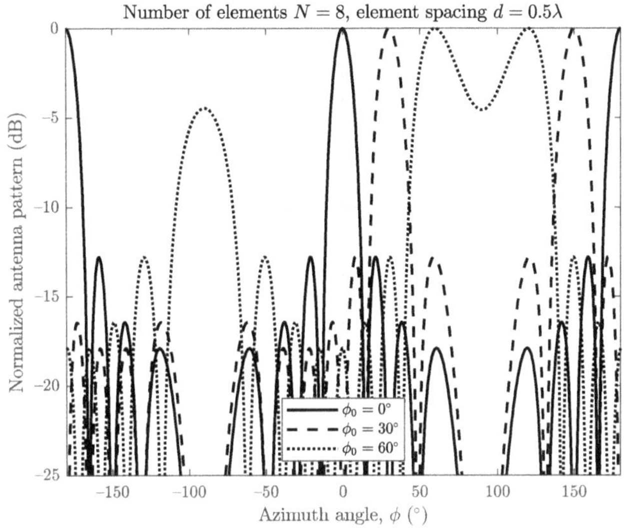

Figure 3.7 Radiation pattern of ULA for beam angles equal to 0°, 30°, and 60°.

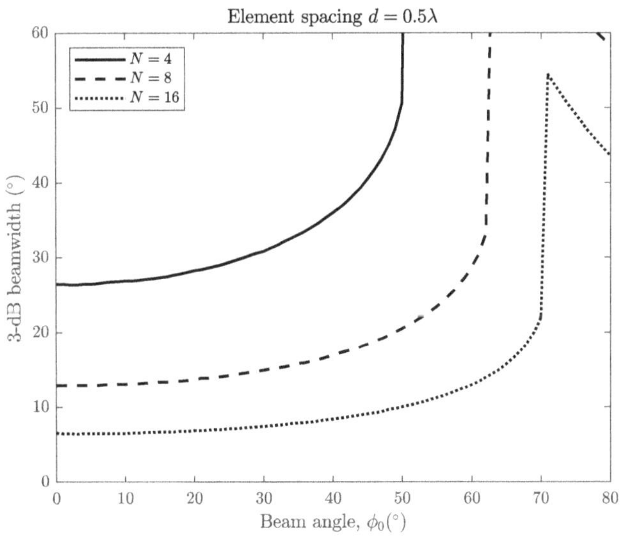

Figure 3.8 Beamwidth for different values of beam steering angles.

3.2.1 Array Tapering

The examples presented so far show that the sidelobe level is at the level of approximately -13 dB below the main beam for the uniform amplitudes $|w_n|$ of the beamforming weights. In order to reduce the sidelobes of the antenna pattern, tapering functions are used. This means that the amplitudes corresponding to individual elements are not equal. Instead, a certain window function is applied, which reduces the amplitudes at the sides of the array. This is the same approach that is used in spectral analysis; a rectangular window in the time domain results in high sidelobes in the spectral domain. Therefore, different windows are applied to the time signal, which reduces the sidelobes in the frequency domain.

Figures 3.9, 3.10, 3.11, and 3.12 show examples of different windows applied for tapering (rectangular as a reference, Taylor, Chebyshev, Gauss). Plots on the left side show the amplitudes of individual elements of the array for $N = 16$. In an untapered array (Figure 3.9), the signals from all elements are summed with the amplitude of 1, which provides the maximum amplitude gain of N. When the tapering is applied, signal amplitudes from certain elements (typically at the edges of the array) are attenuated. The consequence of this attenuation is a reduction in the array gain. Plots on the right side show the array pattern. The pattern values are normalized to the maximum of the untapered pattern. In each of the presented cases, the maximum value is reduced with respect to the uniform-illumination array. Another disadvantage of applying array weighting is widening of the main lobe. However, the desired effect of windowing (i.e., reduction of the sidelobes) can be clearly seen. In the case of the presented windowing functions, the sidelobe level is in the order of -30 dB. Depending on the window type, parameters can be changed in order to adjust the desired sidelobe level.

In Table 3.1 the characteristics of the analyzed patterns are summarized. The first row lists the parameters of the unweighted pattern as a reference. The main lobe widening factor is the ratio of the 3-dB beamwidth of the analyzed pattern to the width of the unweighted pattern. In the presented example the widening is in

Table 3.1
Parameters of Different Tapering Windows ($N = 16$)

Window Name	*Main Lobe Widening Factor*	*Amplitude Loss (dB)*	*SNR Loss (dB)*	*Sidelobe Level (dB)*
Rectangular	1.00	0.00	0.00	13.15
Taylor	1.27	−3.85	−0.69	30.00
Chebyshev	1.25	−3.70	−0.62	30.00
Gaussian	1.29	−4.12	−0.78	30.36

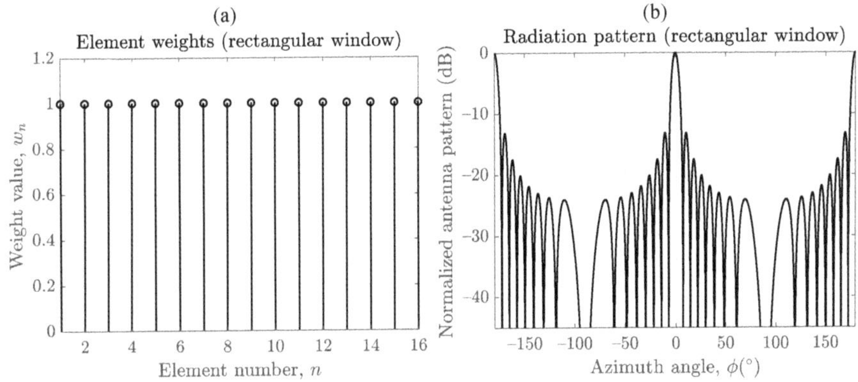

Figure 3.9 (a) Tapering window and (b) array pattern, rectangular window.

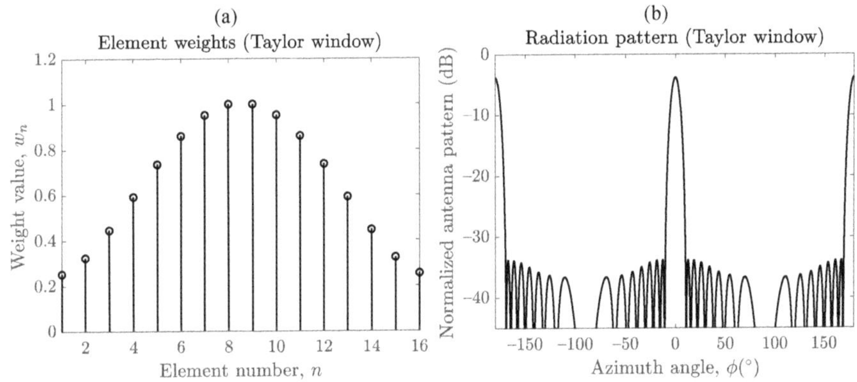

Figure 3.10 (a) Tapering window and (b) array pattern, Taylor window.

the range of 25–30%. The losses in the amplitude gain from the third column of the table are defined as the difference between the maximum value of the analyzed pattern and the maximum value of the reference (unweighted) pattern. As can be observed, array weighting can introduce significant gain losses, in the order of several decibels (approximately 4 dB in the presented case). However, in the case of digital beamforming, the weighting reduces not only the signal but the noise component as well. The parameter that is important in practice is the signal-to-noise ratio loss, which is shown in the third column of Table 3.1. As can be seen, signal-to-noise ratio loss below 1 dB can be expected. The last column shows the sidelobe level, defined as the difference (calculated in decibels) between the main beam amplitude and the maximum amplitude of the sidelobes. The sidelobe level for the

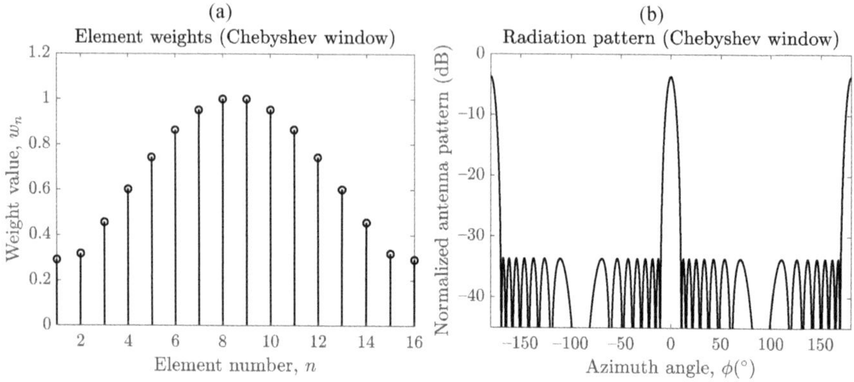

Figure 3.11 (a) Tapering window and (b) array pattern, Chebyshev window.

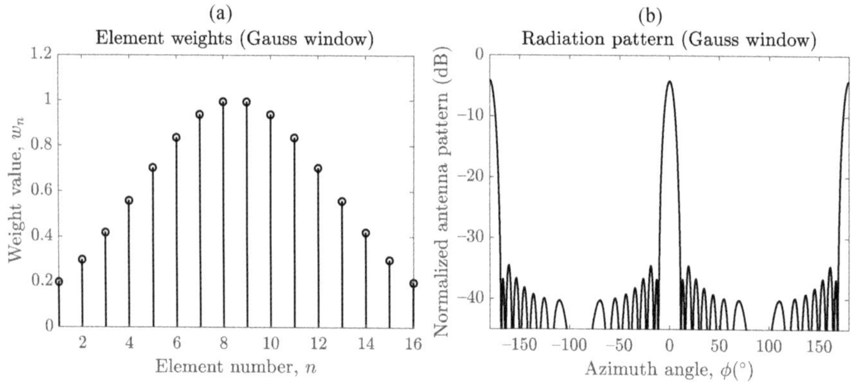

Figure 3.12 (a) Tapering window and (b) array pattern, Gauss window.

rectangular window is only -13 dB; the reason why array weighting is applied. In the case of the Taylor window, the sidelobe parameter was set to -30 dB, with 4 dB of sidelobe variation. The sidelobe level for the Chebyshev window was also set to -30 dB, and these same values were obtained. In case of the Gaussian window, the shape parameter was set to 1.8, which results in a sidelobe level close to 30 dB.

In conclusion, array tapering allows the sidelobe level to be substantially reduced; however, it involves two drawbacks. One of them is the widening of the main beam, which reduces azimuth resolution. The second one is the introduction of losses in the array gain and signal-to-noise ratio, which is the greatest disadvantage of using the tapering windows.

3.2.2 Directional Radiating Element

So far, the analysis has assumed that the array element radiates isotropically. Therefore, only the array factor, resulting from array geometry, was analyzed. In practice the effective antenna array pattern $F_{\text{ef}}(\phi)$ depends on the array factor $F(\phi)$ (radiation pattern of the array) and the radiation pattern of individual elements $F_{\text{el}}(\phi)$. In the case where each element has exactly the same radiation pattern, the effective pattern of the antenna array can be calculated as a product of the element and array factor patterns:

$$F_{\text{ef}}(\phi) = F(\phi) \cdot F_{\text{el}}(\phi) \tag{3.8}$$

In the case where the element pattern is directive (i.e., is looking in one direction), the undesirable symmetrical beam of the ULA can be eliminated. This is shown in Figure 3.13, where the element radiation pattern is assumed to be described with:

$$F_{\text{el}}(\phi) = \cos\left(\frac{\phi}{2}\right) \tag{3.9}$$

The radiation pattern of the element is shown on the left side of Figure 3.13. This pattern reflects a situation with a null in the backward direction. The effective radiation pattern of the array, composed of 8 elements with $\lambda/2$ spacing, is shown on the right side. As can be seen, the beam at $\pm 180°$ is eliminated with the radiation pattern of the single element (compare Figure 3.4).

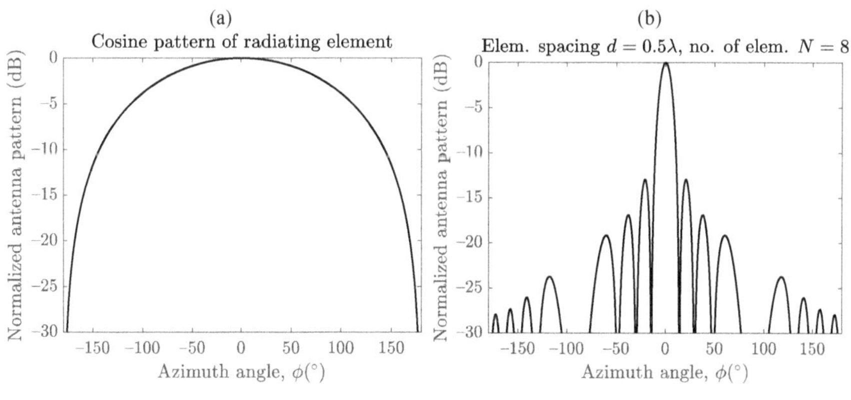

Figure 3.13 (a) Cosine-shaped radiation pattern of a single element and (b) effective radiation pattern of 8-element array with cosine-shaped element pattern.

3.3 UNIFORM CIRCULAR ARRAY (UCA)

Apart from ULAs, uniform circular arrays (UCA) are another common choice for antenna configuration for passive radars. Due to the radial symmetry of the UCA the dependency of beam shape on the beam steering angle is reduced in comparison to a ULA. However, the UCAs have their own drawbacks, such as high sidelobes or relatively wide main beam, as will be investigated in this section. The UCA will also be compared with the ULA taking into account the number of the elements as well as the overall size of the array.

The geometry of a UCA is shown in Figure 3.14 (for 8 elements). The angular spacing between the elements is: $\Delta\phi = 2\pi/N$, where N is the number of elements. The angle of the nth element of the array is $\phi_n = 2\pi n/N = \Delta\phi \cdot n$. The radius of the array is r. Assume that a signal is transmitted from an angle ϕ. The phase shift of the signal received by the nth element with respect to the center of the antenna array can be calculated from:

$$\alpha_n = \frac{2\pi}{\lambda} r \cos(\phi_n - \phi), \tag{3.10}$$

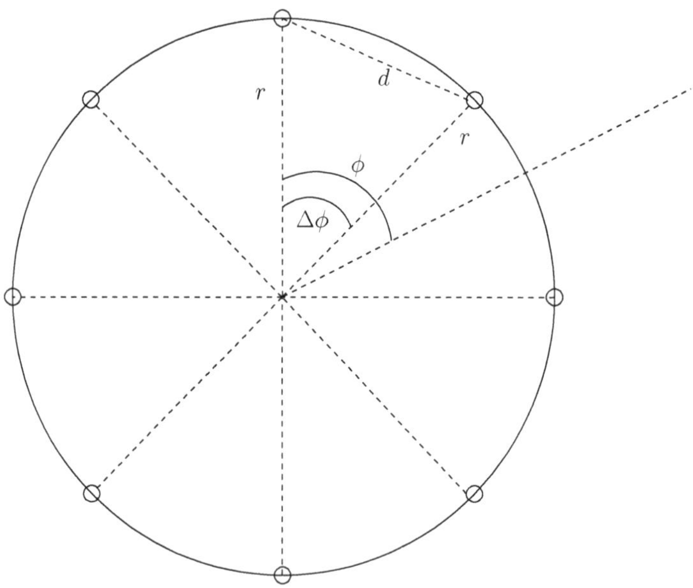

Figure 3.14 Geometry of a uniform circular array with 8 elements.

where λ is the wavelength. The steering vector for the circular array, denoted as $\mathbf{a}(\phi) = [a_1(\phi),\ a_2(\phi),\ \ldots,\ a_N(\phi)]'$ can thus be calculated as:

$$a_n(\phi) = \mathrm{e}^{\mathrm{j}\alpha_n} = \mathrm{e}^{\mathrm{j}\frac{2\pi}{\lambda} r \cos(\phi_n - \phi)}. \tag{3.11}$$

The azimuth antenna pattern in the UCA can be calculated in the same way as for the ULA:

$$F(\phi, \mathbf{w}) = \sum_{n=1}^{N} w_n^* a_n(\phi) = \mathbf{w}^H(\phi)\mathbf{a}, \tag{3.12}$$

where $\mathbf{w} = [w_1, w_2, \ldots, w_N]'$ is the vector of complex beamforming coefficients.

The beam steering in the desired direction ϕ_0 can be obtained by setting phases of the coefficients w_n so that the phases of the steering vector are compensated for in that direction. In a simplified case, amplitudes of the coefficients are unitary, that is $|w_n| = 1$. Therefore, the unweighted beamforming coefficients allowing the beam to be steered in the direction ϕ_0 have the following form:

$$w_n = \mathrm{e}^{\mathrm{j}\frac{2\pi}{\lambda} r \cos(\phi_n - \phi_0)}. \tag{3.13}$$

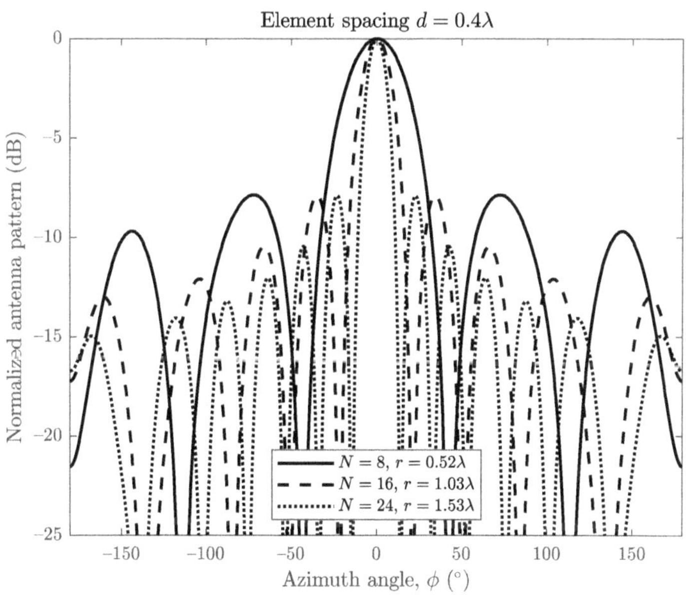

Figure 3.15 UCA radiation pattern for 8, 16, and 24 elements for a constant interelement spacing of 0.4λ.

Similar to the case of the ULA, the properties of the antenna pattern of the UCA depend on the spacing between the elements and their number. In the first example, the influence of the number of the array elements will be investigated. Figure 3.15 shows a UCA array pattern for a different number of elements. The spacing between elements d is constant and equal to 0.4λ. This corresponds to different array radius r, calculated according to:

$$r = \frac{d}{2\sin\left(\frac{\Delta\phi}{2}\right)} \tag{3.14}$$

The sidelobe level is equal to approximately -8 dB for all cases. The number of elements in the array influences the beamwidth. As could be expected, more elements result in a larger array, which in turn results in a narrower beam.

The influence of the array radius on the radiation pattern for a constant number of array elements ($N = 8$) is investigated in Figure 3.16. For the array radius equal to 0.4λ and 0.5λ, the sidelobe level is at the same level, equal to -8 dB. For a larger radius, equal to 0.5λ, grating lobes start to appear at $\pm 140°$.

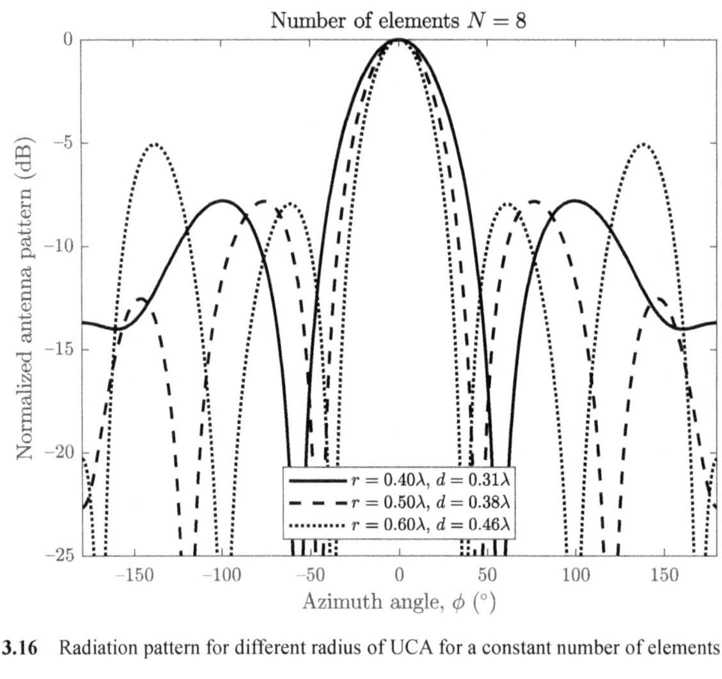

Figure 3.16 Radiation pattern for different radius of UCA for a constant number of elements $N = 8$.

The results presented above reveal an interesting property of the UCA. In contrast to the ULA, there is no second ("mirror") beam in the radiation pattern. This is a serious disadvantage of the ULA, which has to be eliminated with the directional patterns of the radiating element. In the case of the UCA, the array pattern itself eliminates undesirable backward radiation.

Now the beamwidth and sidelobe level versus element spacing will be analyzed. Figure 3.17 shows the results for $N = 8, 12, 16, 32$ elements. Similar to the case of the ULA analysis, the vertical axis on the left side (solid line) shows the beamwidth (measured at -3-dB points), whereas the right side axis (dashed line) shows the sidelobe level. An increase in the array radius leads to a decrease in the beamwidth. This confirms the results presented so far, as well following engineering intuition. For a certain range of values of the element spacing, the sidelobes are at the same level, equal approximately to -8 dB. When the element spacing reaches

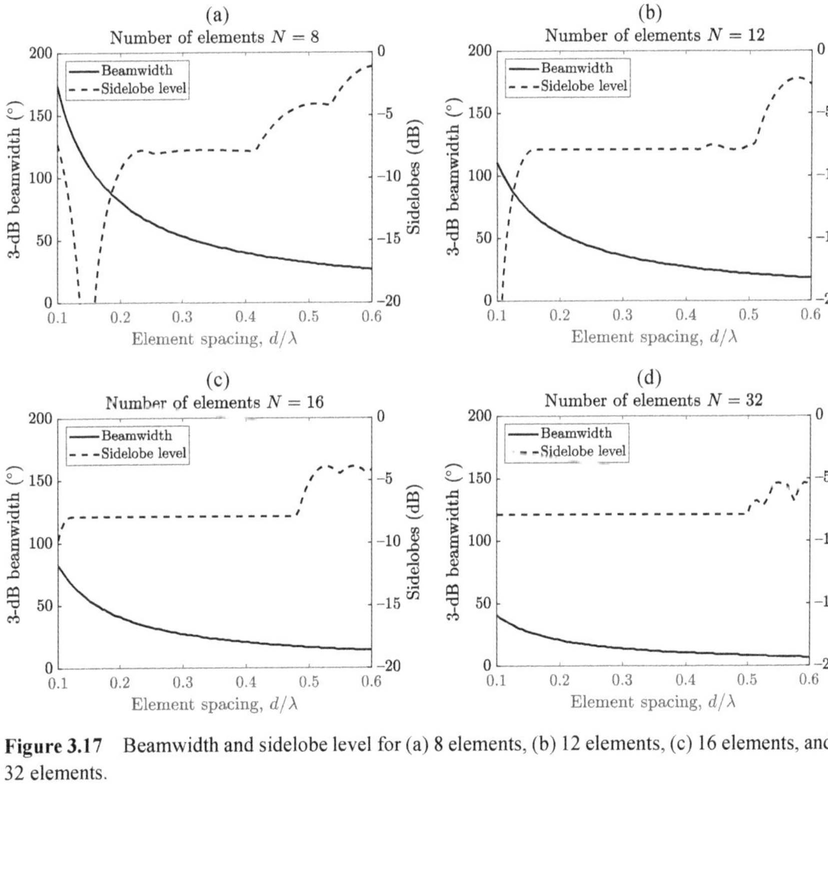

Figure 3.17 Beamwidth and sidelobe level for (a) 8 elements, (b) 12 elements, (c) 16 elements, and (d) 32 elements.

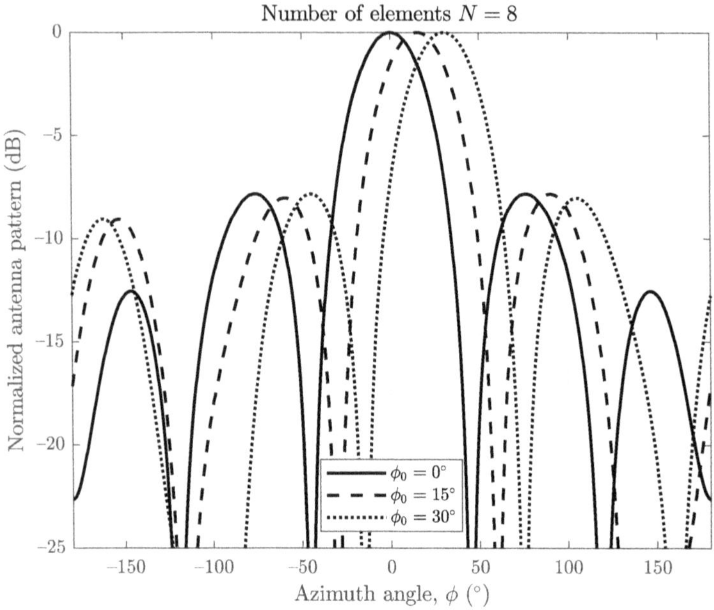

Figure 3.18 Radiation pattern of UCA for different beam steering angles.

approximately 0.5λ, the grating lobes appear. For some values of the element spacing, the sidelobe level reaches very low values, as the main beam is very wide, and no sidelobes are observed.

Due to the radial symmetry of the UCA, beam steering should influence the beam shape to a much lower degree than in the case of the ULA. This is illustrated with an example. The beam is steered to $0°$, $15°$, and $30°$ for an 8-element array, as shown in Figure 3.18. As can be seen, the beamwidth as well as the sidelobe level do not change significantly. Due to array symmetry, the situation will be the same every $2\pi/N$, which in the considered case of 8 elements corresponds to $45°$.

3.3.1 Array Tapering

Similar to the case of the ULA, array tapering can be applied to the UCA in order to reduce the sidelobe level. Under some assumptions, the results obtained for the ULA can be transformed so that they can be applied to the UCA [17–19]. Some numerical optimization methods for calculating the beamforming coefficients have also been proposed for the UCA [20, 21].

Here, an example of a different way of calculating beamforming coefficients is shown [22]. It is based on the numerical optimization of a specially defined cost function. The cost function is chosen so that the desired pattern is obtained. Such an approach ensures versatility, as different array geometries can be used (not only the ULA and UCA) and the design requirements can be changed in a wide range.

Assume that the actual antenna pattern obtained for the beamforming coefficients vector $\mathbf{w}$ is $F(\phi, \mathbf{w})$ and the desired pattern is $F_d(\phi)$. The cost function used in the proposed method is based on the following integral:

$$J_b(\mathbf{w}) = \int_{\phi} |F(\phi, \mathbf{w}) - F_d(\phi)|^2 \mathrm{d}\phi. \tag{3.15}$$

The desired pattern $F_d(\phi)$ is divided into three regions: main beam, sidelobes, and the transition region. The value of the $F_d(\phi)$ pattern in the main beam region is set to a constant value. The value of the pattern in the sidelobe region should be as low as possible (ideally zero). The transition region is neglected (the values of the calculated pattern $F(\phi, \mathbf{w})$ are ignored). The concept of the division of the desired pattern $F_d(\phi)$ into three regions is shown in Figure 3.19. The width of the main beam of the desired pattern should be chosen taking into consideration the physical properties of the array. If N array elements are used, the width of the beam that can be obtained is in the order of $2\pi/N$. The width of the transition region has to be chosen so that the values of the main beam of the pattern does not fall within the sidelobe region.

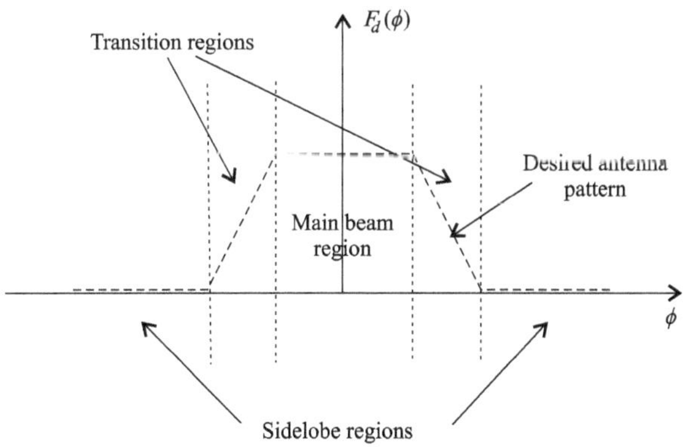

Figure 3.19 The concept of division of the desired antenna pattern into three regions.

The cost function can be rewritten using two integrals corresponding to the main beam (angles ϕ_{main}) and the sidelobes (angles $\phi_{\text{sidelobes}}$) as:

$$J_b(\mathbf{w}) = \int\limits_{\phi_{\text{main}}} |F(\phi,\mathbf{w}) - F_d(\phi)|^2 \mathrm{d}\phi + \int\limits_{\phi_{\text{sidelobes}}} |F(\phi,\mathbf{w}) - F_d(\phi)|^2 \mathrm{d}\phi. \quad (3.16)$$

To put more emphasis on the attenuation of the pattern in the sidelobe region, the second integral is multiplied by a weight w. The larger the value of w, the more the sidelobes will be attenuated. Additionally, taking into account that $F_d(\phi) \equiv 0$ for the sidelobe region (angles $\phi_{\text{sidelobes}}$), the cost function can be written as:

$$J_b(\mathbf{w}) = \int\limits_{\phi_{\text{main}}} |F(\phi,\mathbf{w}) - F_d(\phi)|^2 \mathrm{d}\phi + w \cdot \int\limits_{\phi_{\text{sidelobes}}} |F(\phi,\mathbf{w})|^2 \mathrm{d}\phi. \quad (3.17)$$

In practice, the integrals are substituted by appropriate sums of pattern values calculated for discrete angles ϕ. The minimization of the cost function is carried out numerically [23–26]. An unconstrained optimization procedure is applied with real and imaginary parts of the coefficients **w** tuned during the minimization. The constraints are imposed indirectly by the desired antenna pattern.

Let us consider an example of an 8-element UCA with the radius of $r = 0.36\lambda$, which corresponds to $d = 0.28\lambda$ of element spacing. The normalized radiation patterns created using the unweighted beamforming coefficients calculated according to (3.13) are shown in Figure 3.20 (the values are normalized to the maximum array gain, equal to $N = 8$, which corresponds to 9 dB). Different curves correspond to different steering angles ϕ_0 equal to $0°$, $11.25°$, and $22.5°$, respectively. The patterns for corresponding multiples of angle ranges $(0°, 22.5°)$ will be equivalent, due to the symmetry of the array. As can be seen, the patterns are very similar in terms of shape and sidelobe level. The width of the main lobe is relatively wide (approximately 60° at −3-dB points), and the sidelobe level is high (−8 dB relative to the peak).

An example of the application of the proposed beamforming coefficients optimization method is shown in Figure 3.21. Pattern values are normalized to the maximum pattern value of the unweighted array. The width of the main beam of the desired pattern $F_d(\phi)$ was set to 50°, the width of the transient region was set to 20° (on each side of the main beam), and the sidelobe weight was set to $w = 10$. As can be seen, the sidelobe level has been significantly reduced compared to the result presented in Figure 3.20 (the sidelobe level after optimization is lower than −20 dB with respect to the maximum for all values of the steering angle ϕ_0).

The curves for 0° and 22.5° are symmetrical, as these directions correspond to the direction of one of the array elements (0°) and the direction exactly between two consecutive array elements (22.5°), respectively. The pattern for 11.25° is asymmetrical, with a different sidelobe level on each side of the main beam, the result of the nonsymmetrical position of the beam with respect to the antenna elements. The shape of the main beam has changed as a result of optimization. The obtained shape of the pattern $F(\phi, \mathbf{w})$ is close to the desired one $F_d(\phi)$ (main beamwidth 50°, transient region width 20°). However, as already indicated, the parameters of the desired pattern $F_d(\phi)$ should be related to the physical properties of the array. This is even more important as the number of degrees of freedom in the considered case is relatively small; therefore, the pattern-shaping capabilities are limited.

The reduction of the sidelobe level led to a substantial reduction of the array amplitude gain. Depending on the beam direction, the amplitude losses are in the range of 11.5–12.5 dB. As in the case of tapering of the ULA, the weights reduce not only the signal, but also the noise, therefore the signal-to-noise ratio loss is smaller than the amplitude loss. The signal-to-noise ratio loss is in the order of 9 dB. In many cases, this would be an unacceptable price to be paid for sidelobe attenuation.

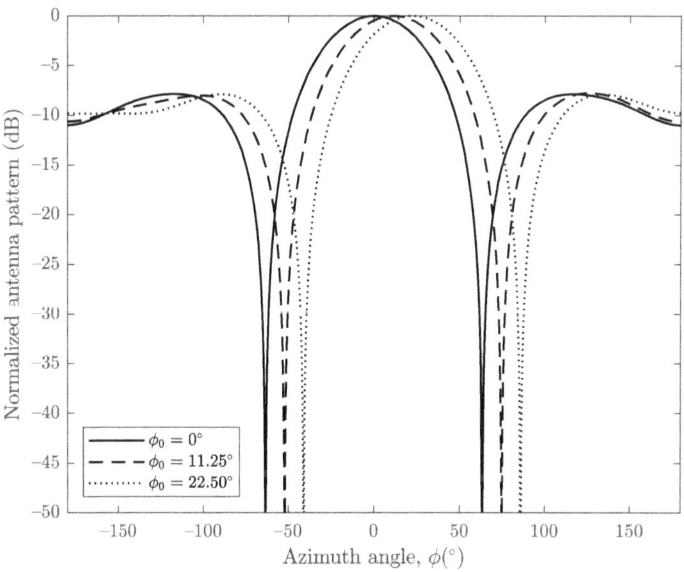

Figure 3.20 Patterns with unweighted beamforming coefficients for different steering angles.

3.4 COMPARISON OF ULA AND UCA

After the presentation of the ULA and UCA a question may arise: which array is better for a passive radar? There is no simple answer to this question, as it depends on numerous aspects (e.g., the application of the passive radar). However, the following analysis can provide some general conclusions on both array types.

One of the main limitations that has to be considered when constructing an antenna array for a passive radar is how many receiving channels are available. All channels have to be coherent, which involves using a common local oscillator for downconversion and a common clock signal for analog-to-digital converters. The number of receiving channels indirectly imposes the size of the array. The more receiving channels, the bigger the array will be. This can be a serious limitation, especially for lower frequency bands, such as FM radio. A comparison of radiation patterns and array geometries for ULA and UCA is presented in Figures 3.22 and 3.23. The figures correspond to a different number of array elements, equal to 8, 12, 16, and 32, respectively. The top plots show the array patterns, whereas the bottom ones depicts the array geometries. In both cases, constant interelement spacing was

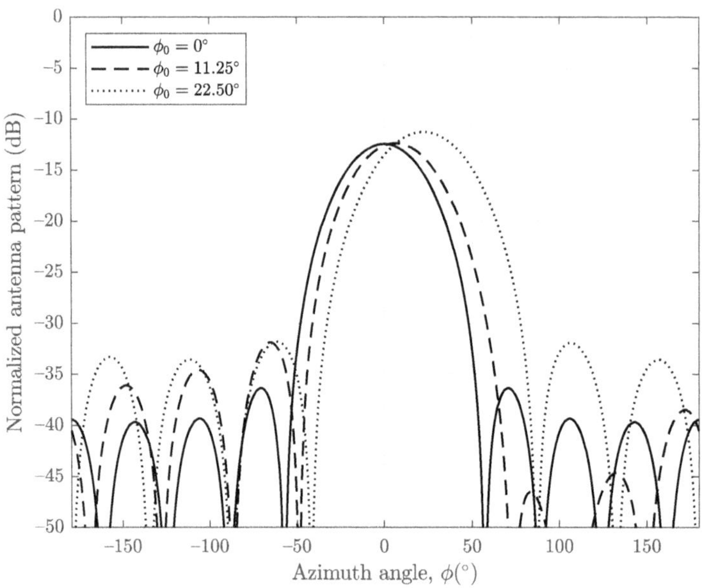

Figure 3.21 Patterns with optimized beamforming coefficients for different steering angles.

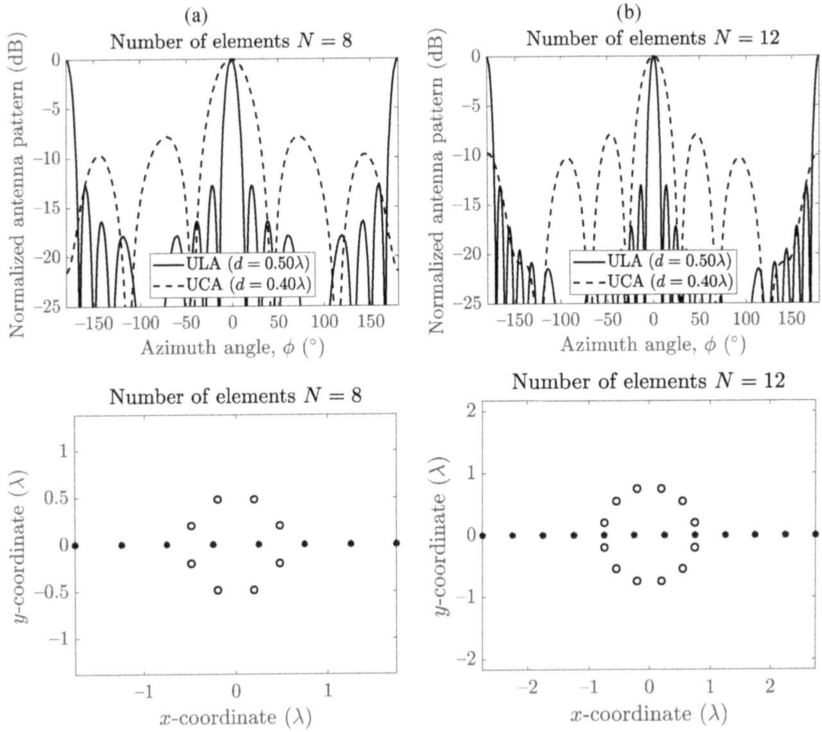

Figure 3.22 Comparison of array radiation patterns and array geometries of ULA and UCA for: (a) 8 elements and (b) 12 elements (uniform weighting used for ULA and UCA).

assumed: 0.5λ for ULA and 0.4λ for UCA. Uniform weighting was applied for both ULA and UCA.

The analysis conducted separately for the ULA and UCA, as well as the radiation patterns shown in Figures 3.22 and 3.23 allow for the following conclusions to be formulated.

- The sidelobe level for an untapered ULA (−13 dB) is lower than for a UCA (−8 dB).
- The main beamwidth for a ULA is lower than for a UCA for the same number of elements (for practical interelement spacing).
- The beam shape for a ULA is highly dependent on the steering angle, whereas the beam shape almost does not change with the angle for a UCA.

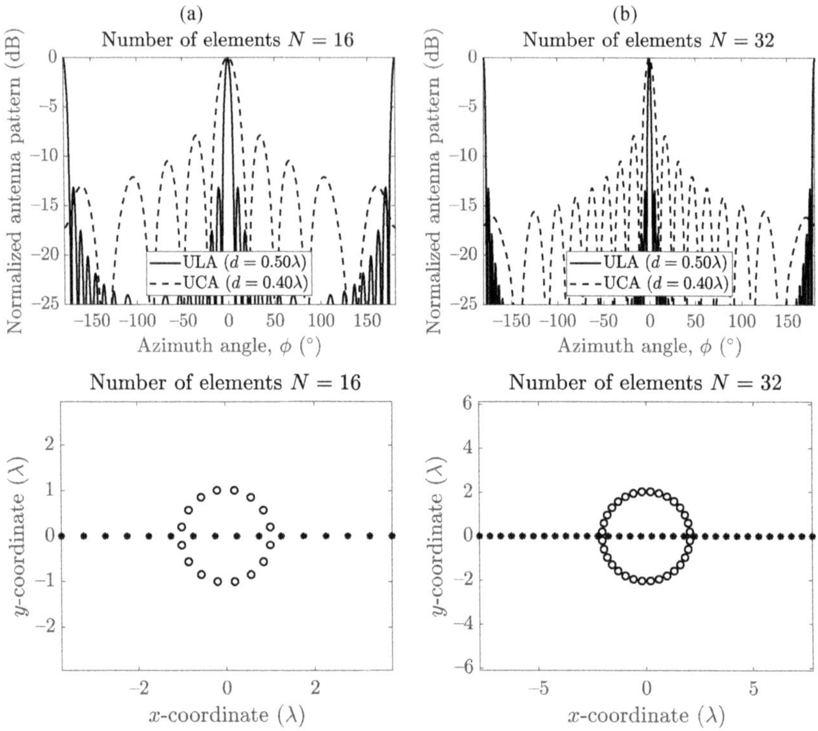

Figure 3.23 Comparison of array radiation patterns and array geometries of ULA and UCA for: (a) 16 elements and (b) 32 elements (uniform weighting used for ULA and UCA).

- A single ULA can be used in practice only for sector observation of 90° to 120° width. In order to obtain 360° coverage, typically 3 or 4 ULAs are used. This entails increase in the number of the receiving channels, as well as the overall size of the antenna system. This is a common practice in active radars, with 3 or 4 facets looking in different directions. In the case of UCA, a single array can provide omnidirectional observation, which is often a preferred solution in passive radar due to the smaller overall size of the array.
- The array radiation pattern of a ULA has a mirrored main beam, which is usually undesirable. In practice this is not a problem, as it is eliminated using directional radiation elements, which attenuate radiation to the back of the array. This feature is connected with the previous one, limiting ULA operation to a sector.

3.5 ARRAYS OPERATING IN A WIDE FREQUENCY BAND

The properties of the array are directly related to the array size with respect to the wavelength. In practice, the physical size of the array is fixed, whereas the wavelength of the received signals can change. As a result, if an array is designed with a certain element spacing (e.g., $\lambda/2$ for one frequency), this relation will be different for another frequency. This poses a problem when an array is designed for the reception of signals in a wide frequency band. Let us review the parameters of typical illuminators of opportunity from the point of view of relative (fractional) bandwidth. Table 3.2 lists the frequency bands, single channel bandwidths, and fractional bandwidths for popular transmitter types. The first fractional bandwidth (channel) is calculated as:

$$\frac{B}{\bar{f}_c} \cdot 100\% \tag{3.18}$$

where B is the single channel bandwidth and $\bar{f}_c$ is the center frequency of the frequency band calculated as $\bar{f}_c = (f_{c\max} + f_{c\min})/2$. The second fractional bandwidth (whole band) is calculated as:

$$\frac{\Delta f_c}{\bar{f}_c} \cdot 100\% \tag{3.19}$$

where Δf_c is the whole band of a particular transmission type $\Delta f_c = f_{c\max} - f_{c\min}$.

The results from the table indicate that when a single channel is considered, the listed signals are fairly narrowband. Even the largest value of 1.14% for DVB-T can be easily classified as narrowband. The situation is rather different when the whole bandwidth of a particular signal type is considered. For ex-

Table 3.2
Fractional Bandwidth for Different Transmitter Types

Transmitter Type	*Frequency Band,* $f_{c\min} - f_{c\max}$ (MHz)	*Signal Bandwidth,* B (MHz)	*Fractional Bandwidth (Channel) (%)*	*Fractional Bandwidth (Whole Band) (%)*
FM	88–108	0.15	0.15	20.4
DAB	174–230	1.5	0.74	27.7
DVB-T	470–860	7.6	1.14	58.6
GSM	935–960	0.2	0.02	2.6
WiFi	2,400–2,500	16	0.65	4.1

ample, the whole bandwidth of FM radio (88–108 MHz) corresponds to the fractional bandwidth of 20.4%. This means that the antenna array size (expressed in wavelengths) will be substantially different when the signal from the beginning (88 MHz) and end (108 MHz) of the band is being received. This entails using different beamforming coefficients for different frequencies in order to match phase shifts to varying spacing (in wavelengths) between array elements. This is mostly pronounced for DVB-T, where the fractional bandwidth for the whole DVB-T band is 58.6%.

Let us consider an example of a ULA and UCA for a wide band. Figure 3.24 shows the array radiation pattern for the approximate frequency range of DVB-T signals (i.e., 450–900 MHz). The horizontal axis corresponds to frequency, and the vertical axis corresponds to the azimuth angle. The intensity in-

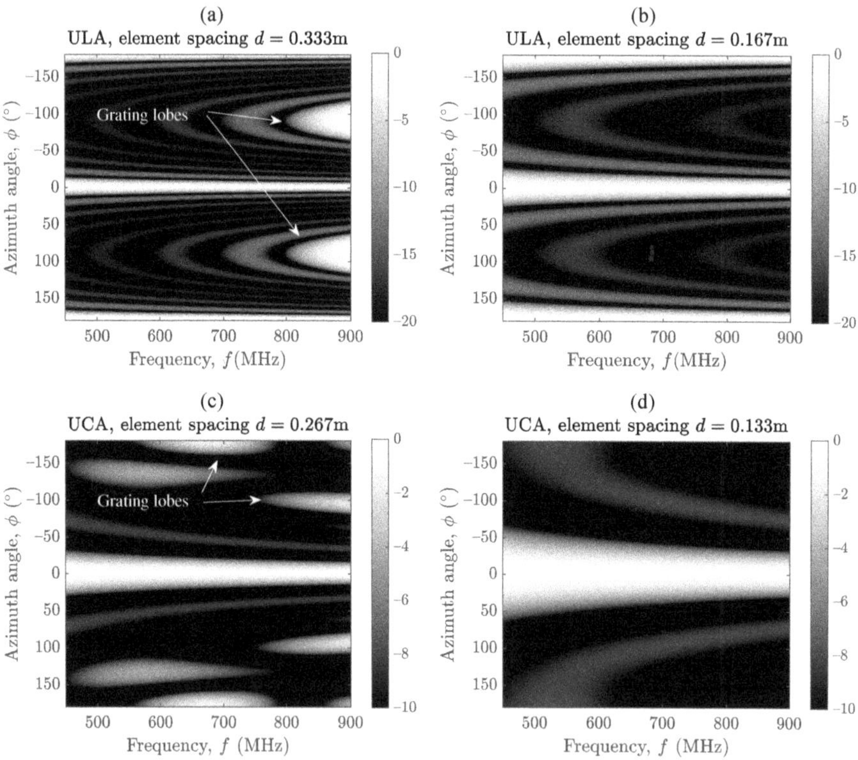

Figure 3.24 Array radiation pattern for a wide band: (a) ULA matched to 450 MHz, (b) ULA matched to 900 MHz, (c) UCA matched to 450 MHz, and (d) UCA matched to 900 MHz.

dicates the value of the radiation pattern (the values are normalized to 0 dB, and saturated at -20 dB for ULA and -10 dB for UCA). An 8-element array was simulated in both cases (ULA and UCA). Two upper plots (Figure 3.24(a, b)) are for the ULA, whereas the two bottom ones (Figure 3.24(c, d)) are for the UCA. The plots on the left and right correspond to element spacing matched to the lowest (450 MHz) and highest (900 MHz), respectively. In the case of the ULA, matching of the element spacing meant using $\lambda/2$ distance for the corresponding frequency. For the UCA, the element spacing was set to 0.4λ. The beam is pointing at $0°$, which is visible as a horizontal region across all frequencies. The main beamwidth changes for different frequencies, as the relation of the wavelength to the element spacing changes. For higher frequencies the beam gets narrower because the array size, with respect to the wavelength, is larger. The arrays from the left side plots are matched to the lowest frequency. This provides a narrow main beam, however, for higher frequencies grating lobes appear. If the array spacing is matched to the highest frequency, no grating lobes are present, but the main lobe is much wider. In the case of the ULA, apart from the main beam at $0°$, a symmetrical beam appears at $\pm 180°$, an effect already investigated during the analysis of the ULA.

In addition to the discussion above, concerning the array geometry with respect to the wavelength, the radiation pattern of individual elements also has to be taken into consideration. For example, one of the simplest forms of a radiating element, a half-wave dipole, is relatively narrowband. This results from the fact that the resonance connected with the half-wavelength of the antenna occurs only at a certain frequency. If the frequency, and thus the wavelength, is changed, the dipole is no longer half-wave, and is not very effective at radiating electromagnetic energy. For this reason, wideband array elements, such as discone antennas [27, 28], are used.

3.6 ADAPTIVE BEAMFORMING

The beamforming approach presented in this chapter so far is often referred to as *data independent* [29]. In this approach the beamforming weights are fixed and do not depend on the received signal. An alternative approach is the so-called *data-dependent* or *statistically optimum* beamforming, where the weights are chosen in order to provide optimum (in certain sense) performance in a specific scenario with desired signal and interference. Statistically optimum beamforming requires the knowledge of statistical parameters of the received signals (typically second-order statistics), which in practice are usually not known. Moreover, the statistics can change in time; therefore, in practice, the statistics are estimated, and the beamforming parameters are adapted to the incoming signals. This approach is called *adaptive beamforming.*

A typical application of adaptive beamforming is steering the beam in the direction of interest, and/or creating nulls in the directions of source of interference/ unwanted signals [30, 31]. This seems to be perfectly suited to passive radar. For example, the surveillance beam could be steered in the direction of targets, while nulls could be created in the direction the transmitter and sources of strong clutter. Usually the direction of the transmitter is known, as the positions of the transmitter and receiver are known. The clutter sources directions, are normally not known. Even if they were known, creating nulls in those directions would make the radar blind in those directions. Moreover, the number of antenna array elements limits the number of null directions: only $N - 1$ nulls can be created (where N is the number of array elements). As can be seen, adaptive beamforming has some limitations.

One of the possible solutions to these limitations of adaptive beamforming, often applied in passive radar, is to apply conventional, data-independent beamforming with known directions of interest [22]. Then the removal of interfering signals (i.e., DPI and clutter) is realized by the *adaptive filtering* of the received signals. The adaptive filtering operates in time (or the frequency) domain, whereas adaptive beamforming performs spatial filtering. As a result, a combination of the conventional beamforming and adaptive filtering in the time domain leads to desired cancellation of interfering signals. Yet another approach is a combination of adaptive spatial and temporal filtering [32].

Let us consider an example of adaptive beamforming. A 16-element ULA with $\lambda/2$ spacing will be assumed. The adaptive beamforming applied is MVDR (minimum variance distortionless response), one of the most popular algorithms. It will be compared against the conventional beamforming. The plots in Figure 3.25 show the radiation patterns for the conventional and MVDR beamformers for different angles of the desired signal, equal to 10°, 40°, and 60°. In each case the angle of the interfering signal is 0°. In the first case (Figure 3.25(a)), the angle of the desired signal (10°) is close to the interference signal angle (0°). This results in the serious distortion of the radiation pattern. In two other cases (Figure 3.25(b, c)), the separation between the desired and interfering directions is greater. As a consequence, the influence of creating a null in the interfering direction has a lower impact on the overall pattern.

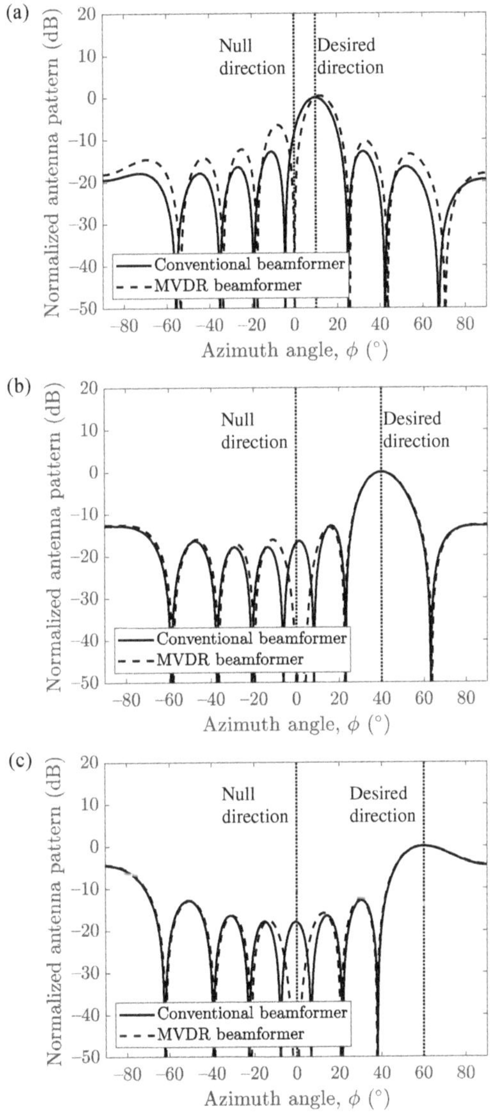

Figure 3.25 Comparison of conventional and MVDR beamformer radiation patterns for desired direction equal to: (a) 10°, (b) 40°, and (c) 60°. Null direction is 0° in all cases.

3.7 DIGITAL SIGNAL RECONSTRUCTION

A topic that is connected with digital beamforming is the reconstruction of the reference signal. The beamforming approach that is usually applied is to produce a beam that provides high gain in the direction of interest and low sidelobes in the direction of the interference. In the case of using beamforming for the reference channel, the desire is to obtain a clean copy of the transmitted signal. Clean means that there should be as little as possible of:

- Multipath;
- Noise;
- Target echoes;
- Interference.

This can be achieved to some extent by generating a very directional beam pattern. However, an alternative approach is to try to reconstruct the signal to the form in which it has been transmitted [33–37]. This is possible in the case of digital modulations, such as WiFi, GSM, or DVB-T.The idea is as follows: the analog copy of the reference signal is received. Next, it is decoded to the level of bits. Error correction algorithms can be applied, when possible. If the signal-to-noise ratio is moderately high, a relatively low bit-error rate can be achieved. In the next step, an analog copy of the signal is reconstructed from the decoded bit stream. In theory this signal should represent the ideal copy of the transmitted signal, without multipath, noise, target echoes, and interference. In practice, the situation is more complicated.

In many cases the standard defining a particular signal imposes only boundary conditions that the signal should fulfill. Because of this, the same bit stream can produce a different analog signal, depending on the manufacturer or even a particular piece of equipment, even though they all comply with the standard defined for the signal.

Another aspect of signal reconstruction is that the exact copy of the transmitted signal is not always the best reference signal. In some cases, the reconstructed signal is modified on purpose in order to improve the shape of the ambiguity function. An example of such an approach is shown in [38]. In the paper, a DVB-T signal is reconstructed and modified. The aim of the signal modification is to reduce spurious peaks on the ambiguity function originating from, for example, pilot frequencies or cyclic prefixes.

At the beginning of the previous chapter, it was stated that a passive radar needs at least two receiving channels to operate: reference and surveillance. This requirement can be alleviated when signal reconstruction is used: a passive radar can be built with only a single receiving channel. In such a case, the received signal is used for the reconstruction of the transmitted signal template. The same analog signal is then correlated with the reconstructed reference signal.

3.8 NUMERICAL RESULTS

Now a practical example of a UCA for passive radar will be analyzed based on the *PaRaDe* system, developed at the Warsaw University of Technology [22]. The antenna array under consideration is shown in Figure 3.26. It consists of 8 omnidirectional elements (half-wave dipoles) placed in a circle. The radius r of the array is equal to 1.2m, which corresponds to 0.4λ for a carrier frequency of 100 MHz. The spacing between two consecutive elements is equal to 0.97m, which is equal to 0.32λ. Due to such close spacing between the elements, strong coupling can be expected. As will be presented, the coupling strongly affects the procedure of beamforming, and has to be compensated for in order to be able to obtain desired results.

The concept of beamforming used in the *PaRaDe* system involves providing maximum coverage. To achieve this aim, 8 beams are formed, separated in angle by $45°$. This provides $360°$ in angular coverage, with relatively small losses at angles between neighboring beam directions. One of the eight beams serves as the reference beam; it points towards the transmitter. The other seven beams are used as the echo beams.

Figure 3.26 Circular 8-element antenna array of the *PaRaDe* system.

The beams created using the tapering method presented in Section 3.3.1 are presented in Figure 3.27 (values are normalized to 0 dB). The beams are shown in Cartesian and polar coordinates. The neighboring beams intersect at the level of approximately −2.5 dB, which is acceptable in most practical applications.

In the analysis carried out so far, it has been assumed that no coupling between the array elements exists. In reality, however, the currents induced in an array

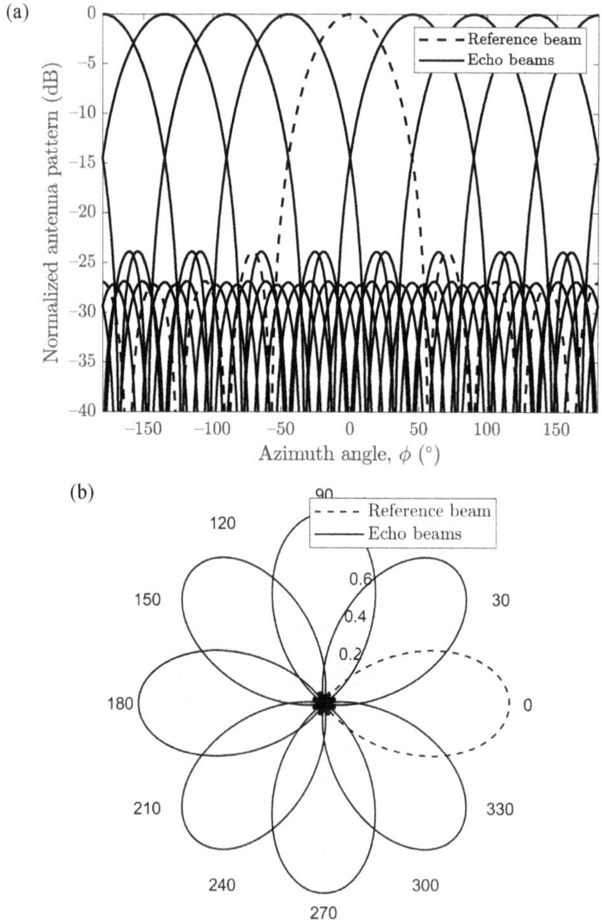

Figure 3.27 Digitally formed beams using the tapering method based on optimization of cost function: (a) in Cartesian coordinates and (b) in polar coordinates. (©2008 IEEE Reprinted, with permission, from [22].)

element by the received electromagnetic wave cause induction of currents in other elements. This phenomenon is known as mutual coupling of array elements [12]. Mutual coupling can significantly influence the radiation pattern, especially when the interelement spacing is small. In this section, the calibration of mutual coupling will be considered.

Consider the vector of complex amplitudes of signals received by an antenna array in an ideal case (i.e., when no coupling between the array elements exists):

$$\mathbf{X}(\phi) = \mathbf{a}(\phi)S, \tag{3.20}$$

where $\mathbf{a}(\phi)$ is the steering vector, and S is the complex amplitude of the transmitted signal. In reality coupling exists, modifying the received signal values. Moreover, transmission lines from antennas to receivers can introduce nonequal amplitude and phase distortions. A more realistic signal model, taking into the account these two effects, can be written as:

$$\widetilde{\mathbf{X}}(\phi) = \mathbf{CTa}(\phi)S = \hat{\mathbf{C}}\mathbf{a}(\phi)S, \tag{3.21}$$

where $\mathbf{C}$ is the antenna element coupling matrix, and $\mathbf{T} = \text{diag}\left([T_1, T_2, \ldots, T_N]\right)$ is a diagonal matrix representing the transmission lines coefficients. The matrix $\hat{\mathbf{C}} = \mathbf{CT}$ represents the overall influence of the nonidealities due to the coupling and transmission lines. As can be seen, the nonidealities are modeled by matrix multiplication. If the matrix $\hat{\mathbf{C}}$ were known, the ideal signals $\mathbf{X}(\phi)$ could be obtained by the multiplication by $\hat{\mathbf{C}}^{-1}$:

$$\mathbf{X}(\phi) = \hat{\mathbf{C}}^{-1}\widetilde{\mathbf{X}}(\phi). \tag{3.22}$$

The calibration method used in this case is based on using several calibration measurements with a signal transmitter placed at different positions with respect to the array [8, 22]. The measured data are then used to estimate the matrix $\hat{\mathbf{C}}$ by minimizing a certain cost function.

Assume that M calibration measurements have been performed. In each calibration measurement, a signal transmitter at a known position with respect to the array is used. The signals recorded for each measurement is given by:

$$\begin{aligned}
\widetilde{\mathbf{X}}_1(\phi_1) &= \hat{\mathbf{C}}\mathbf{a}_1(\phi_1)S_1, \\
\widetilde{\mathbf{X}}_2(\phi_2) &= \hat{\mathbf{C}}\mathbf{a}_2(\phi_2)S_2, \\
&\vdots \\
\widetilde{\mathbf{X}}_M(\phi_M) &= \hat{\mathbf{C}}\mathbf{a}_M(\phi_M)S_M.
\end{aligned} \tag{3.23}$$

The known parameters are the measured signals $\widetilde{\mathbf{X}}_m(\phi_m)$ and the steering vectors $\mathbf{a}_m(\phi_m)$ (which can be calculated from the relative position of the receiver and transmitter). The unknowns are the matrix $\hat{\mathbf{C}}$ (the same matrix for all measurements) and complex amplitude of the transmitted signals S_m.

Rewriting (3.23), error vectors $\mathbf{V}_m(\phi_m)$ can be defined as:

$$\begin{aligned}
\mathbf{V}_1(\phi_1) &= \widetilde{\mathbf{X}}_1(\phi_1) - \hat{\mathbf{C}}\mathbf{a}_1(\phi_1)S_1, \\
\mathbf{V}_2(\phi_2) &= \widetilde{\mathbf{X}}_2(\phi_2) - \hat{\mathbf{C}}\mathbf{a}_2(\phi_2)S_2, \\
&\vdots \\
\mathbf{V}_M(\phi_M) &= \widetilde{\mathbf{X}}_M(\phi_M) - \hat{\mathbf{C}}\mathbf{a}_M(\phi_M)S_M.
\end{aligned} \tag{3.24}$$

Ideally vectors $\mathbf{V}_m(\phi_m)$ should be zero. This fact can be used to calculate unknown matrix $\hat{\mathbf{C}}$ and complex amplitudes S_m based on the measured signals $\widetilde{\mathbf{X}}_m(\phi_m)$. To obtain this, the following cost function is defined:

$$J_c(\hat{\mathbf{C}}, S_1, S_2, \ldots .S_M) = \sum_{m=1}^{M} \| \mathbf{V}_m(\phi_m) \|^2 . \tag{3.25}$$

This function is minimized numerically by adjusting the values of matrix $\hat{\mathbf{C}}$ and amplitudes S_m [39, 40]. The real and imaginary parts of the optimized complex parameters are tuned during minimization. Box constraints are used to keep the parameters values within reasonable bounds during optimization.

Let us consider the application of the presented calibration methods to the real UCA of the *PaRaDe* system. The calibration effectiveness will be investigated by measuring the radiation pattern. To measure the antenna pattern, two approaches can be used. In the first one, the array is stationary, and the source of the radiation is moved around it. The second approach involves turning the antenna array in the presence of a stationary transmitter. In the experiment, a method based on the second approach was used. Both the array and the transmitter were stationary; however, the beam was swept digitally through 360°. Assuming that the transmitter is the dominant source of the radiation and that the antenna pattern does not change significantly for different beam directions, the measurement of the output signal power yields the approximate antenna pattern.

In Figure 3.28 the theoretical and measured antenna patterns are shown. In the case of the theoretical pattern an ideal situation has been assumed (i.e., without coupling). The measured pattern was obtained with the method described above, that is, with a stationary FM radio transmitter and digital beam steering. Both

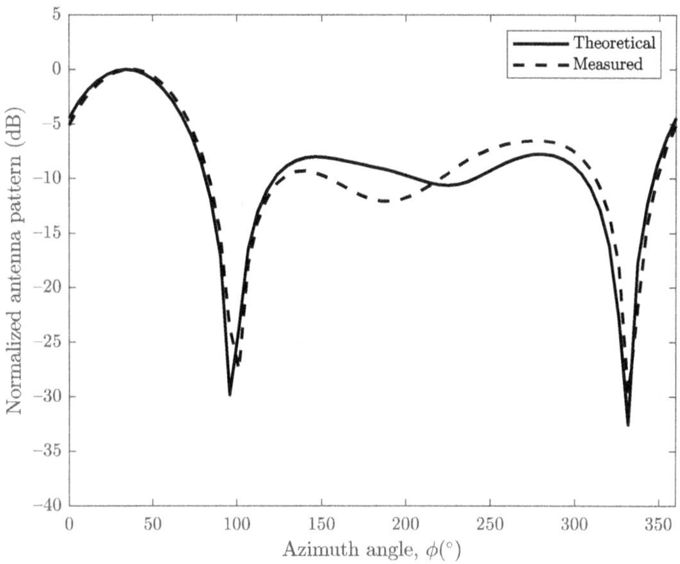

Figure 3.28 The theoretical and measured antenna patterns without beamforming coefficients' optimization, without coupling calibration.

patterns were obtained without the optimization of the beamforming coefficients. The measured pattern is relatively close to the theoretical one; however, some differences are visible. These differences are caused by the mutual coupling, as well as reflections from the obstacles in the vicinity of the array.

The difference between the measurement and theory is much greater when a more demanding antenna pattern is required. An example of such a result is shown in Figure 3.29. In this case, the described tapering approach based on the optimization was applied. The theoretical pattern has low sidelobes, compared with the unoptimized case. However, the measured pattern has much worse characteristics. The sidelobe level is at -12 dB, whereas the sidelobe level of the theoretical pattern is -24 dB. This difference can be attributed to uncompensated coupling between the antenna array elements.

The coupling was compensated for by the proposed calibration procedure. First, the $\hat{\mathbf{C}}$ matrix was estimated. To perform this, a single-tone signal was transmitted from a dipole antenna from different positions surrounding the antenna array. The carrier frequency used in the calibration should not interfere with the broadcast signal, as it may distort the measurements. For this reason, the frequency of the calibration signal was set to 91.20 MHz. This frequency corresponds to the next FM

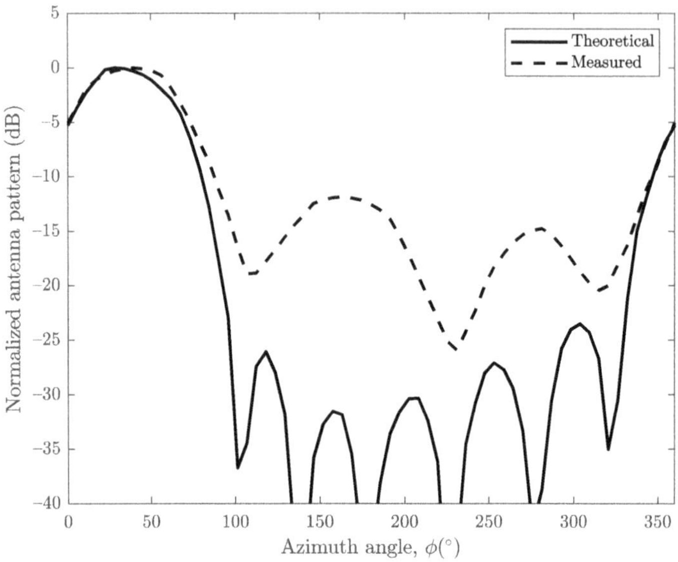

Figure 3.29 The theoretical and measured antenna patterns after beamforming coefficients' optimization, without coupling calibration.

radio channel relative to the broadcast at 91 MHz used in the actual beamforming experiment. The value of matrix $\hat{\mathbf{C}}$ was calculated by finding the minimum of the cost function (3.25) using the calibration measurements. To simplify the solution of the calibration procedure, an assumption was made that only five neighboring elements of the array are coupled (i.e., an element is coupled only with its two adjacent elements on each side). This assumption was based on the performed measurements of the array parameters, which showed that two adjacent elements cause the most significant coupling. In such a case, the coupling matrix has the following form:

$$\hat{\mathbf{C}} = \begin{bmatrix} \hat{C}_{11} & \hat{C}_{12} & \hat{C}_{13} & 0 & 0 & 0 & \hat{C}_{71} & \hat{C}_{81} \\ \hat{C}_{12} & \hat{C}_{22} & \hat{C}_{23} & \hat{C}_{24} & 0 & 0 & 0 & \hat{C}_{82} \\ \hat{C}_{13} & \hat{C}_{23} & \hat{C}_{33} & \hat{C}_{34} & \hat{C}_{35} & 0 & 0 & 0 \\ 0 & \hat{C}_{24} & \hat{C}_{34} & \hat{C}_{44} & \hat{C}_{45} & \hat{C}_{46} & 0 & 0 \\ 0 & 0 & \hat{C}_{35} & \hat{C}_{45} & \hat{C}_{55} & \hat{C}_{56} & \hat{C}_{67} & 0 \\ 0 & 0 & 0 & \hat{C}_{46} & \hat{C}_{56} & \hat{C}_{66} & \hat{C}_{67} & \hat{C}_{68} \\ \hat{C}_{71} & 0 & 0 & 0 & \hat{C}_{57} & \hat{C}_{67} & \hat{C}_{77} & \hat{C}_{78} \\ \hat{C}_{81} & \hat{C}_{82} & 0 & 0 & 0 & \hat{C}_{68} & \hat{C}_{78} & \hat{C}_{88} \end{bmatrix}. \quad (3.26)$$

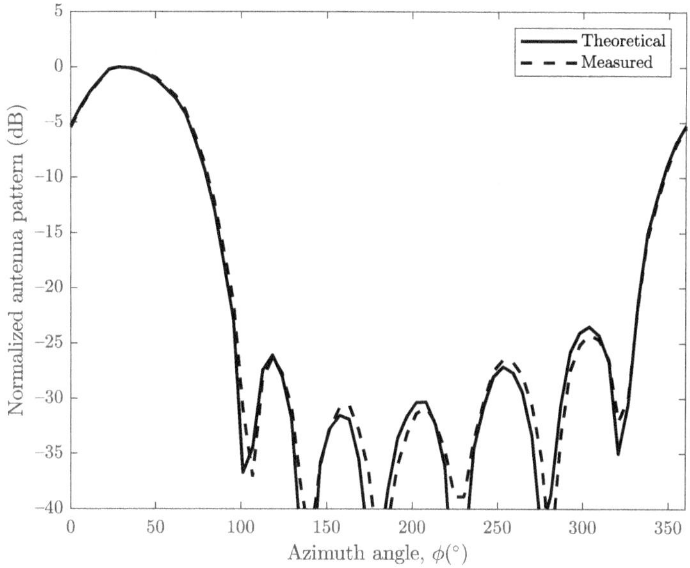

Figure 3.30 The theoretical and measured antenna patterns after beamforming coefficients' optimization and after coupling calibration. (©2008 IEEE Reprinted, with permission, from [22].)

The estimated value of $\hat{\mathbf{C}}$ was then used for signal correction according to (3.22) before the beamforming was carried out. The results of the application of the calibration method are shown in Figure 3.30. As can be seen, the calibration resulted in substantial reduction of the differences between the measured and theoretical patterns. In this case, the measured pattern is almost identical to the theoretical one. The mutual coupling introduces losses, as the signals from individual elements are not independent; thus, the gain of the array is reduced. In the presented example, the losses due to the mutual coupling were in the order of 1–2 dB.

3.9 SOME PRACTICAL REMARKS

Advanced antenna systems have been successfully constructed for radar for decades. These include active electronically scanned arrays (AESA) with thousands of transmit/receive modules. In comparison to this, it might seem that building a receive-only array with just a few elements for a passive radar is a trivial task. This is only partially true. One of the aspects that make it easier to construct an antenna array for passive radar than an array for classical active radar is long wavelength. Typical active radars operate in the microwave region, with centimeter wavelengths. This puts high requirements on the accuracy of array manufacturing. In the case of passive radar, which typically operate in VHF or UHF bands, those requirements are less strict. However, a long wavelength is also what makes it more difficult to build a passive radar antenna array. The reason for this is that measuring and calibrating antennas in a meter-wave region in an anechoic chamber is a challenging task. Because the attenuation of electromagnetic waves by absorbers at those frequencies is not very effective, high interference levels can be expected. Even more so because the frequencies used by passive radars are continually used by the illuminators of opportunity, and most probably no one is going to switch off all radio or TV transmitters in a 100-km radius so that the calibration of a passive radar antenna array can be performed without interference.

3.10 SUMMARY

In this chapter beamforming methods have been reviewed. Two basic types of arrays have been investigated: uniform linear and circular arrays. The presented results show important differences between the ULA and UCA. The ULA provides a narrower beamwidth and a lower sidelobe level for a given number of elements. The disadvantages of the ULA include dependency of the beamwidth on the beam angle and the mirror main beam. This means that in practice the ULA can be used for observing a sector of angles. If omnidirectional coverage is required, multiple ULAs have to be used or a single UCA.

A practical example of a UCA for an FM radio-based radar was analyzed. The importance of the mutual coupling of the array elements was demonstrated. A method for the decoupling of the array was proposed and verified.

References

[1] M. Malanowski, *Signal Processing Algorithms and Target Localisation Methods for Continuous-Wave Passive Bistatic Radar*. No. 179 in Prace Naukowe Politechniki Warszawskiej. Elektronika, Warszawa, ul. Polna 50: Oficyna Wydawnicza Politechniki Warszawskiej, 2012.

[2] P. E. Howland, D. Maksimiuk, and G. Reitsma, "FM Radio Based Bistatic Radar," *IEE Proc. Radar, Sonar and Navigation*, vol. 152, pp. 107–115, June 2005.

[3] A. D. Lallo, A. Farina, R. Fulcoli, P. Genovesi, R. Lalli, and R. Mancinelli, "Design, Development and Test on Real Data of an FM Based Prototypical Passive Radar," in *Proc. RadarCon 2008*, (Rome, Italy), p. CD, May 2008.

[4] H. Kuschel, M. Glende, J. Heckenbach, S. Müller, J. Schell, and C. Schumacher, "Experimental Passive Radar Systems Using Digital Illuminators (DAB/DVB)," in *Proc. International Radar Symposium 2007*, (Cologne, Germany), pp. 411–417, September 2007.

[5] B. D. Nordwall, " 'Silent Sentry' A New Type of Radar," *Aviation Week and Space Technology*, vol. 149, pp. 70–71, November 1998.

[6] D. Poullin, M. Flécheux, and M. Klein, "New Challenges for PCL System: 3D Requirements and 'Optimal' Resources Management for PCL Air Traffic Survey," in *Proc. International Radar Symposium 2010*, (Vilnius, Lithuania), pp. 8–11, June 2010.

[7] A. Schroeder and M. Edrich, "CASSIDIAN Multiband Mobile Passive Radar System," in *Proc. International Radar Symposium 2010*, (Vilnius, Lithuania), pp. 286–291, June 2010.

[8] F. Belfiori, S. Monni, W. V. Rossum, and P. Hoogeboom, "Mutual Coupling Compensation Applied to a Uniform Circular Array," in *Proc. International Radar Symposium 2010*, (Vilnius, Lithuania), pp. 100–103, June 2010.

[9] M. Malanowski, K. Kulpa, and J. Misiurewicz, "PaRaDe – Passive Radar Demonstrator Family Development at Warsaw University of Technology," in *Proc. Microwaves, Radar and Remote Sensing Symposium – MRRS*, (Kiev, Ukraine), pp. 75–78, September 2008.

[10] D. W. O'Hagan, C. J. Baker, and H. D. Griffiths, "Signal and Interference Analysis: Proposed Analogue Signal Suppression Techniques for PCL Radar," in *Proc. European Radar Conference – EURAD*, (Manchester, UK), pp. 296–298, September 2006.

[11] R. Saini, M. Cherniakov, and V. Lenive, "Direct Path Interference Suppression in Bistatic System: DTV Based Radar," in *Proc. International Conference on Radar*, (Adelaide, Australia), pp. 309–314, September 2003.

[12] S. J. Orfanidis, *Electromagnetic Waves and Antennas*. Sophocles J. Orfanidis, 2008.

[13] R. Cardinali and E. Anniballi, "ARGUS 3D System: Operational Requirements and Architecture," tech. rep., European Commission, 2012.

[14] A. Farina, "Selex Sistemi Integrati Passive Covert Location Radar: AULOS," in *Proc. Joint military meeting of SET-152/RTG*, (Rome, Italy), p. CD, February 2012.

[15] P. Piasecki and M. Gasztold, "Dual Polarized Circular Array Antenna for PCL System and Possibility of Digital Beamforming of an Antenna Pattern," in *2015 European Microwave Conference (EuMC)*, pp. 718–721, September 2015.

[16] K. Kulpa and M. Malanowski, "The Concept of Simple MIMO PCL Radar," in *Proc. European Radar Conference – EURAD*, (Amsterdam, The Netherlands), pp. 240–243, October 2008.

[17] D. E. N. Davies, "A Transformation Between the Phasing Techniques Required for Linear and Circular Aerial Arrays," *Proc. IEE*, vol. 112, pp. 2041–2045, November 1965.

[18] M. Al-Husseini, E. Yaacoub, K. Y. Kabalan, and A. El-Hajj, "Pattern Synthesis with Uniform Circular Arrays for the Reduction of WCDMA Intercell Interference," *Turkish Journal of Electrical Engineering and Computer Science*, vol. 16, pp. 207–215, November 2008.

[19] F. Belfiori, S. Monni, W. V. Rossum, and P. Hoogeboom, "Side-Lobe Suppression Techniques for a Uniform Circular Array," in *Proc. European Radar Conference – EURAD*, (Paris, France), pp. 113–116, September – October 2010.

[20] N. H. Noordin, V. Zuniga, A. O. El-Rayisa, N. Haridas, A. T. Erdogan, and T. Arslan, "Uniform Circular Arrays for Phased Array Antenna," in *Loughborough Antennas and Propagation Conference*, (Loughborough, UK), pp. 1–4, November 2011.

[21] B. Basu and G. K. Mahanti, "Evolutionary Algorithms for Synthesis of Uniform Circular Array with Minimum Side Lobe Level and Maximum Directivity," in *Annual IEEE India Conference – INDICON*, (Kolkata, India), pp. 1–4, December 2010.

[22] M. Malanowski and K. Kulpa, "Digital Beamforming for Passive Coherent Location Radar," in *Proc. RadarCon 2008*, (Rome, Italy), p. CD, May 2008.

[23] C. G. Broyden, "The Convergence of a Class of Double-Rank Minimization Algorithms," *Journal Inst. Math. Applic.*, vol. 6, pp. 76–90, 1970.

[24] R. Fletcher, "A New Approach to Variable Metric Algorithms," *Computer Journal*, vol. 13, pp. 317–322, 1970.

[25] D. Goldfarb, "A Family of Variable Metric Updates Derived by Variational Means," *Mathematics of Computing*, vol. 24, pp. 23–26, 1970.

[26] D. F. Shanno, "Conditioning of Quasi-Newton Methods for Function Minimization," *Mathematics of Computing*, vol. 24, pp. 647–656, 1970.

[27] P. Knott, T. Nowicki, and H. Kuschel, "Design of a Disc-Cone Antenna for Passive Radar in the DVB-T Frequency Range," in *2011 German Microwave Conference*, pp. 1–4, March 2011.

[28] H. Kuschel, J. Heckenbach, and M. Ummenhofer, "Passive Radar Collision Warning System PARASOL," *IEEE Aerospace and Electronic Systems Magazine*, vol. 32, pp. 40–52, February 2017.

[29] B. D. V. Veen and K. M. Buckley, "Beamforming: a Versatile Approach to Spatial Filtering," *IEEE ASSP Magazine*, vol. 5, pp. 4–24, April 1988.

[30] Z. Jiabing, H. Yi, and T. Liang, "Adaptive Beamforming Passive Radar Based on FM Radio Transmitter," in *2006 CIE International Conference on Radar*, pp. 1–4, October 2006.

[31] T. Peto and R. Seller, "Quad Channel DVB-T Based Passive Radar," in *2016 17th International Radar Symposium (IRS)*, pp. 1–4, May 2016.

[32] J.-H. Deng, J.-K. Hwang, C.-Y. Lin, and S.-M. Liao, "Adaptive Space-Time Beamforming Technique for Passive Radar System with Ultra Low Signal to Interference Ratio," in *2010 IEEE International Conference on Wireless Information Technology and Systems*, pp. 1–4, August 2010.

[33] P. Krysik, K. Kulpa, M. Malanowski, and P. Samczyński, "Reconstruction of GSM Reference Signals for PCL Processing," in *Proc. 3rd FHR Focus Days on PCL*, (Wachtberg, Germany), pp. 1–19, May 2011.

[34] M. Bączyk and M. Malanowski, "Reconstruction of the Reference Signal in DVB-T-Based Passive Radar," *International Journal of Electronics and Telecommunications, PAN*, vol. 57, no. 1, pp. 43–48, 2011.

[35] D. W. O'Hagan, M. Setsubi, and S. Paine, "Signal Reconstruction of DVB-T2 Signals in Passive Radar," in *2018 IEEE Radar Conference (RadarConf18)*, pp. 1111–1116, April 2018.

[36] D. Bok, "Reconstruction and Reciprocal Filter of OFDM Waveforms for DVB-T2 Based Passive Radar," in *2018 International Conference on Radar (RADAR)*, pp. 1–6, August 2018.

[37] W. Xianrong, W. Junfang, H. Sheng, and T. Hui, "Reconstruction of Reference Signal for DTMB-Based Passive Radar Systems," in *Proceedings of 2011 IEEE CIE International Conference on Radar*, vol. 1, pp. 165–168, October 2011.

[38] R. Saini and M. Cherniakov, "DTV Signal Ambiguity Function Analysis for Radar Application," *IEE Proc. Radar, Sonar and Navigation*, vol. 152, pp. 133–142, June 2005.

[39] S. Han, "A Globally Convergent Method for Nonlinear Programming," *Journal of Optimization Theory and Applications*, vol. 22, p. 297, 1977.

[40] M. J. D. Powell, "A Fast Algorithm for Nonlinearly Constrained Optimization Calculations," *Numerical Analysis, Lecture Notes in Mathematics*, vol. 630, pp. 144–175, 1978.

Chapter 4

Correlation Processing

4.1 INTRODUCTION

In this chapter, correlation processing in passive radar is analyzed, which is the basis for detection and bistatic parameters estimation [1]. In practice, correlation is preceded by clutter filtering in a typical processing scheme. In this book, however, clutter filtering will be described after the correlation processing. The reason for this is that the effects of clutter presence in the echo signal can be understood only on the basis of correlation.

The detection and estimation in passive radar are carried out using of the correlation of the reference and echo signals [2, 3] by way of the *cross-ambiguity function* (CAF), closely related to the ambiguity function analyzed in Chapter 2. In this chapter, the signal model used in the calculation of the classical form of the CAF is introduced first. Next, practical methods for the calculation of the CAF are shown. As indicated in Chapter 2, when (pseudo-)random signals are correlated, two types of sidelobes are present: random and deterministic. The random sidelobe level depends on the signal bandwidth and integration time, and influencing these sidelobes in a classical way (i.e., by windowing) is not effective. The deterministic sidelobes can be shaped by applying windowing in the time or frequency domains, similar to the approach used in spectral analysis. The corresponding influence on the sidelobes in the range and velocity dimensions is analyzed. Next, the processing gain obtained thanks to the correlation is investigated. At the end of the theoretical part, extensions of the CAF are introduced, including compensation of the range and velocity cell migration. In the numerical section, some real-life cases are analyzed, such as the application of a long integration time or observation of maneuvering targets.

4.2 SIGNAL MODEL AND CORRELATION PROCESSING

In this section, the signal model of a target echo that is usually used in passive radar is introduced. Based on this model, the cross-ambiguity function is defined. The signal model, and consequently, the cross-ambiguity function are derived using some simplifying assumptions. These assumptions are that the bistatic motion can be modeled by a first-order polynomial and that the influence of the bistatic velocity on the time scale stretch of the baseband signal is negligible. The aforementioned assumptions are generally accepted in literature [4–6]. They can be justified by the fact that the simplifications can be applied without loss of detection performance in conditions typical for passive radar. However, it will be shown that in some situations these assumptions are not valid (which leads to range and velocity cell migration), and more complicated signal processing is required in order to prevent performance degradation. Two typical situations are the application of a long integration time and the detection of highly maneuvering targets. In the last part of this chapter, the extended motion model will also be presented together with the extended signal processing scheme based on this model.

Assume that $x_r(t)$ is the baseband complex information signal. The modulated radio frequency signal can be written as:

$$x_r^{RF}(t) = \Re\{x_r(t) \cdot \exp(\mathrm{j}2\pi f_c t)\}, \tag{4.1}$$

where $\Re\{\}$ is the real-value operator, and $f_c = c/\lambda$ is the carrier frequency, c is the speed of light, and λ is the wavelength. The baseband signal $x_r(t)$ can usually be modeled as a band-limited complex noise signal (as in Chapter 2). Even in cases when the analyzed signal contains deterministic components, such as cyclic prefix and pilots in DVB-T standard, band-limited noise is a good model for studying the correlation properties of the signal in passive radar.

Often the motion model of a target is defined in Cartesian coordinates. Passive radar, however, performs measurements in bistatic coordinates. The relationship between the Cartesian and bistatic parameters is nonlinear (compare (2.1)–(2.4) for bistatic range and (2.6) for bistatic velocity). Therefore, even for a simple linear motion model in Cartesian coordinates, the bistatic motion model is nonlinear. Moreover, the relationship between Cartesian and bistatic coordinates is relatively complicated. For this reason, it is convenient to use a generic model for the bistatic range, which does not directly correspond to the Cartesian parameters.

Consider a model of the instantaneous bistatic range, defined as the difference between the transmitter-target-receiver and transmitter-receiver paths, expressed as an infinite-order polynomial:

$$r(t) = \sum_{k=0}^{\infty} \frac{r^{(k)}(t)}{k!} t^k = R + Vt + \frac{At^2}{2} + \ldots, \tag{4.2}$$

where R is the mean bistatic range (later called simply bistatic range), V is the bistatic velocity, and A is the bistatic acceleration. The signal reflected from a target with instantaneous bistatic range $r(t)$ is (at the radio frequency):

$$x_e^{RF}(t) = \Re\left\{C \cdot x_r\left(t - \frac{r(t)}{c}\right) \cdot \exp\left(\mathrm{j}2\pi f_c\left(t - \frac{r(t)}{c}\right)\right)\right\}, \tag{4.3}$$

where C is the target echo amplitude. During the reception, the signal is downconverted to the form:

$$x_e(t) = C' \cdot x_r\left(t - \frac{r(t)}{c}\right) \cdot \exp\left(\mathrm{j}2\pi f_c \frac{r(t)}{c}\right), \tag{4.4}$$

where C' is the complex target echo amplitude. In most cases, some simplifications are made to this model. First of all, it is assumed that during the integration time or coherent processing interval (CPI),[1] the infinite-order polynomial representing instantaneous bistatic range $r(t)$ can be sufficiently approximated by:

$$r(t) \approx R + Vt. \tag{4.5}$$

It means that only the bistatic range R and bistatic velocity V are considered in the signal model. Higher-order motion model derivatives (bistatic acceleration, derivative of bistatic acceleration) are neglected.

An additional assumption is that the time-varying bistatic range (4.5) has negligible influence on the baseband information signal $x_r(t)$, apart from a constant delay. This can be expressed as:

$$x_r\left(t - \frac{R + Vt}{c}\right) \approx x_r\left(t - \frac{R}{c}\right). \tag{4.6}$$

Taking into account the above assumptions, the simplified model of the received echo signal can be written as:

$$x_e(t) = C'' \cdot x_r\left(t - \frac{R}{c}\right) \cdot \exp\left(\mathrm{j}\frac{2\pi}{\lambda} Vt\right), \tag{4.7}$$

1 The terms "integration time" and "CPI" will be used interchangeably.

where $C'' = C' \cdot \exp(\mathrm{j}2\pi R/\lambda)$. According to (4.7), the received echo signal is a scaled (attenuated), time-delayed and Doppler-shifted copy of the reference signal.

The signal model (4.7) is widely accepted in passive radar literature [4–6], as it is suitable in most cases. Using this model leads to the classical cross-ambiguity function used for target detection [2].

To detect a target in passive radar, the correlation function of the reference and echo signals is processed. Due to the long CPI typically used in passive radar, in the order of a fraction of a second, or several seconds, the Doppler resolution is very fine. As a result, the Doppler shift of the target echo has to be taken into consideration while calculating the correlation. Based on the model of the received signal (4.7), a correlation function can be defined:

$$\psi(R, V) = \int_{-T/2}^{T/2} x_e(t) \cdot x_r^*\left(t - \frac{R}{c}\right) \cdot \exp\left(-\mathrm{j}\frac{2\pi}{\lambda}Vt\right) \mathrm{d}t, \tag{4.8}$$

where T is the integration time (CPI), and * denotes complex conjugation. Function $\psi(R, V)$ in (4.8) is usually referred to as the cross-ambiguity function (CAF) [7–9]. In practice, the values of the bistatic range R and bistatic velocity V of the target are unknown. For this reason, the CAF is calculated for a certain span of bistatic ranges and velocities, where the target echoes are expected. As a result, a range-velocity map is created, on which target echoes are searched for. If the delay and Doppler shift applied to the reference signal are the same as those of a target echo, a correlation peak appears indicating target bistatic range and velocity. The CAF is then processed by the detection algorithm, which decides on the presence or absence of a target echo by thresholding the absolute values of CAF.

It is worth noting the very close resemblance of the cross-ambiguity function (4.8) with the ambiguity function (2.32) from Chapter 2. In the ambiguity function used for signal evaluation, the signal is compared to itself with respect to the delay and Doppler shift. In the cross-ambiguity function, two signals are compared: the delayed and Doppler-shifted reference signal and the echo signal.

The calculation of the CAF in passive radar corresponds to two operations typically carried out in active pulse radar: matched filtering and Doppler processing. As a result of matched filtering, a range profile is obtained. By processing several range profiles via Fourier transform, the Doppler separation of target echoes can be obtained. In this way, with a two-stage process, range and velocity dimensions are created. In the CAF calculated in passive radar, both dimensions are created simultaneously.[2]

It is worth noting that a typical CPI in passive radar is much longer than in active radar. In classical active radar the illumination time is usually limited to tens of microseconds for a single pulse and tens of millisecond for a burst of pulses. Because of this relatively short observation time, the Doppler resolution is rather coarse, and the integration gain is limited. Moreover, certain characteristic phenomena, such as helicopter rotor blade flashes, can be missed, as they happen with the period of tens of millisecond. In passive radar, the transmitter is illuminating the airspace continuously. This provides the possibility to observe the target for a long time, thus enabling very fine Doppler resolution to be obtained, as well as large integration gain, which can exceed 50 dB. This enables target imaging or micro-Doppler analysis to be effectively performed, including helicopter rotor blade flashes analysis [10, 11].

One of the fundamental problems in active pulse radar is delay and Doppler frequency measurement ambiguities. Delay measurement ambiguity originates from the fact that when a series of identical pulses is sent, a late echo from a far target can be mistaken for an echo of a close target corresponding to an earlier pulse. Theoretically, the pulses could be sent so rarely that no ambiguity can practically occur. This, however, contradicts the requirement connected with the second ambiguity, that of the Doppler frequency measurement. Sending pulses by radar can be interpreted as a sampling of the space. As well known from the sampling theory, the signal is represented unambiguously when its maximum frequency is in the $(-f_s/2, +f_s/2)$ region, where f_s is the sampling frequency. For this reason, if a target echo with a high Doppler shift is to be observed unambiguously, a high sampling frequency, known as the *pulse repetition frequency* (PRF) in radar, is required. In practice, three operating regimes are defied for pulse radar:

1. High PRF: range is measured ambiguously, velocity is measured unambiguously;
2. Medium PRF: range is measured ambiguously, velocity is measured ambiguously;
3. Low PRF: range is measured unambiguously, velocity is measured ambiguously.

2 In practice, a particular way of calculating of the CAF can involve the separation of the range and velocity dimensions. An example of this is Method 3 of calculating the CAF, shown in the following part of this chapter.

If dealing with the ambiguities of range/velocity measurement were impossible or very difficult, radar would not be as popular as it is now. These ambiguities can be solved, which is done routinely in most radars [12]. However, it is an additional problem that has to be overcome in the processing chain of a radar.

A continuous-wave radar with unmodulated carrier frequency can be viewed as the fourth, extreme case of the operating regime. In such a case, the velocity is measured unambiguously, and the range is not measured at all.

As opposed to pulsed radar, range and velocity measurements in passive radar are in general unambiguous (apart from some special cases, e.g., when the repetitive nature of the signal applied introduces ambiguities) due to the continuous-wave operation and noise-like properties of the utilized signals. This is a very important practical aspect of passive radar, which allows very fast and very far targets to be detected without the need for applying algorithms to solve measurement ambiguities.

4.3 PRACTICAL CALCULATION OF THE CROSS-AMBIGUITY FUNCTION

In real passive radar, the cross-ambiguity function is calculated in the digital domain (i.e., on sampled signals). It can be written as:

$$\psi(m,k) = \sum_{n=0}^{N-1} x_e(n) \cdot x_r^*(n-m) \cdot \exp\left(-\mathrm{j}\frac{2\pi}{N}kn\right), \tag{4.9}$$

where m is an integer number corresponding to a signal delay in samples, and k is an integer number corresponding to a frequency shift expressed in frequency bins. N is the number of samples corresponding to the CPI (i.e., $T = N/f_s$), where f_s is the sampling frequency. It is worth mentioning that the CPI in passive radar can be chosen arbitrarily. Certain limits exist in practice, some of which will be discussed in this chapter; however, CPI is one of the few parameters that the passive radar designer can control.

In theory, (4.9) can be directly used for calculating the CAF for a desired span of time delays and frequency shifts. This, however, would be very time-consuming, as the multiplication and summation of very long signals are required for each pair of delay m and frequency shift k. In this section, some of the practical approaches for effective calculation of the CAF will be outlined. In order to convey the idea of each method, symbolic rather than strict mathematical notation will be used.

4.3.1 Method 1

One of the approaches used for effective calculation of the CAF is based on an observation that (4.9) can be viewed as a discrete Fourier transform (DFT), which can be calculated using the fast Fourier transform (FFT). This method is sometimes referred to as the *Fourier transform of lag product*. If a new signal is defined:

$$y^m(n) = x_e(n) \cdot x_r^*(n - m) \tag{4.10}$$

the calculation of the CAF (4.9) can be written as:

$$\psi(m, k) = \text{FFT}\{y^m(n)\} \tag{4.11}$$

where FFT$\{\cdot\}$ is the fast Fourier transform operator, and k is the frequency bin index resulting from the FFT. By calculating the FFT, the whole span of normalized frequency is obtained (from -0.5 to $+0.5$). In the case of passive radar, only a narrow span of Doppler frequencies in comparison with the sampling frequency is of interest. This means that it is sufficient to calculate only the spectrum in the close vicinity of zero frequency. This can be obtained in the following way: first, the $y^m(n)$ signal is lowpass-filtered; thus high frequency components are rejected. Then the signal is decimated by taking every Rth sample from the filtered signal. In order to avoid spectrum aliasing, the (normalized) cutoff frequency of the lowpass filter should be at most $1/R$. Usually R is in the order of tens or hundreds. At this point, the signal contains only interesting spectrum components. From a computational point of view, the calculation of the FFT on a signal that is R times shorter provides a great advantage.

The calculation of the CAF in the way explained above can be written as:

$$\psi(m, k) = \text{FFT}\{\text{DEC}_R\{\text{LPF}_{1/R}\{y^m(n)\}\}\}, \tag{4.12}$$

where DEC$_R\{\cdot\}$ is the operator of decimation by R samples and LPF$_{1/R}\{\cdot\}$ is the lowpass filtering operator with a cutoff frequency equal to $1/R$.

For large values of R, the filtering and decimation can be done in several steps by decomposing R into $R_1 \cdot R_2 \cdot \ldots$ so that $R = R_1 \cdot R_2 \cdot \ldots$, which reduces the requirements for lowpass filters. In the process of decimation the cascaded integrator-comb (CIC) filters can also be used, which provide very effective means of realizing lowpass filtering and decimation.

A block diagram outlying the presented algorithm is shown in Figure 4.1. The two signals, echo and reference, are shown in the top left corner (the reference signal

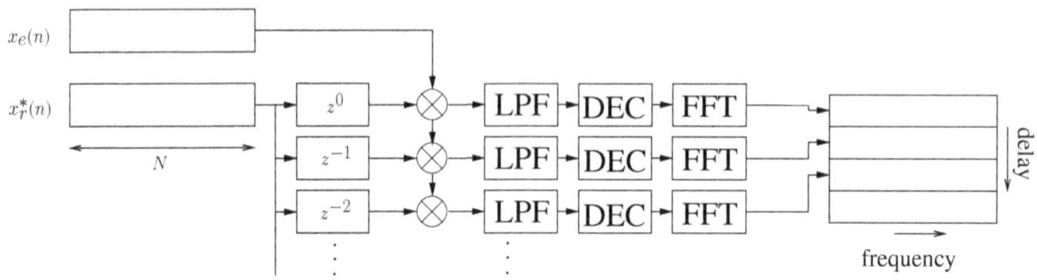

Figure 4.1 Block diagram representing the calculation of CAF using Method 1.

is conjugated). The block denoted with z^{-d} represents the signal delay by d samples. The delayed reference signal is multiplied with the echo signal. The multiplied signal is then lowpass-filtered and decimated. The spectrum of this signal, calculated using FFT, is one row in the CAF matrix, which corresponds to a single delay. Consecutive rows are obtained by introducing varying delay m to the reference signal.

4.3.2 Method 2

Another approach to the calculation of the CAF is based on the realization that (4.9) can be viewed as a correlation of two signals. This method is sometimes called the *filter method/filter bank method* [13, 14]. With the definition of the two signals:

$$y_1(n) = x_e(n) \tag{4.13}$$

$$y_2^k(n) = x_r(n) \cdot \exp\left(\mathrm{j}\frac{2\pi}{N}kn\right) \tag{4.14}$$

the calculation of the CAF can be written as:

$$\psi(m,k) = \mathrm{COR}\{y_1(n), y_2^k(n)\} \tag{4.15}$$

where COR$\{\cdot\}$ is the cross-correlation operator. This interpretation has the following rationale: the correlation is a means of comparing two signals with respect to the delay. A large value of cross-correlation means high similarity of two signals. Therefore, cross-correlating the surveillance channel signal ($y_1(n)$) with the frequency-shifted reference signal ($y_2(n)$) shows the similarity of the reference signal to the potential echo signal.

Because of the close relationship of cross-correlation with the convolution, the properties of the Fourier transform can be used to speed up the computations. According to the *convolution theorem*, the Fourier transform of the convolution of two signals is equivalent to the multiplication of the corresponding Fourier transforms. As a result of this, the convolution can be calculated as the inverse Fourier transform of the product of the signal spectra, which can be written in a generic form as: $x * y = \mathcal{F}^{-1}\{\mathcal{F}\{x\} \cdot \mathcal{F}\{y\}\}$. Therefore, the cross-correlation can be realized by calculating the FFT of the two signals, multiplying the results and performing inverse FFT. This can be symbolically written as:

$$\psi(m,k) = \text{IFFT}\{\text{FFT}\{y_1(n)\} \cdot \text{FFT}^*\{y_2^k(n)\}\} \tag{4.16}$$

where IFFT$\{\cdot\}$ is the inverse fast Fourier transform operator. The calculation of the whole cross-correlation will provide a full span of delays, that is, from $-N$ to $+N$ (where N is the number of signal samples). From the point of view of passive radar, only a small portion of the span of delays is interesting, for which the target echoes are expected. Usually the maximum delays reach hundreds or thousands of samples, compared to N equal to hundreds of thousands of samples. Moreover, only positive delays are of interest, as the echo is always delayed with respect to the direct signal. For this reason, the sizes of the FFTs can be chosen so that the only valuable result is not distorted by the cyclic convolution.

A visual representation of the analyzed algorithm is shown in Figure 4.2. The reference signal is multiplied by the $\exp(j2\pi/Nkn)$ function, which introduces Doppler shift. Then the echo signal is correlated with the frequency-shifted reference

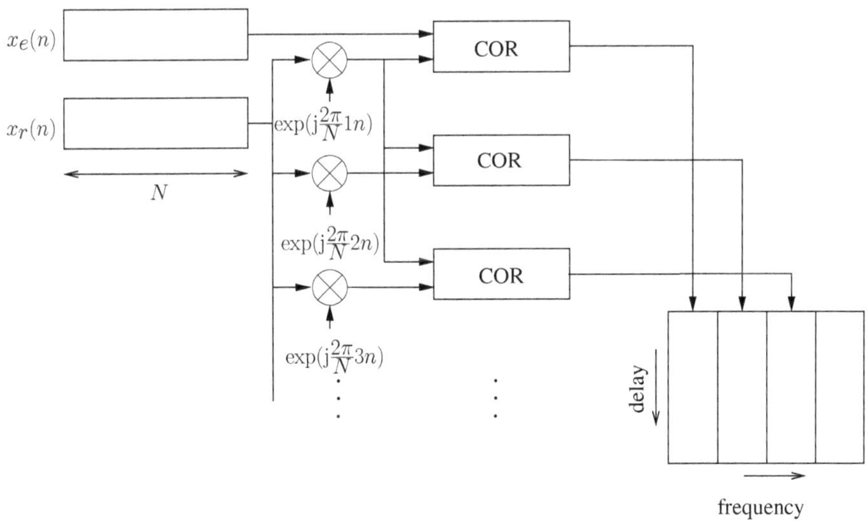

Figure 4.2 Block diagram representing the calculation of CAF using Method 2.

signal. The result is a single column in the CAF matrix, corresponding to a single frequency bin. In order to obtain the required span of frequencies, the reference signal has to be shifted correspondingly.

4.3.3 Method 3

The third method that is often applied for the calculation of the CAF, usually referred to as the *batch algorithm* [15], utilizes the fact that for short time intervals the Doppler resolution is very poor, therefore the Doppler shift can be neglected. This allows partial correlations to be calculated first (this step corresponds to matched filtering in pulse radar resulting in a range profile), followed by the Fourier transform (this step corresponds to Doppler processing in pulse radar).

Let us consider definition of the cross-ambiguity function (4.9) again. It can be reformulated in the following way:

$$\begin{aligned}\psi(m,k) &= \sum_{q=0}^{Q-1}\sum_{p=0}^{P-1} x_e(qP+p)x_r^*(qP+p-m)\exp\left(-\mathrm{j}\frac{2\pi}{N}k(qP+p)\right) \approx \\ &\sum_{q=0}^{Q-1}\left\{\sum_{p=0}^{P-1} x_e(qP+p)x_r^*(qP+p-m)\right\}\exp\left(-\mathrm{j}\frac{2\pi}{N}kq\right) = \\ &\mathrm{FFT}\left\{\sum_{p=0}^{P-1} x_e(qP+p)x_r^*(qP+p-m)\right\} = \\ &\mathrm{FFT}\left\{\mathrm{COR}\left\{x_e(qP+p), x_r(qP+p)\right\}\right\}\end{aligned} \tag{4.17}$$

In the first line of the above equation, the discrete time n is merely substituted by $n = qP + p$, where $p = 0, \ldots, P-1$ and $q = 0, \ldots, Q-1$. This means that the signal is divided into Q blocks, each P samples long. The p index denotes the sample index within the block, whereas q is the block index. The transition from the first to the second line is the main point of the presented method for calculation of the CAF. It is assumed that within one block (i.e., for $p = 0, \ldots, P-1$) the phase change of the exp$(\cdot)$ function is negligible. This allows for the exp$(\cdot)$ function to be excluded from the internal sum (over p). For this reason, the calculation of the CAF can be viewed as an FFT of the internal sum (third line of the above equation). The internal sum, however, has to be calculated for each of the Q blocks. It is a simple correlation of two signals (without Doppler shift) performed for a short block. This fact is indicated in the last line of the above equation.

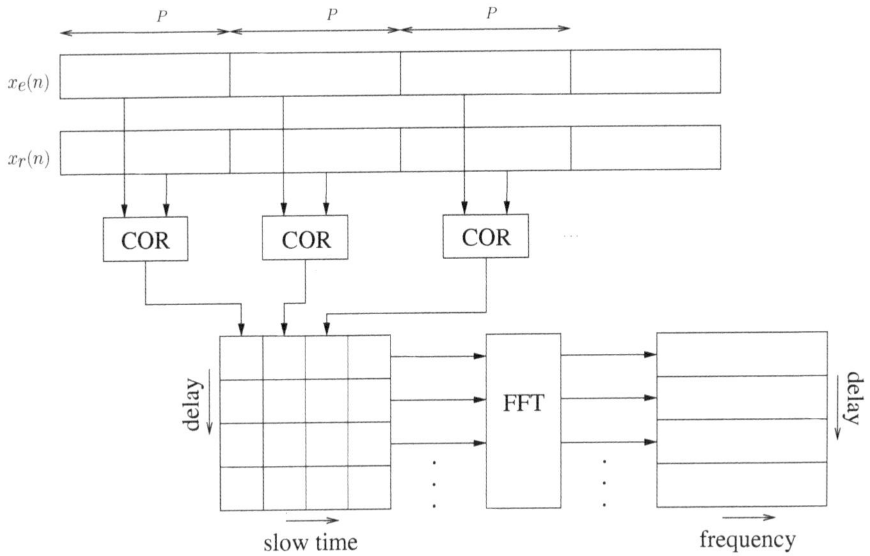

Figure 4.3 Block diagram representing the calculation of CAF using Method 3.

A block diagram of the third algorithm is depicted in Figure 4.3. The reference and echo signals are divided into blocks, each P samples long. The signals from each block are correlated (usually using FFT/multiply/IFFT approach, similar as in (4.16)). The result can be formed into a matrix, whose columns are the output of partial correlations. In this matrix the row index corresponds to the time delay between the reference and echo signals. The column index, corresponds to *slow time*. In pulse radar slow time corresponds to consecutive pulses, whereas *fast time* denotes samples within one pulse. In this case slow time means the time index decimated by the division of the signal into blocks. Calculation of the spectrum of the signal in each row, using FFT, results in a CAF matrix.

4.4 REDUCTION OF DETERMINISTIC SIDELOBES

As shown in Chapter 2, at first glance it seems that a white noise signal has a perfect ambiguity function. Admittedly, there are random sidelobes spanning across all ranges and velocities, but there is only a single peak at zero range and zero velocity, seemingly without any additional sidelobes. This perfect impression was destroyed when the band-limited noise signal was investigated. The ambiguity function of such a signal reveals sidelobes in the range direction, which are directly connected to the shape of the signal spectrum. This effect is in fact related to the sampling of the ambiguity function. If the noise is white (where a discrete noise signal occupies the whole spectrum), the autocorrelation function has a sinc$(\cdot)$ (i.e., $\sin(x)/x$) shape. Assume that an cross-ambiguity function is calculated; two identical signals are correlated with zero delay. When the correlation is calculated with one sample spacing, the sinc$(\cdot)$ function is sampled at nulls (except the argument equal to 0, where the function reaches the value of 1). Even when the delay between two identical signals is nonzero, but is a multiple of a sampling interval, the sidelobes would be invisible. The sidelobes would become apparent when the delay between the correlated signal was noninteger. In such a case, the sinc$(\cdot)$ function would be sampled at points different than nulls.

The phenomenon of sampling the sinc$(\cdot)$ function is shown in Figure 4.4. The horizontal axis represents the normalized delay of the cross-ambiguity function. The vertical axis corresponds to the amplitude of the cross-ambiguity function (in linear scale). The dashed line shows the envelope of the true sinc-shaped cross-ambiguity function. In the first case, shown in Figure 4.4(a), the function is sampled with spacing equal to 1. In this case the samples fall into nulls of sinc$(\cdot)$, except for a delay equal to 0, where the maximum value of 1 is obtained. In the second case, shown in Figure 4.4(b), where the sampling spacing is equal to 0.1, the samples reveal the shape of a sinc$(\cdot)$ function in much greater detail. In this case, the sidelobes are clearly visible. Denser sampling, equal to 0.1, corresponds to a situation when a signal with a rectangular-shaped spectrum is oversampled 10 times. The conclusion of the presented discussion is valid not only for the range dimension in the cross-ambiguity function, but also for the velocity dimension, as will be presented in the following part of this section.

The fact that sampling the sinc$(\cdot)$ function at nulls gives an apparently ideal result is well known in spectral analysis. When a finite-time sinusoidal signal is sampled in such a way that there is an integer number of periods in the observation interval, the discrete Fourier transform (DFT) provides an ideal single peak. When the frequency of the signal is shifted, the sidelobes (previously not visible) appear in the spectrum. This phenomenon is known as spectral leakage, as the power of the signal leaks from the peak to the sidelobes. The sidelobes are also revealed when the DFT is oversampled; the spectrum is calculated denser than with $1/N$ spacing (in normalized frequency). A common practice to limit the leakage is to apply signal windowing in the time domain. The same approach can be used when calculating the cross-ambiguity function.

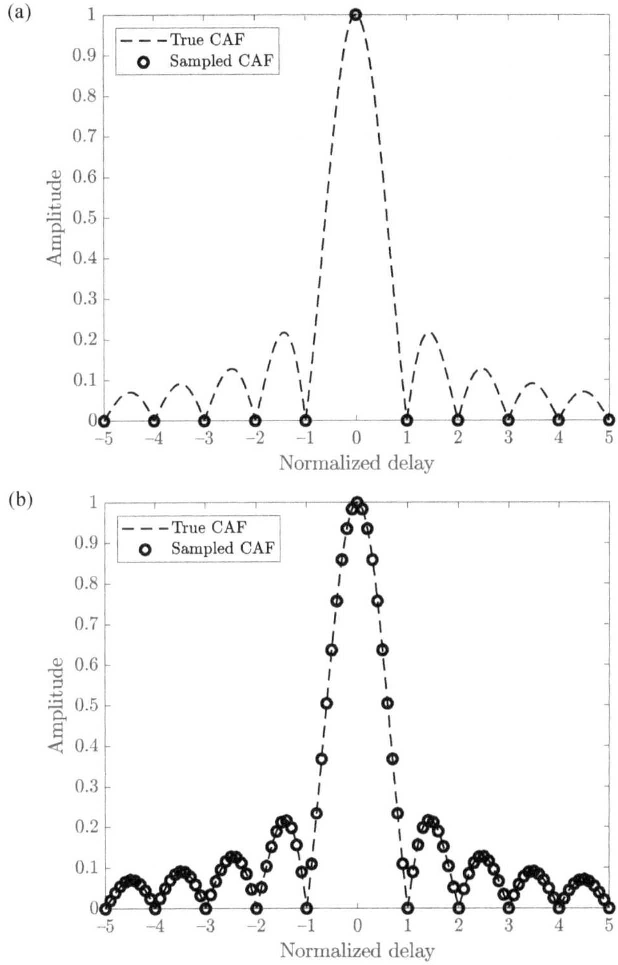

Figure 4.4 Cross-section of sinc-shaped cross-ambiguity function sampled with (a) spacing equal to 1 and (b) spacing equal to 0.1.

An example of a white noise signal analysis is shown in Figure 4.5. The noise signal occupies a bandwidth of 200 kHz, and the sampling frequency is also 200 kHz, which means that the discrete signal has a character of white noise. The cross-ambiguity function is calculated without oversampling in the range nor velocity dimension. As already mentioned, in such a case, the cross-ambiguity function has

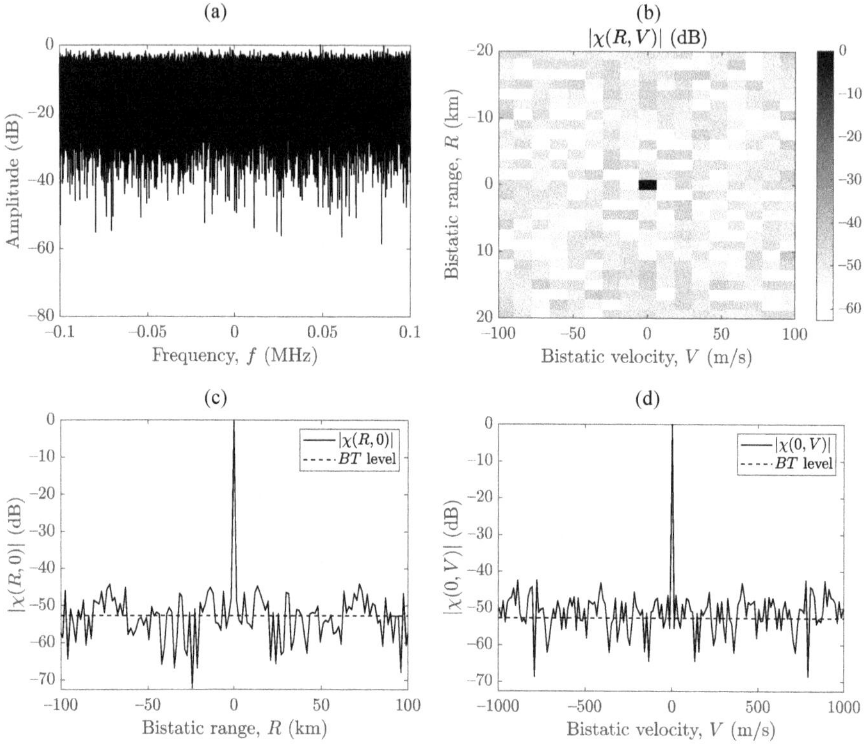

Figure 4.5 White noise signal without oversampling. (a) Signal spectrum. (b) Cross-ambiguity function. (c) Range cross-section of the cross-ambiguity function. (d) Velocity cross-section of the cross-ambiguity function.

only a single peak at zero range and velocity. The residual fluctuations at the BT level below the peak are also present, but are not the focus of this section.

In Figure 4.6 the same signal with the bandwidth of 200 kHz is presented, but this time the sampling frequency is 2 MHz. The discrete signal is therefore a band-limited noise with the normalized bandwidth of 0.1. In addition to oversampling in the time domain, the cross-ambiguity function was oversampled in the velocity dimension. As a result of oversampling in the time domain, range sidelobes appear in the CAF. They have a shape similar to the sinc($\cdot$) function, with the first sidelobe at the level of -13 dB below the main peak. This is usually unacceptable, as a target echo could mask weaker surrounding echoes. A similar situation is observed in the velocity dimension. The sidelobes also have a sinc($\cdot$) shape, and can cause target echo masking.

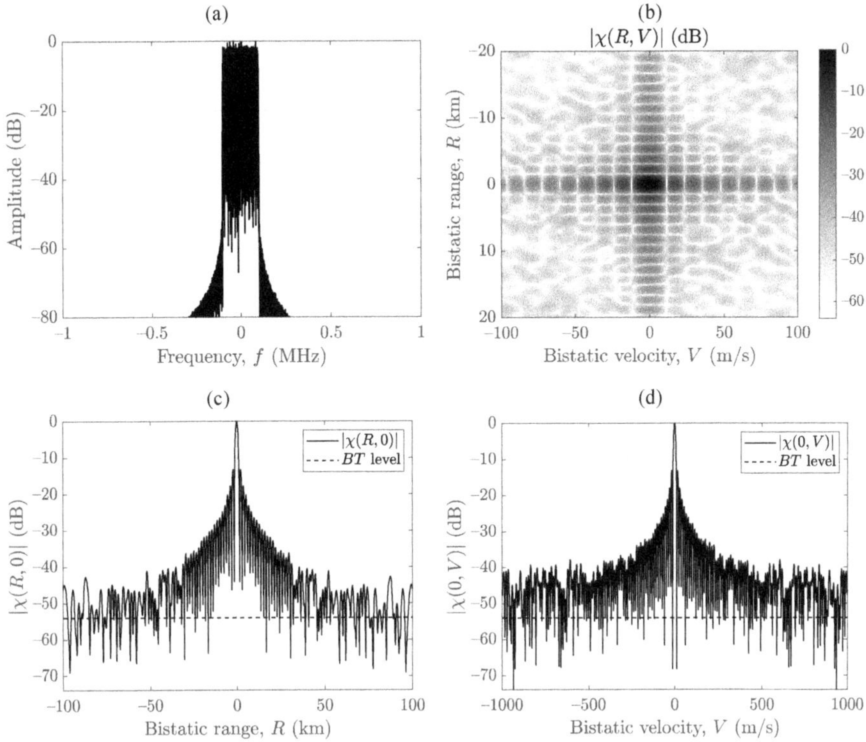

Figure 4.6 Oversampled noise signal. (a) Signal spectrum. (b) Cross-ambiguity function. (c) Range cross-section of the cross-ambiguity function. (d) Velocity cross-section of the cross-ambiguity function.

In order to reduce the sidelobes in the velocity dimension of the CAF, a time windowing function can be introduced into the equation in the following way:

$$\psi^w(R, V) = \int_{-T/2}^{T/2} x_e(t) \cdot x_r^* \left(t - \frac{R}{c}\right) \cdot w(t) \cdot \exp\left(-\mathrm{j}\frac{2\pi}{\lambda}Vt\right) \mathrm{d}t, \qquad (4.18)$$

where $w(t)$ is the time windowing function. This has a straightforward interpretation. The calculation of the CAF can be interpreted as the Fourier transform of signal product $x_e(t) \cdot x_r^* \left(t - R/c\right)$. In order to reduce spectral sidelobes, the windowing function has to be applied in the time domain. The windowing function can be one typically applied in spectral analysis, such as Hamming, Blackman, or Gauss windows.

An example of the application of a time domain Hamming window is shown in Figure 4.7. In this case, the same signal with the same oversampling in the frequency domain as in the previous example was considered. As can be seen, deterministic sidelobes in the velocity dimension dissipated almost completely. However, this comes at the cost of a widening of the main peak, a behavior known from spectral analysis. Another disadvantage of applying a windowing function is the reduction of the main peak amplitude (by approximately 5 dB in the analyzed example). This results from the fact that the window function reduces signal power, as the values at the edges of the window are substantially reduced. However, similar as in the case of digital beamforming, windowing reduces not only the useful signal, but also the noise. Therefore, the parameter that is important is not the amplitude loss, but the

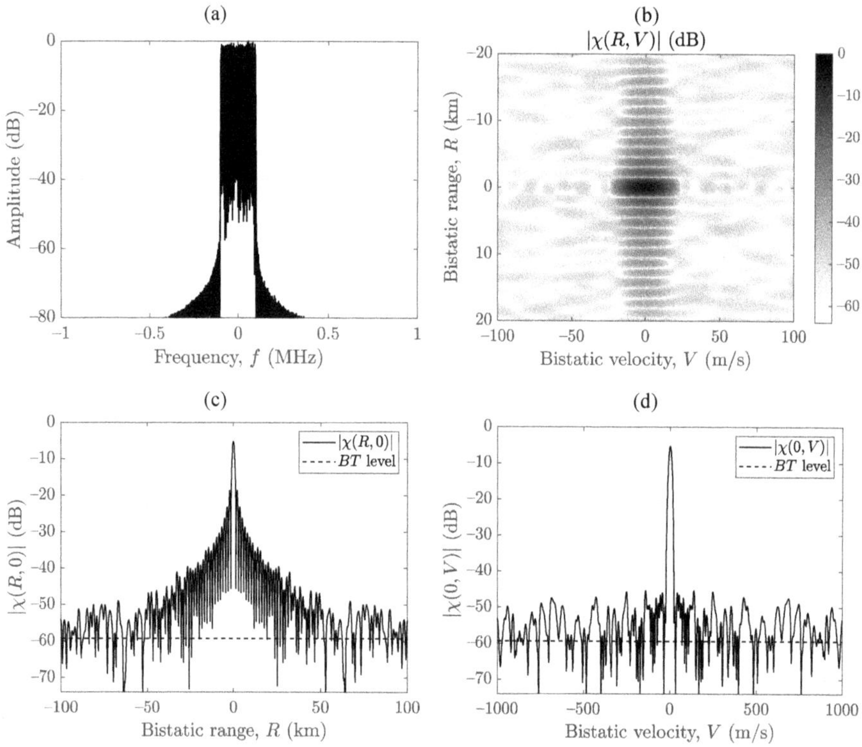

Figure 4.7 Oversampled noise signal. (a) Signal spectrum. (b) Cross-ambiguity function calculated with time windowing. (c) Range cross-section of the cross-ambiguity function. (d) Velocity cross-section of the cross-ambiguity function.

signal-to-noise ratio loss. In this example, the signal-to-noise ratio loss is 1.2 dB. The time windowing has not changed the structure of the sidelobes in the range dimension, which still remain to be attenuated.

Reduction of the range sidelobes can be achieved by the modification of the spectrum of the analyzed signal. More specifically, a windowing function has to be applied in the frequency domain in order to shape the spectrum:

$$x_r^w(t) = \mathcal{F}^{-1}\{\mathcal{F}\{x_r(t)\} \cdot W(f)\} \tag{4.19}$$

The above equation means that a windowing function $W(f)$ in the frequency domain is applied to the spectrum of the analyzed signal $\mathcal{F}\{x_r(t)\}$. Similar to time domain windowing, a spectral windowing function attenuates the signal at the edges, and has a maximum in the middle. This can be observed in Figure 4.8. Figure 4.8(a) shows the spectrum of the signal, which is no longer rectangular. An envelope caused by the Gaussian windowing function applied in this example is clearly visible. The CAF calculated using both time-domain windowing and frequency-domain windowing (shaping of spectrum) can be seen in Figure 4.8(b–d). This time, deterministic sidelobes in both the range and velocity dimensions were attenuated. Reduction of the range sidelobes is achieved due to the spectrum shaping. Attenuation of the sidelobes in the velocity dimension, can be attributed to the time-domain windowing. In both cases widening of the main peak is clearly visible. In addition, the amplitude of the main peak was reduced by approximately 12 dB, which corresponds to a loss in the signal-to-noise ratio of 3.2 dB, compared to the case of lack of time and frequency windowing.

Summarizing the discussion on deterministic sidelobes and their reduction, the following conclusions can be drawn: when a white noise signal is investigated, it seems that there are no sidelobes present (apart from the random noise floor). This is only an illusion. The deterministic sidelobes are present, and are clearly visible when oversampling in range or velocity is applied. In order to attenuate these sidelobes, signal windowing can be applied. The application of time domain windowing reduces sidelobes in the velocity dimension. Windowing in the frequency domain, in turn, mitigates range sidelobes. In both cases main peak broadening occurs (in respective dimensions). More importantly, however, the application of windowing is connected to the reduction of the peak amplitude, which in practice means a lower signal-to-noise ratio, thus a poorer detection performance. This is a very serious disadvantage of using windowing in the time or frequency domains. As the sensitivity is one of the main parameters of radar, limiting the losses introduced in the processing should be one of the main priorities. For this reason, any windowing should be applied with care, only after analysis of its effects. Finally, it is also worth

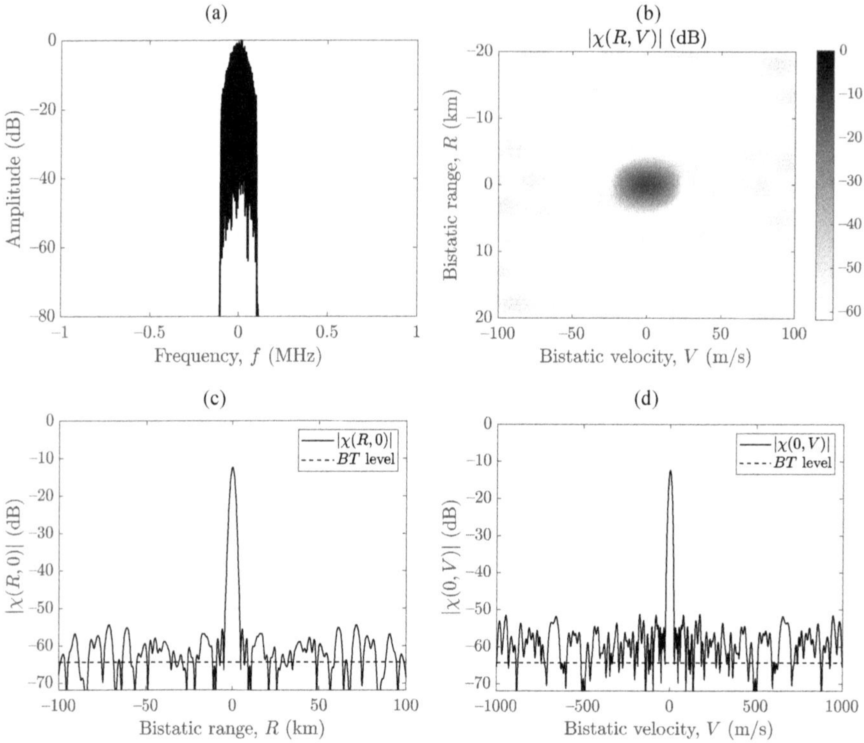

Figure 4.8 Oversampled noise signal (with frequency windowing). (a) Signal spectrum. (b) Cross-ambiguity function calculated with time and frequency windowing. (c) Range cross-section of the cross-ambiguity function. (d) Velocity cross-section of the cross-ambiguity function.

noting that the described windowing operations modify only deterministic sidelobes, and they have no influence on the random fluctuations of the cross-ambiguity function.

4.5 INTEGRATION GAIN

Calculation of the CAF realizes the coherent integration of the signal. Indeed, when the delay m and frequency shift k of the reference signal in (4.9) match the ones of the target echo, the signal in summed coherently, whereas the noise component in the echo signal is summed incoherently. As a result, the CAF provides integration gain. The output signal-to-noise ratio is increased with regards to the input as:

$$\mathrm{SNR} = \mathrm{SNR_{in}} BT, \tag{4.20}$$

therefore, the integration gain is:

$$G_{\mathrm{int}} = BT. \tag{4.21}$$

The gain depends on the signal bandwidth B and integration time T, or CPI. The bandwidth[3] of the signal is related to the type of the illuminator of opportunity used by the passive radar, and is beyond the control of the radar designer. The CPI, on the other hand, can be adjusted in order to change radar performance.

If the signal bandwidth B is equal to (in practice usually slightly less than) the sampling frequency f_s, the integration gain can be calculated as:

$$G_{\mathrm{int}} \approx f_s T = \frac{T}{t_s} = N \tag{4.22}$$

where $t_s = 1/f_s$ is the sampling interval, and N is the number of signal samples corresponding to T, the CPI.

In practice the output SNR, which directly influences the capability of target detection, can be measured by investigating the CAF. The peak on the CAF corresponding to the target echo represents the signal power after coherent integration. The noise level, can be estimated by investigating the surrounding values of the CAF. Often the CAF values outside the target echo region will represent the level corresponding to noise present in the signal; however, this does not always have to be the case. Sometimes the constant level of the CAF may correspond to residual fluctuations of some strong target echo; as previously discussed, each correlation peak has related residual fluctuations at the level of BT below the maximum.

The described method of measuring the SNR after correlation is called the *peak-to-noise floor ratio* (PNFR) [16–18]. The PNFR can be calculated from the number of samples of the signal N and the input SNR as:

$$\mathrm{PNFR} = \frac{N}{1 + \mathrm{SNR_{in}^{-1}}}, \tag{4.23}$$

An example of a comparison of a theoretical output SNR calculated using (4.20) and a PNFR calculated using (4.23) is shown in Figure 4.9. It was assumed that the number of samples is $N = 10^6$, which provides the integration gain

3 The bandwidth B is the signal's effective bandwidth rather than nominal bandwidth (compare discussion in Section 2.5.1).

$G_{\text{int}} = 60$ dB. As can be seen the SNR is simply the value of SNR_{in} shifted by 60 dB of integration gain. The PNFR corresponds to the SNR up to the value of 60 dB. At this point the values of the PNFR saturate, whereas values of the SNR increase linearly. This behavior results from the aforementioned origin of noise floor measured with the PNFR parameter. For the values of the SNR below 60 dB the noise floor actually corresponds to the noise level. When this level is exceeded, the noise floor corresponds to the residual fluctuations related to the correlation peak of the target echo. For this reason, for larger SNR the values of the PNFR do not change. This is one of the limitation of assessing SNR using PNFR. The second limitation will become apparent during the analysis of the next example.

In Figure 4.10 a comparison of theoretical and measured PNFRs is presented. The calculations were performed for a number of samples equal to $N = 10^4, 10^5, 10^6$. The theoretical values of the PNFR were calculated using (4.23). The measured values of the PNFR were obtained as a ratio of the maximum value of the CAF to the mean value of the CAF. The mean value was calculated from a range-velocity region of the CAF that did not contain a simulated target echo. As can be seen, for higher values of the SNR the theoretical and measured values of the PNFR are almost identical. For lower values of the SNR, theory and measurement diverge.

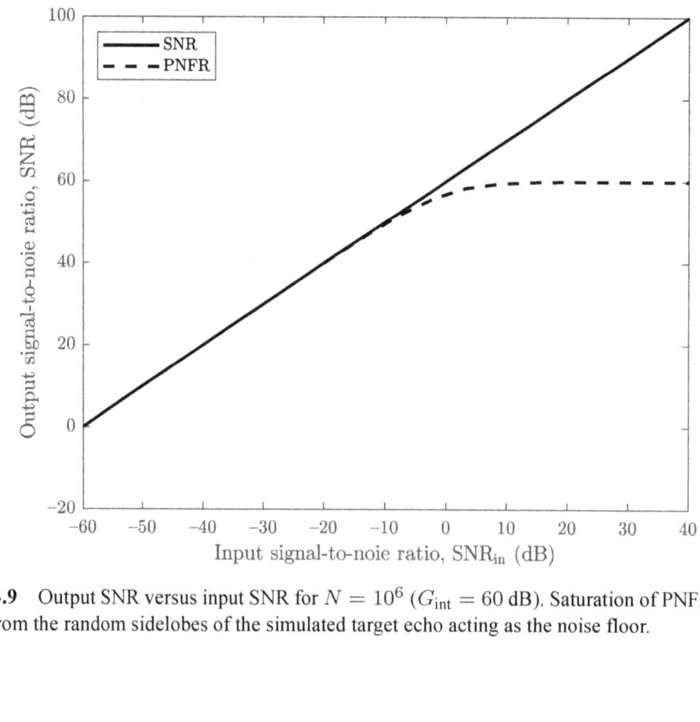

Figure 4.9 Output SNR versus input SNR for $N = 10^6$ ($G_{\text{int}} = 60$ dB). Saturation of PNFR at 60 dB results from the random sidelobes of the simulated target echo acting as the noise floor.

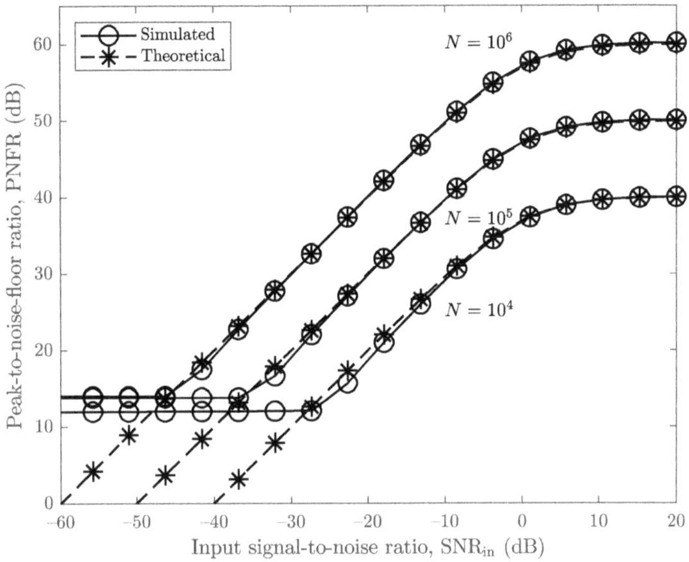

Figure 4.10 Peak-to-noise floor ratio for $N = 10^4, 10^5, 10^6$. Saturation of the PNFR for the low values of SNR_{in} results from the fact that the peak value disappears in the noise for SNR = 10–12 dB.

This is the second mentioned limitation of the PNFR. The values of the PNFR are saturated at the level of approximately 12 dB. The reason for this is that the maximum value of the CAF taken during the calculation of the PNFR no longer corresponds to the target echo, but to some random value of the noise floor.

4.6 EXTENDED SIGNAL MODEL AND CORRELATION PROCESSING

In this section an extended signal model (i.e., a model with higher-order derivatives) will be considered. From (4.3) it can be seen that the instantaneous bistatic range $r(t)$ influences two components of the signal: the baseband signal $x_r(t)$ and the carrier frequency. In both cases some simplifying assumptions have been used. As already mentioned, these simplifications may lead to a significant degradation of the detection performance and the parameter estimation accuracy. The extensions to the signal model will be presented. First, the effects of the stretching of the envelope of the reference signal $x_r(t)$ will be considered. This is connected with the range cell migration. Next, the effects of the influence of the motion model on the carrier frequency will be analyzed. This, in turn, is associated with the velocity cell migration.

4.6.1 Range Cell Migration and Its Reduction

Once again consider the model of the received signal with a generic form of the instantaneous bistatic range $r(t)$:

$$x_e(t) = C' \cdot x_r\left(t - \frac{r(t)}{c}\right) \cdot \exp\left(\mathrm{j}\frac{2\pi}{\lambda}r(t)\right). \tag{4.24}$$

Assume that the actual instantaneous bistatic range is $r(t) = R + Vt$. Then the exact form of the received signal is [19–21]:

$$x_e(t) = C'' \cdot x_r\left(t - \frac{R + Vt}{c}\right) \cdot \exp\left(\mathrm{j}\frac{2\pi}{\lambda}Vt\right). \tag{4.25}$$

The difference between the (4.25) model and (4.7) is the variable delay of the complex envelope $x_r(t)$. If the correlation scheme is built on this signal model, the extended cross-ambiguity function is derived [19]:

$$\psi_s(R, V) = \int_{-T/2}^{T/2} x_e(t) \cdot x_r^*\left(t - \frac{R + Vt}{c}\right) \cdot \exp\left(-\mathrm{j}\frac{2\pi}{\lambda}Vt\right) \mathrm{d}t. \tag{4.26}$$

where $\psi_s(R, V)$ denotes CAF with signal stretch.

A close analysis of the variable delay of the reference signal is required. The signal with variable delay can be rewritten as:

$$x_r\left(t - \frac{R + Vt}{c}\right) = x_r\left(t\left(1 - \frac{V}{c}\right) - \frac{R}{c}\right) = x_r\left(t \cdot \alpha - \frac{R}{c}\right), \tag{4.27}$$

where $\alpha = 1 - V/c$. It can be seen that, apart from a constant delay, equal to R/c, the time scale is changed by the α factor. If a significant change of the time scale is not compensated during the correlation CPI, SNR losses can be expected. The losses will depend on the ratio V/c and on the length of the correlated signal. The larger the ratio V/c and/or the longer the correlated signal, the larger the loss of the SNR. Because in practice the correlation of signals is calculated using an appropriate sum of signal samples, the length of the correlated signals is measured by the number of samples. Consider a case where N samples are correlated. If the time scale coefficient is equal to $\alpha = 1 - 1/N$, it means that at the end of the CPI, the signals will be shifted by 1 sample. This remark is important from the point of view of analyzing the results of the next experiment.

The effects of the time scale changes have been analyzed by performing computer simulations. In Figure 4.11 the SNR losses versus the number of samples integrated in CAF for different values of the bistatic velocity is shown. The losses are greater for a larger number of samples and larger bistatic velocity. It is worth analyzing the results quantitatively and comparing them with typical parameters used in passive radars. For the bistatic velocity $V = 300$ m/s, the ratio V/c is equal to 10^{-6} and the value of the time scale coefficients is equal to $\alpha = 0.999999$. It means that if $N = 10^6$ samples are correlated, the signals will be shifted by 1 sample at the end of the CPI. From Figure 4.11 it can be seen that the losses for $V = 300$ m/s start to appear for $N = 10^5$, that is, when the time shift corresponds to a 0.1 of the sampling interval. A similar situation can be observed for $V = 3000$ m/s, for which $V/c = 10^{-5}$, and $\alpha = 0.99999$. The losses start to increase noticeably for $N = 10^4$, which corresponds to 0.1 sampling interval shift. If larger

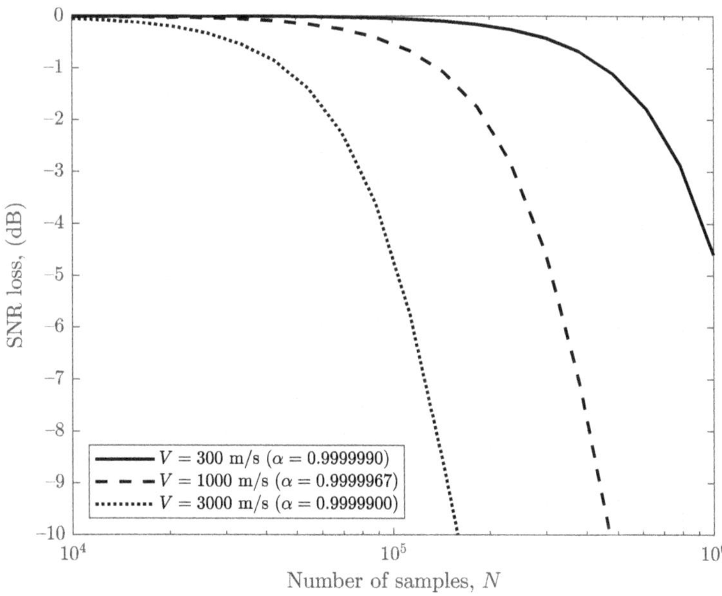

Figure 4.11 Signal-to-noise ratio loss versus number of samples integrated in CAF for different values of the bistatic velocity.

losses are acceptable (e.g., 1 dB), a shift of approximately 0.5 sample per CPI can remain uncompensated. These observations allow for the formulation of a practical rule, according to which the time scale stretch has to be compensated for if the shift exceeds a certain part (later this fraction will be denoted with C_{mr}) of the sampling interval during the CPI.

Let us analyze the phenomenon of the range cell migration from a different angle. A target with the bistatic velocity equal to V shifts during the CPI of T in bistatic range by VT. This shift should be small compared to the bistatic range resolution cell c/B; otherwise, range cell migration occurs. Assume that the signal bandwidth B is almost equal to the sampling frequency $B \approx f_s$, which means that the bistatic range resolution can be approximated as c/f_s. Then the following condition can be formulated:

$$VT < C_{mr}\frac{c}{f_s}, \tag{4.28}$$

which means that the target echo shift VT should not exceed a certain fraction C_{mr} of the range resolution cell c/f_s. Because the product of the CPI and the sampling frequency yields the number of correlated samples, $Tf_s = N$, the above relation can be reformulated as:

$$N_{\max} < C_{mr}\frac{c}{V} \tag{4.29}$$

This can be interpreted in the following way. The maximum number of samples in the CPI is limited by the maximum expected velocity of the target. For example, for $V = 1,000$ m/s, the maximum number of correlated samples is $3 \cdot 10^5 = 300,000$ (assuming $C_{mr} = 1.0$). Another way of looking at this limitation is by calculating the maximum allowable velocity $V_{\max}$ for a given number of samples N:

$$V_{\max} < C_{mr}\frac{c}{N} \tag{4.30}$$

This rule can be related to typical parameters used in passive radars based on, for example, FM radio and DVB-T television. The nominal bandwidth of an FM signal is 150 kHz, and a typical sampling frequency f_s of an FM signal is approximately 200 kHz. The CPI for FM radio is in the order of 1s; therefore the number of samples is $N = 200 \cdot 10^3$. The maximum bistatic velocity for such a situation, which will not introduce significant losses without compensation, is $V = 1,500$ m/s (calculated using (4.30) for $C_{mr} = 1.0$). Such a value is rarely encountered in practice; thus, no compensation of the range cell migration is required for the typical integration time in the FM radio case.

For the case of a DVB-T signal, the bandwidth of the signal depends on the version of the standard, but is usually close to 8 MHz, and the sampling frequency f_s is in the order of 10 MHz. A typical CPI for DVB-T is 0.1s; therefore, $N = 10^6$ samples are processed. The maximum bistatic velocity value in this case is $V = 300$ m/s, a value not uncommon in typical scenarios.

From these considerations, the following conclusion can be drawn: the range cell migration is more significant for longer signals and larger bistatic velocities. For typical parameters of FM-based radar, the signal stretch is insignificant. In the case of DVB-T-based radar, the change of the time scale may become relevant if a longer-than-usual CPI is used, or fast targets are observed.

4.6.1.1 Stretch Processing

There are several ways in which the range cell migration correction can be implemented in practice. Let us consider again (4.27). It suggests that for each desired velocity, the scaling factor is slightly different; therefore, the reference signal should be resampled separately for each velocity. That would not be practical. However, it can be noticed that the bistatic velocity assumed in the resampling does not have to exactly match the target velocity; the reference signal resampled with one velocity can be used for a span of velocities, as suggested by (4.30). The span can be calculated as $C_{mr}c/N$, where N is the number of correlated samples, c is the speed of light, and C_{mr} is the coefficient adjusted depending on the acceptable losses.

This leads us to an algorithm, which can be called the *stretch processing* [19–21]. First, the reference signal is stretched (or squeezed, depending on the sign of the velocity) by the following resampling:

$$x_r(t) \rightarrow x_r(t \cdot \alpha) \tag{4.31}$$

for a particular value of the bistatic velocity, which influences the scale coefficient $\alpha = 1 - V/c$. The resampling can be realized in several ways, for example, by spline interpolation, or using the FFT/Chirp-Z transform. Next, the CAF is calculated in a classical way, but the resampled reference signal is used instead of the original one. The calculated CAF is matched to a certain velocity or span of velocities. By performing stretching matched to different velocities, calculating CAFs, and extracting only appropriate segments, a final CAF matrix can be built.

Let us consider a situation where $N = 10^6$ samples are correlated, and the desired span of velocities is equal to $1{,}500$ m/s (from -750 to $+750$ m/s). The span of velocity for a single resampling of the reference signal is $C_{mr}c/N = 300$ m/s (for $C_{mr} = 1.0$). Therefore, the whole CAF can be divided into 5 segments, each 300 m/s wide. For each of the segments, apart from $V = 0$ m/s, the reference signal is resampled with a velocity equal to the middle of the segment, that is, $-600, -300, +300, +600$ m/s. Then the CAF is calculated, but only the part corresponding to the current velocity segment is used in the final CAF. In this example, the resampling would have to be carried out 4 times and calculation of the CAF 5 times.

4.6.1.2 The Keystone Transform

The correction method described above is not very efficient, as it requires a separate resampling and CAF calculations for multiple values of the velocity. Another approach involves using the *Keystone Transform* [15, 22–27]. The approach is usually applied in Method 3 of calculation of the CAF. As already mentioned, relatively short blocks of the reference and echo signals are first correlated. This is usually done by performing FFT on the two corresponding blocks, multiplying the results, and performing IFFT. Next, another FFT is applied in the slow-time dimension, which results in calculated CAF matrix.

The range cell migration with the use of the Keystone Transform consists in resampling of the data matrix after the first FFT and multiplication (i.e., in the fast frequency-slow time domain). After the resampling the data matrix has trapezoid shape, hence the name of the Keystone Transform. The resampling, as in the previous method, is relatively time-consuming; however, it is performed only once for all velocities.

4.6.1.3 FIR Filtering

Yet another approach involves calculating the CAF only once, and then post-processing it using a two-dimensional finite impulse response (FIR) filter in order to mitigate range cell migration, as shown in [28]. The filter operates in the bistatic range-bistatic velocity domain and acts as a matched filter, which focuses spread target echoes. The impulse response of the filter is precomputed by simulating a target echo with range cell migration present and calculating the CAF. In this way, a spread target echo is obtained. This echo is cropped from the CAF and used as an impulse response of the two-dimensional filter. Even with relatively small impulse response, 5×5 or 7×7, satisfactory results can be obtained. Similar as with the stretch processing, the CAF can be divided into several velocity segments, and the compensation is performed separately for each segment.

4.6.2 Velocity Cell Migration and Its Reduction

The effects of the target motion on the carrier wave of the reference signal will now be considered. In the derivation of the signal model presented at the beginning of this chapter, it was assumed that the instantaneous bistatic range could be approximated by a first-order polynomial. It is equivalent to the assumption that the bistatic velocity does not change significantly during the CPI. In practice this condition does not have to be satisfied. One of the examples is the detection of highly maneuvering targets, which have significant bistatic acceleration. Another example is when a long CPI is applied, for which the velocity cannot be considered constant.

Consider an extended bistatic motion model of a target. Instead of using the first-order polynomial (4.5), a second-order polynomial will be applied, with an additional term corresponding to the bistatic acceleration A [29, 30]:

$$r(t) \approx R + Vt + \frac{At^2}{2}. \tag{4.32}$$

If the extended motion model is applied to the received signal equation (4.4), the following result is obtained:

$$x_e(t) = C'' \cdot x_r\left(t - \frac{R}{c}\right) \cdot \exp\left(\mathrm{j}\frac{2\pi}{\lambda}\left(Vt + \frac{At^2}{2}\right)\right). \tag{4.33}$$

As can be seen, an additional, quadratic phase component appeared in the equation, connected with the bistatic acceleration A. The delay of the baseband signal $x_r(t)$ is still treated as constant (equal to R/c). In such a case, the correlation of the reference and echo signals results in a modified cross-ambiguity function:

$$\begin{aligned} &\psi_A(R, V, A) = \\ &\int_{-T/2}^{T/2} x_e(t)x_r^*\left(t - \frac{R}{c}\right) \exp\left(-\mathrm{j}\frac{2\pi}{\lambda}\left(Vt + \frac{At^2}{2}\right)\right) \mathrm{d}t = \\ &\int_{-T/2}^{T/2} x_e(t)x_r^*\left(t - \frac{R}{c}\right) \exp\left(-\mathrm{j}\frac{2\pi}{\lambda}Vt\right) \exp\left(-\mathrm{j}\frac{2\pi}{\lambda}\frac{At^2}{2}\right) \mathrm{d}t \end{aligned} \tag{4.34}$$

where $\psi_A(R, V, A)$ denotes CAF with bistatic acceleration. The cross-ambiguity function (4.34) depends not only on the bistatic range R and the bistatic velocity

V, but also on the bistatic acceleration A. The extended cross-ambiguity can be calculated by performing the following substitution:

$$x_r\left(t-\frac{R}{c}\right) \rightarrow x_r\left(t-\frac{R}{c}\right)\cdot \exp\left(-\mathrm{j}\frac{2\pi}{\lambda}\frac{At^2}{2}\right), \tag{4.35}$$

and using the obtained signal in the classical formula (4.8). This substitution has to be applied for each of the desired values of the acceleration A. It means that a bank of cross-ambiguity functions can be created, each for a different value of A. In this way, a three-dimensional data matrix is obtained (R, V, A). The location of a target echo peak provides an estimate of the bistatic range R, the bistatic velocity V, and the bistatic acceleration A.

The effects of the bistatic acceleration on the SNR losses for the classical cross-ambiguity (4.8), where the acceleration is not taken into account, are investigated with the use of computer simulations. The SNR losses versus the bistatic acceleration for the classical cross-ambiguity are plotted in Figure 4.12. In the simulations typical parameters of an FM-based radar have been assumed (λ=3m, f_s=200 kHz). Each curve corresponds to a different value of the CPI. The losses increase with

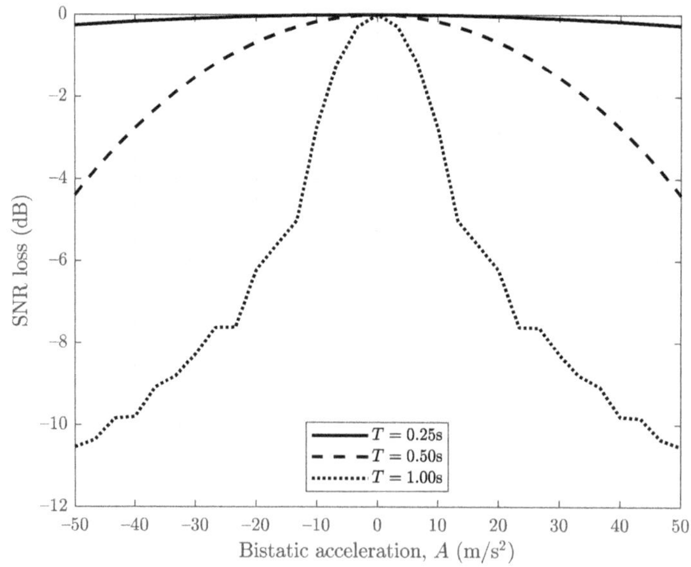

Figure 4.12 Signal-to-noise ratio loss versus bistatic acceleration for different values of the CPI (λ = 3m).

the value of acceleration, as greater acceleration causes target echo energy to spread over different velocity resolution cells. The effect is stronger for longer CPIs, as the velocity resolution is finer in such cases. These results indicate that if a target with significant bistatic acceleration is observed, the extended processing scheme has to be applied to avoid deterioration of the detection performance. The same conclusion refers to long CPIs, when the bistatic velocity cannot be regarded as constant.

The limits on the velocity cell migration can be formulated in the following way: the target with the bistatic acceleration A changes velocity during the CPI of T by AT. However, the bistatic velocity resolution cell size is λ/T. Therefore, the following condition can be formulated:

$$AT < C_{mv}\frac{\lambda}{T} \tag{4.36}$$

which means that the shift of the target velocity should not exceed a fraction C_{mv} of the velocity resolution cell. This condition can be reformulated as:

$$A_{\max} < C_{mv}\frac{\lambda}{T^2} \tag{4.37}$$

which allows the maximum acceleration $A_{\max}$ to be calculated assuming a certain value of the CPI. Or alternatively, the maximum CPI $T_{\max}$ can be calculated based on the expected bistatic target acceleration:

$$T_{\max} < \sqrt{C_{mv}\frac{\lambda}{A}} \tag{4.38}$$

The extension of the motion model from the first to the second-order polynomial has been considered so far. An analogous procedure can be carried out for higher derivatives of the motion model. An example of such an extension is shown in a real-data example in the subsequent part of this chapter [31].

As in the case of the practical calculation of the CAF in order to mitigate range cell migration, some practical approaches are worth considering in the case of velocity cell migration. As already mentioned, the CAF for a certain bistatic acceleration A can be obtained by substituting the reference signal according to (4.35), and calculating the CAF in a standard way. This operation would have to be repeated for each desired value of the bistatic acceleration A. Another approach can be used, where a single CAF is calculated, and different accelerations are obtained by one-dimensional FIR filtering, as presented in [30, 32]. The filtering approach results from the following considerations: one of the interpretations of the CAF is the Fourier transform of the product of two signals: the echo signal and delayed reference signal. Applying this reasoning to (4.34) yields the following result:

$$\psi_A(R,V,A) = \mathcal{F}\left\{x_e(t)\cdot x_r^*\left(t-\frac{R}{c}\right)\cdot \exp\left(-\mathrm{j}\frac{2\pi}{\lambda}\frac{At^2}{2}\right)\right\} = \mathcal{F}\left\{x_e(t)\cdot x_r^*\left(t-\frac{R}{c}\right)\right\} * \mathcal{F}\left\{\exp\left(-\mathrm{j}\frac{2\pi}{\lambda}\frac{At^2}{2}\right)\right\} \tag{4.39}$$

As can be seen, the multiplication by the component connected with bistatic acceleration A in the CAF can be interpreted as convolution with a certain function. This function is the Fourier transform of the $\exp\left(-\mathrm{j}\frac{2\pi}{\lambda}\frac{At^2}{2}\right)$ component, whose argument is the bistatic velocity V. In practice, the values of this function decay relatively quickly with bistatic velocity; thus, the Fourier transform can be truncated without having a significant impact on the results. For this reason, the following algorithm can be applied: first, the classical CAF is calculated. Then the function for a specific value of A is computed:

$$h_A(V) = \mathcal{F}\left\{\exp\left(-\mathrm{j}\frac{2\pi}{\lambda}\frac{At^2}{2}\right)\right\} \tag{4.40}$$

This function will be convolved with the CAF, therefore can be treated as an impulse response of a certain filter. In practice, the computations are carried out in the digital domain, therefore the impulse response represents a discrete FIR filter. The length of the impulse response can be limited to a few samples (e.g., 7 or 9 [32]). Next, the CAF for a selected acceleration value A is calculated by the following convolution in the bistatic velocity V dimension:

$$\psi_A(R,V,A) = \psi(R,V) * h_A(V) \tag{4.41}$$

In this way, the time-consuming calculation of the CAF is carried out only once. Then the CAF values for different values of A are obtained by simple one-dimensional convolution, where the length of the impulse response is just a few samples.

4.7 NUMERICAL RESULTS

The presented extended processing schemes (i.e., including higher-order motion derivatives) have been verified on real data in two typical situations where their application is advantageous: long CPI and a highly maneuvering target. First, the results for the extended CPI will be considered. Two cases will be presented: FM radio and DVB-T television signals. Due to the different bandwidths and carrier frequencies of these signals, the effects of the increased CPI are different. Next, the real-data result of the detection of a maneuvering target will be shown for FM radio.

4.7.1 Extension of the Integration Time (FM Radio)

The integration time in passive radar can be chosen arbitrarily, as the continuous stream of signal samples can be divided into blocks of any length for correlation processing. There are, however, typical CPIs that are usually applied in practice. These typical lengths are, related to the simplified signal model and the processing based on it. The values of the CPIs usually used in practice are short enough for the simplified processing to provide satisfactory results [21, 31]. The reason for using a longer-than-typical CPI is to improve detection performance, as larger integration gain provides higher detection probability.

Typical values of the CPI vary with illuminators of opportunity, which is connected with different signal bandwidth and carrier frequency. For example, the CPI typically used in FM-based radar is 1s. However, in the case of DVB-T, the CPI is in the order of 0.1s. Two examples of extending the integration time above the typical values will be considered next. In the first case, the data from FM-based radar will be presented. In the next case, the DVB-T-radar signals will be investigated.

Consider an example of extending the CPI for a real target echo with the use of FM radio signals. The signals have been acquired with the *PaRaDe* system [33, 34]. The experiment has been performed by choosing a single target echo and measuring the SNR after the correlation for different CPIs. The measurement of the SNR has been carried out by the PNFR (compare Section 4.5). The measurement results are shown in Figure 4.13. In theory, the SNR should increase proportionally to the CPI, according to (4.20). This theoretical curve is marked in the figure as a solid line with * markers. The other curves correspond to different motion models assumed when constructing the cross-ambiguity function. In all cases, the extensions of the motion model only influenced the carrier frequency component of the signal (no stretch of the baseband signal has been considered). In the first case, the classical first-order polynomial has been assumed ($r(t) = R + Vt$). It can be seen that up to approximately 1s of CPI, the obtained SNR follows the theoretical curve. At an integration time of 2s the SNR values start to decrease. This is the result of the uncompensated bistatic acceleration and higher motion derivatives, which leads to velocity cell migration. If the extended motion model is used, including the bistatic

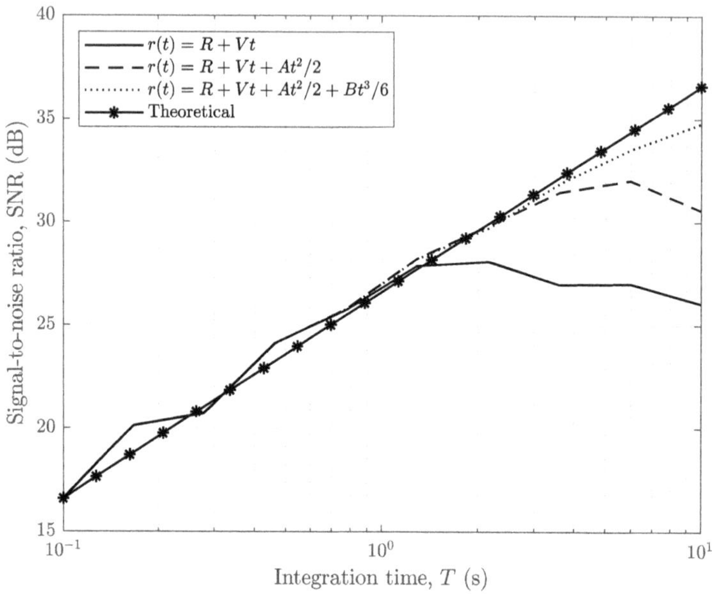

Figure 4.13 Signal-to-noise ratio versus the CPI for a real target echo (FM radio signal); different curves correspond to different motion models.

acceleration ($r(t) = R + Vt + At^2/2$), the SNR values start to decrease for an integration time equal to 6s. The motion model can be extended further to include higher derivatives. The last curve has been obtained by incorporating the derivative of bistatic acceleration B, referred to as the jerk ($r(t) = R + Vt + At^2/2 + Bt^3/6$). In this case the measured SNR follows the theoretical curve almost to 10s. As can be seen, some of the measured values of the SNR in Figure 4.13 are higher than the theoretical ones. This difference can be attributed to the imperfect measurement of the SNR, especially for shorter integration times.

In the presented example, the bistatic velocity of the target was relatively low and the size of the range resolution cell was large; therefore, the effects associated with the signal envelope stretch, connected to the range cell migration, were negligible. It is important to note that the motion model used in the derivation of the signal processing is not the only limitation of the obtainable integration gain. Another limitation may arise from the fluctuation of the target echo amplitude; even when the target motion model is perfectly compensated, the fact that the target echo amplitude is not constant may be the limiting factor. The problem has been analyzed in [31].

4.7.2 Extension of the Integration Time (DVB-T Television)

Now consider an example of extending the CPI for DVB-T signals. The signals used in the experiment have been obtained courtesy of Dr. Karl Erik Olsen from the Norwegian Defense Research Establishment (FFI). The system used to capture the data is capable of recording four separate DVB-T channels simultaneously. Two directional antennas were used to provide spatial separation between the transmitter and observed target [21, 35].

In Figure 4.14 a part of the cross-ambiguity function zoomed on the target of interest echo is shown for consecutive time instants. The cross-ambiguity was calculated in the classical way, according to (4.8). Each of the crossambiguities was calculated with a CPI of 0.25s, and each of the snapshots is separated by 0.25s.

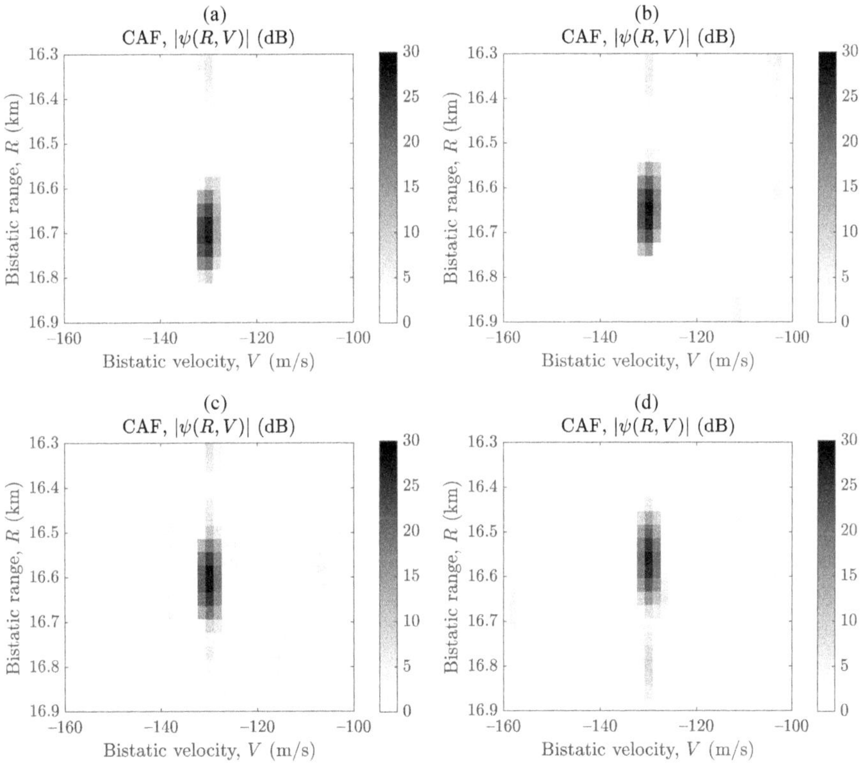

Figure 4.14 Cross-ambiguity function of a target under consideration. Each snapshot corresponds to 0.25s CPI. Snapshots are separated by 0.25s. (a) t=0s, (b) t=0.25s, (c) t=0.5s, and (d) t=0.75s.

Therefore, the overall time of 1s is illustrated in Figure 4.14. It can clearly be seen that during the 1s interval the bistatic range of the target changes considerably. If the cross-ambiguity was calculated for the whole interval of 1s, without the compensation of the time-variable signal delay, integration losses would be expected due to the range cell migration. The application of the extended cross-ambiguity scheme defined by (4.26) should compensate the change of the target range during the CPI.

The comparison of using the classical (4.8) and extended (4.26) cross-ambiguity functions is shown in Figure 4.15. The dotted curve shows the theoretical SNR value; increasing proportionally to the CPI. The use of the classical cross-ambiguity function (4.8), with the constant delay of the reference signal R/c, leads to the decrease of the SNR for the CPI exceeding 0.6s. This is the effect of the uncompensated time-variable delay of the signal, which leads to the range cell migration. If the extended cross-ambiguity (4.26) is applied, with the delay of the reference signal equal to $(R+Vt)/c$, the SNR increases almost identically with the theoretical curve.

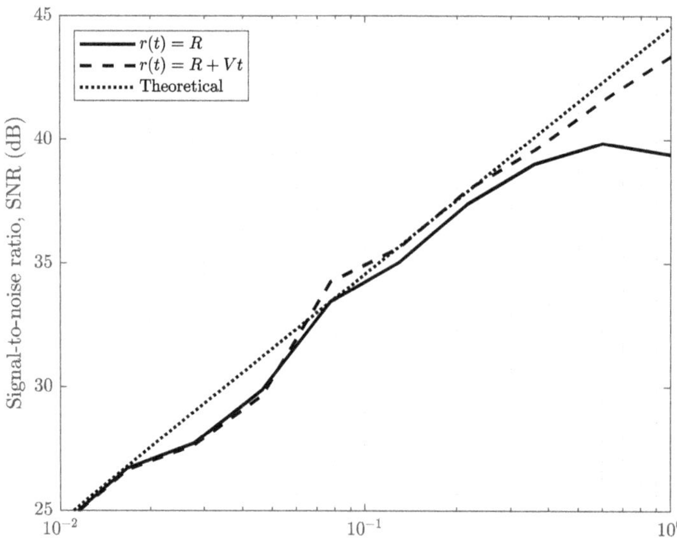

Figure 4.15 Signal-to-noise ratio versus the CPI for a real target echo (DVB-T signal); different curves correspond to different motion models.

4.7.3 Detection of Maneuvering Targets (FM Radio)

As an example of another application of the extended processing schemes, the detection of a maneuvering target will be presented. The results shown here have been acquired with the *PaRaDe* system in the FM radio band. The use of a typical CPI of 1s is considered; however, the observed target was maneuvering fast, which resulted in a performance degradation when using standard processing schemes.

Consider the snapshots of the cross-ambiguity function presented in Figure 4.16. The figure shows a zoomed part of the cross-ambiguity function of a maneuvering target. The cross-ambiguity was calculated in the classical way, that is, according to (4.8). The CPI was equal to 1s, and the separation between consecutive

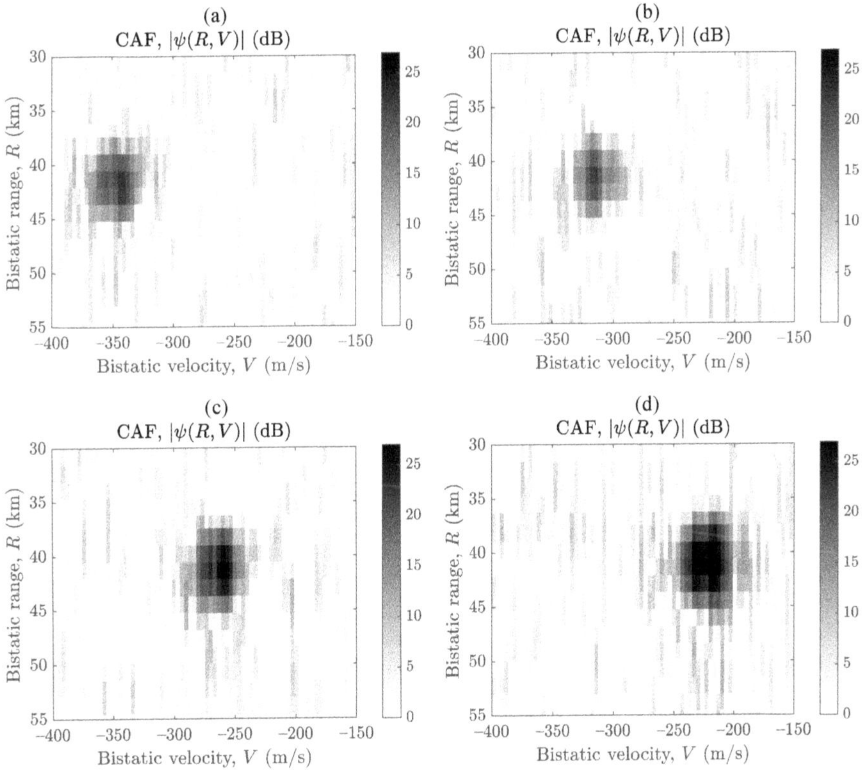

Figure 4.16 Cross-ambiguity function of a maneuvering target. Each snapshot corresponds to 1s CPI. Snapshots are separated by 1s. (a) t=0s, (b) t=1s, (c) t=2s, (d) t=3s.

snapshots was also 1s. It can be observed that the target rapidly changes bistatic velocity, from approximately -350 m/s to -220 m/s for the whole time interval. The bistatic acceleration can be roughly estimated as $(-220\text{ m/s}+350\text{ m/s})/3\text{s}\approx 40\text{ m/s}^2$. Such a high value of acceleration leads to a detection performance degradation when standard processing is applied, which results from the fact that the echo energy is spread across different velocity resolution cells. This leads to the decrease of the SNR, which degrades the detection capability of a radar.

To investigate the problem of SNR losses when the standard cross-ambiguity function is used, consider the results shown in Figure 4.17. The two curves correspond to bistatic velocity slices of the classical cross-ambiguity (4.8) and the extended cross-ambiguity (4.34) for the selected maneuvering target. In the case of the extended cross-ambiguity function, the acceleration value for the signal template was chosen to compensate for the actual acceleration of the target (approximately 40 m/s^2). Two results can be clearly seen.

First, the compensation of the acceleration leads to a significant increase in the SNR in comparison with the classical processing scheme, which is equal to approximately 5 dB. In the considered case, the target echo was strong enough, so that the detection could be performed even with significant SNR loss (SNR with

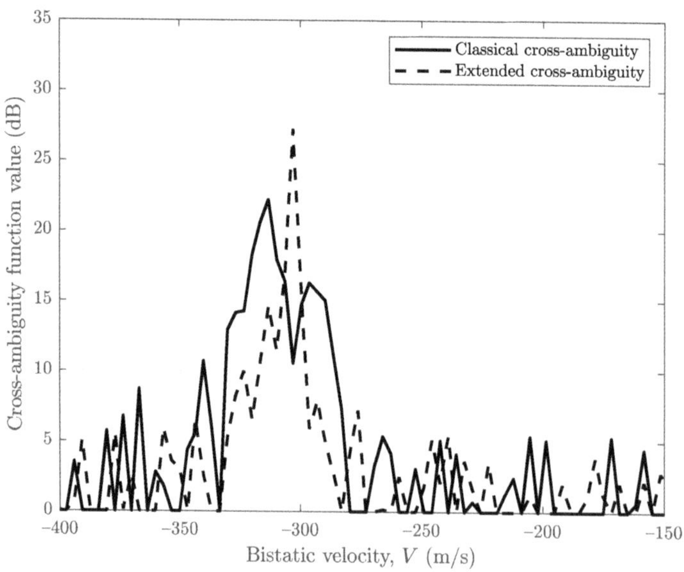

Figure 4.17 Comparison of the bistatic velocity slice of the classical and extended cross-ambiguity function of a maneuvering target.

losses equal to approximately 23 dB). However, the effect of the energy degradation due to the uncompensated target acceleration is the same for weaker target echoes. Therefore, in the case of weaker target echoes, the SNR loss can decide target detection performance.

Second, the energy is distributed across different velocity resolution cells in the case when the classical cross-ambiguity function is applied, as visible in Figure 4.17. If the acceleration is taken into account in the extended processing scheme, the target echo is much more focused. This leads not only to the aforementioned increase of the SNR, but also allows for the velocity measurement to be performed with higher accuracy.

In the next experiment the estimation of the acceleration using the extended cross-ambiguity function has been investigated. In each integration interval the acceleration has been estimated directly from the extended cross-ambiguity function (4.34). In addition, the acceleration has been approximated by comparing the values of the bistatic velocity from two consecutive integration intervals. The plots obtained in this way are shown in Figure 4.18. As can be seen, the two curves show consistent values of approximately 40 m/s^2.

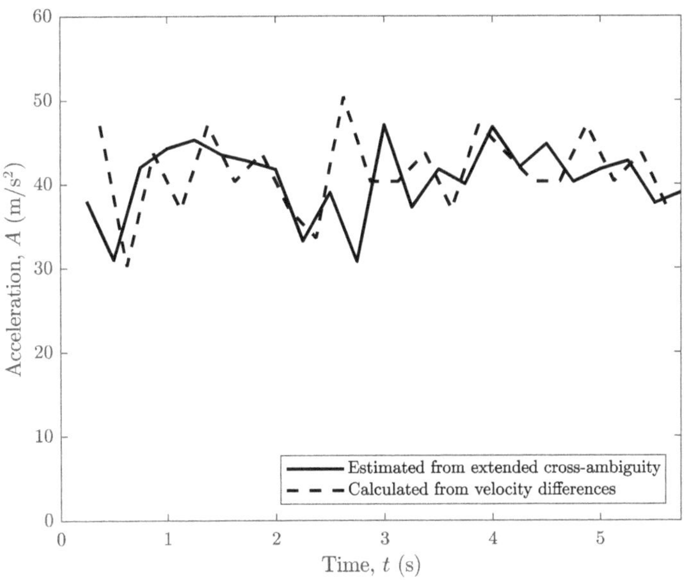

Figure 4.18 Comparison of the bistatic acceleration of a target versus time estimated directly (using extended cross-ambiguity function) and calculated by differentiating the consecutive bistatic velocity values.

This result shows that the compensation of the acceleration using the extended cross-ambiguity function (4.34) leads not only to an improved detection performance due to the increased SNR, but a reliable estimate of the bistatic acceleration is also provided. This estimate can be used by the tracking algorithm to improve its robustness against losing tracked targets.

4.8 SUMMARY

In this chapter, correlation processing has been investigated. The correlation is realized by computing the cross-ambiguity function (CAF). Its calculation directly from the definition would be very time-consuming. For this reason, different methods are applied that allow the computation time to be significantly reduced. Three such methods have been presented.

As mentioned in Chapter 2, there are two types of sidelobes present in the ambiguity function: random and deterministic. The random sidelobe level depends on the BT product, and cannot be influenced by the traditional approach of windowing. The deterministic sidelobes can be shaped by applying a window in the time or frequency domains. The price that has to be paid is the widening and decreasing of the amplitude of the correlation peak, effects typical for windowing (e.g., in spectral analysis).

An important feature of correlation processing is the processing gain, which results from the coherent integration of the signal. Typical values of the gain for passive radar reach 40–60 dB. The output SNR results from the input SNR being increased by the integration gain. The output SNR can be measured as a ratio of the correlation maximum to the average level of the noise floor, the PNFR (peak-to-noise-floor ratio). This is appropriate as long as the noise floor actually represent noise level and not random correlation sidelobes.

The classical cross-ambiguity function calculation is based on a simplified signal model. Usually, the simplifying assumptions are satisfied; therefore, classical processing can be successfully applied. However, in some cases the signal model and the processing scheme based on it have to be extended to prevent performance degradation.

One typical situation where extended processing is required is the use of longer-than-usual integration times. Another situation is the detection of highly maneuvering targets. In such cases, the target echo has a more complicated structure than usually assumed, and a lack of appropriate compensation in the processing leads to losses in the signal-to-noise ratio caused by the range or velocity cell migration.

Two types of extensions to the correlation scheme have been investigated. In the first case, range cell migration has been considered. Typically, the delay of the baseband signal is assumed to be constant. In reality, this delay is time-dependent, which has been taken into consideration during the processing.

In the second case, the velocity cell migration has been analyzed. Usually, only a constant Doppler shift is taken into account in the processing. If the target changes its velocity during the integration interval, the use of the classical cross-ambiguity calculation leads to performance deterioration. The extension of the correlation involves considering higher derivatives in the motion model.

The extensions of classical correlation have been tested on real data for FM radio and DVB-T signals. It has been demonstrated that a significant gain in the signal-to-noise ratio can be obtained when more complicated processing is applied. Moreover, the estimate of the bistatic acceleration can also be obtained, which can facilitate target tracking.

References

[1] M. Malanowski, *Signal Processing Algorithms and Target Localisation Methods for Continuous-Wave Passive Bistatic Radar*. No. 179 in Prace Naukowe Politechniki Warszawskiej. Elektronika, Warszawa, ul. Polna 50: Oficyna Wydawnicza Politechniki Warszawskiej, 2012.

[2] P. E. Howland, D. Maksimiuk, and G. Reitsma, "FM Radio Based Bistatic Radar," *IEE Proc. Radar, Sonar and Navigation*, vol. 152, pp. 107–115, June 2005.

[3] A. D. Lallo, A. Farina, R. Fulcoli, P. Genovesi, R. Lalli, and R. Mancinelli, "Design, Development and Test on Real Data of an FM Based Prototypical Passive Radar," in *Proc. RadarCon 2008*, (Rome, Italy), p. CD, May 2008.

[4] F. Colone, D. O'Hagan, P. Lombardo, and C. Baker, "A Multistage Processing Algorithm for Disturbance Removal and Target Detection in Passive Bistatic Radar," *IEEE Trans. Aerospace and Electronic Systems*, vol. 45, pp. 698 – 722, April 2009.

[5] K. Wang, R. Tao, Y. Ma, and T. Shan, "Adaptive Multipath Cancellation Algorithm in Passive Radar," in *Proc. IEEE CIE International Conference on Radar*, (Shanghai, China), p. CD, October 2006.

[6] K. Kulpa, "Adaptive Clutter Rejection in Bi-Static CW Radar," in *Proc. International Radar Symposium 2004*, (Warsaw, Poland), pp. 61–66, May 2004.

[7] S. Stein, "Algorithms for Ambiguity Function Processing," *IEEE Trans. Acoustics, Speech and Signal Processing*, vol. ASSP-29, pp. 588–599, June 1981.

[8] D. Gould, R. Pollard, C. Sarno, and P. Tittenso, "A Multiband Passive Radar Demonstrator," in *Proc. International Radar Symposium 2006*, (Cracow, Poland), pp. 657–660, May 2006.

[9] P. Liu and J. Liu, "Analysis of Passive Targets Detection Using CDMA Signal," in *Proc. 2005 IEEE International Workshop on VLSI Design and Video Technology*, (Suzhou, China), pp. 408–410, May 2005.

[10] M. K. Bączyk, P. Krysik, P. Samczyński, J. Misiurewicz, and K. Kulpa, "Identification of Helicopter Rotor Parameters Using Multistatic Passive Radar," in *2015 IEEE Radar Conference (RadarCon)*, pp. 1464–1467, May 2015.

[11] M. K. Bączyk, P. Samczyński, K. Kulpa, and J. Misiurewicz, "Micro-Doppler Signatures of Helicopters in Multistatic Passive Radars," *IET Radar, Sonar Navigation*, vol. 9, no. 9, pp. 1276–1283, 2015.

[12] M. A. Richards, *Fundamentals of Radar Signal Processing*. New York: McGraw-Hill, 2005.

[13] R. Tolimieri and S. Winograd, "Computing the Ambiguity Surface," *IEEE Trans. Acoustics, Speech, and Signal Processing*, vol. 33, pp. 1239–1245, October 1985.

[14] L. J. Jie, H. You, and S. Jie, "The Algorithms and Performance Analysis of Cross Ambiguity Function," in *2009 IET International Radar Conference*, pp. 1–4, April 2009.

[15] F. Pignol, F. Colone, and T. Martelli, "Lagrange-Polynomial-Interpolation-Based Keystone Transform for a Passive Radar," *IEEE Trans. Aerospace and Electronic Systems*, vol. 54, pp. 1151–1167, June 2018.

[16] S. R. J. Axelsson, "Noise Radar for Range/Doppler Processing and Digital Beamforming Using Low-Bit ADC," *IEEE Trans. Geoscience and Remote Sensing*, vol. 41, pp. 2703–2720, December 2003.

[17] S. R. J. Axelsson, "Noise Radar Using Random Phase and Frequency Modulation," *IEEE Trans. Geoscience and Remote Sensing*, vol. 42, pp. 2370–2384, November 2004.

[18] S. R. J. Axelsson, "Analysis of Random Step Frequency Radar and Comparison with Experiments," *IEEE Trans. Geoscience and Remote Sensing*, vol. 45, pp. 890–904, April 2007.

[19] K. Kulpa and J. Misiurewicz, "Stretch Processing for Long Integration Time Passive Covert Radar," in *Proc. IEEE CIE International Conference on Radar*, (Shanghai, China), pp. 1–4, October 2006.

[20] J. Misiurewicz and K. Kulpa, "Stretch Processing for Masking Effect Removal in Noise Radar," *IET Proc. Radar, Sonar and Navigation*, vol. 2, pp. 274–283, August 2008.

[21] M. Malanowski, K. Kulpa, and K. E. Olsen, "Extending the Integration Time in DVB-T-Based Passive Radar," in *2011 8th European Radar Conference*, pp. 190–193, October 2011.

[22] L. Liu, R. Tao, and N. Zhang, "The CAF-DFRFT-KT Algorithm for High-Speed Target Detection in Passive Radar," in *2011 First International Conference on Instrumentation, Measurement, Computer, Communication and Control*, pp. 748–751, October 2011.

[23] Y. Feng, T. Shan, Z. Zhuo, and R. Tao, "The Migration Compensation Methods for DTV Based Passive Radar," in *2013 IEEE Radar Conference (RadarCon13)*, pp. 1–4, April 2013.

[24] W. C. Barott and J. Engle, "Single-Antenna ATSC Passive Radar Observations with Remodulation and Keystone Formatting," in *2014 IEEE Radar Conference*, pp. 0159–0163, May 2014.

[25] K. M. Scott, W. C. Barott, and B. Himed, "The Keystone Transform: Practical Limits and Extension to Second Order Corrections," in *2015 IEEE Radar Conference (RadarCon)*, pp. 1264–1269, May 2015.

[26] T. Shan, S. Liu, Y. D. Zhang, M. G. Amin, R. Tao, and Y. Feng, "Efficient Architecture and Hardware Implementation of Coherent Integration Processor for Digital Video Broadcast-Based Passive Bistatic Radar," *IET Radar, Sonar Navigation*, vol. 10, no. 1, pp. 97–106, 2016.

[27] T. Martelli, F. Filippini, F. Pignol, F. Colone, and R. Cardinali, "Computationally Effective Range Migration Compensation in PCL Systems for Maritime Surveillance," in *2018 IEEE Radar Conference (RadarConf18)*, pp. 1406–1411, April 2018.

[28] M. Malanowski and K. Kulpa, "Correction of Range Cell Migration with FIR Filter for Passive Radar," in *2018 IEEE Radar Conference (RadarConf18)*, pp. 1123–1128, April 2018.

[29] K. Kulpa, "Target Acceleration Estimation for Continuous Wave Noise Radar," in *Proc. International Radar Symposium 2005*, (Berlin, Germany), pp. 177–183, September 2005.

[30] M. Malanowski and K. Kulpa, "Acceleration Estimation for Passive Coherent Location Radar," in *Proc. RadarCon 2008*, (Rome, Italy), p. CD, May 2008.

[31] M. Malanowski and K. Kulpa, "Analysis of Integration Gain in Passive Radar," in *Proc. International Conference on Radar*, (Adelaide, Australia), p. CD, September 2008.

[32] M. Malanowski, "Detection and Parameter Estimation of Manoeuvring Targets with Passive Bistatic Radar," *IET Radar, Sonar Navigation*, vol. 6, pp. 739–745, October 2012.

[33] M. Malanowski, K. Kulpa, and J. Misiurewicz, "PaRaDe – Passive Radar Demonstrator Family Development at Warsaw University of Technology," in *Proc. Microwaves, Radar and Remote Sensing Symposium – MRRS*, (Kiev, Ukraine), pp. 75–78, September 2008.

[34] M. Malanowski, K. Kulpa, P. Samczyński, J. Misiurewicz, J. Kulpa, P. Roszkowski, and L. Podkalicki, "Experimental Results of the PaRaDe Passive Radar Field Trials," in *Proc. International Radar Symposium 2012*, (Warsaw, Poland), May 2012.

[35] K. Olsen, *Investigation of Bandwidth Utilisation Methods to Optimise Performance in Passive Bistatic Radar*. PhD thesis, University College London, 2011.

Chapter 5

Clutter Filtering

5.1 INTRODUCTION

In this chapter, methods for canceling clutter returns, as well as the direct-path interference (DPI), are considered [1]. As clutter may mask moving target echoes, its suppression is crucial for the operation of passive radar. Two classes of filtering methods are analyzed: iterative and block. Different approaches are compared with regards to the convergence rate, computational complexity, and frequency selectivity. The results of processing real signals are also presented.

Passive radars considered in this book are continuous-wave and receive all signals (i.e., target echoes, direct signal from the transmitter) at the same time. To separate the signals in time and frequency, the cross-ambiguity function (CAF) of the reference and echo signals, introduced in Chapter 4, is used. As already mentioned, two types of sidelobes in the CAF are present: deterministic and random (residual fluctuations). Deterministic sidelobes can be shaped by appropriate windowing in the time and frequency domains. The random sidelobes are at the level of BT below the correlation peak, and their level cannot be changed by traditional windowing; this results from the finite integration time. The residual fluctuations may pose (and they usually do) a problem when attempting to detect weak target echoes in passive radar. For practical values of the signal bandwidth B and integration time T used in passive radar, the product BT is usually within the range of 40–70 dB. This value is usually insufficient, as the overall dynamic range of a passive radar often has to exceed 100 dB to reach the system noise level.

The concept of masking targets by the correlation sidelobes is demonstrated in Figure 5.1. The vertical axis represents the cross-ambiguity function amplitude (in decibels). The horizontal axis corresponds to the bistatic range or velocity (the

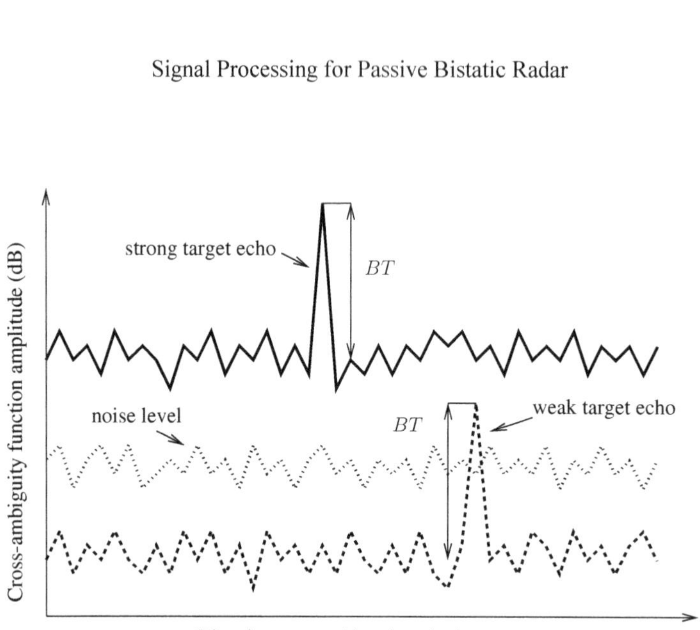

Figure 5.1 The principle of masking weak target echoes by strong reflections in passive radar.

principle of target masking is the same in these two dimensions, as the random sidelobes span across all ranges and velocities). For each target echo, a correlation peak can be observed with associated residual fluctuations at the level of BT below the maximum. Usually, the strongest components in the received signal in passive radar are the DPI and clutter (unwanted reflections from stationary objects). The correlation sidelobes mask all target echoes that are below the BT level with respect to the strongest peak. In practice, the echoes of moving targets are very often masked by the correlation sidelobes of the DPI and clutter, even though the echoes are above the receiver noise level. This prevents them from being detected without additional processing.

In practice, the correlation peak resulting from the DPI is most often the strongest one. The clutter reflections are usually 20–40 dB below the DPI level, but still tens of decibels above the receiver noise level, which can lead to the residual fluctuations originating from clutter being the limiting factor. Typically, the required DPI attenuation should be in the order of 30–50 dB to reach the limits of the receiver sensitivity.

Several methods can be used to minimize the problem of target echo masking by the direct signal and clutter. Examples of such methods are physical shielding of the reference and echo antennas, or analog cancelation of the direct-path signal [2, 3]. These methods have limited effectiveness, and virtually all passive radars resort to clutter cancelation in the digital domain. An approach already mentioned in Chapter 3 is the adaptive beamforming, which realizes filtering in the spatial domain. Another class of approach performs iterative adaptive filtering in the time domain, which is the topic of this chapter [4]. Filters such as least mean squares (LMS), normalized least mean squares (NLMS), recursive least squares (RLS), gradient adaptive lattice (GAL), least square lattice (LSL), and others are used in this application [5–8]. One of the disadvantages of the iterative methods is the problem of their convergence and stability. Another approach to the problem of clutter filtering is to use block adaptive methods [9–12]. A slightly different method, but closely related to clutter filtering, is the CLEAN algorithm [13, 14].

In this chapter, a signal model is introduced that will be used in the formulation of the adaptive filtering problem. Next, the results of the application of selected adaptive filtering algorithms for clutter cancellation in passive radar will be presented. The problem of filters' convergence and frequency selectivity will be analyzed.

The echo signal received by the radar in a real scenario consists of [6, 12]:

- **Direct-path interference (DPI).** Despite the use of directional antennas, or digital beamforming, the direct signal leaks into the surveillance channel. Usually, it is the strongest contribution in the received signal. The direct-path signal has no delay and no Doppler shift.
- **Clutter.** The clutter reflections under consideration are the echoes from stationary objects. In principle, the clutter will therefore consist of the delayed copies of the reference signal with no Doppler shift. In practice, however, it may occur that due to the slow movement of the clutter sources (e.g., branches of trees or sea waves), some spread in the Doppler domain will be present. This topic will be discussed in the following part of this book. Nevertheless, the clutter will be treated as delayed copies of the reference signal with little or no Doppler shift.
- **Target echoes.** Target echoes of interest originate from moving targets, such as aircraft, ships, and vehicles. Since they are moving, the echoes are Doppler-shifted (unless the target is moving tangentially to the bistatic ellipsoid, the Doppler shift is equal to zero in such a case) and delayed copies of the transmitted signal.
- **Noise.** Usually in radar, it is assumed that noise is dominated by the thermal noise of the receiver. This assumption is true for radars operating at microwave frequencies (usually S, C, and X bands). However, at lower frequencies, such as UHF or VHF, which are typical for passive radars utilizing com-

mercial illuminators of opportunity, such as FM radio or DVB-T television, the dominant noise component may be man-made [15–19].

Using the signal model (4.7) for a single target derived in Chapter 4, the aforementioned four components of the received signal can be expressed as:

$$x_e(t) = C_0^s \cdot x_r(t) + \sum_{i=1}^{N_s} C_i^s \cdot x_r\left(t - \frac{R_i}{c}\right) + \sum_{i=1}^{N_m} C_i^m \cdot x_r\left(t - \frac{R_i}{c}\right) \exp\left(\mathrm{j}\frac{2\pi}{\lambda} V_i t\right) + w(t), \tag{5.1}$$

where C_i^s are the amplitudes of the stationary object echoes, C_i^m are the amplitudes of moving target echoes, R_i are the bistatic ranges, V_i are the bistatic velocities, N_s is the number of stationary objects, and N_m is the number of moving targets. The first component on the right side of (5.1) represents the DPI (reference signal[1] with no delay and no Doppler shift). The first sum in the equation represents clutter reflections (delayed copies of the reference signal with no Doppler shift). The second sum corresponds to the moving targets (both delays and Doppler shifts are present). The last component of (5.1) is the noise. It can be seen that the DPI and clutter can be modeled together by a single sum, by defining bistatic range as equal to zero $R_0 = 0$ for the reference signal. In this way a simpler model can be written:

$$x_e(t) = \sum_{i=0}^{N_s} C_i^s \cdot x_r\left(t - \frac{R_i}{c}\right) + \sum_{i=1}^{N_m} C_i^m \cdot x_r\left(t - \frac{R_i}{c}\right) \exp\left(\mathrm{j}\frac{2\pi}{\lambda} V_i t\right) + w(t). \tag{5.2}$$

The first sum in (5.2) is the unwanted component, which the clutter filtering method should remove. It can be distinguished from the desired component (i.e., target echoes) by the Doppler shift.

A general idea of the operation of an adaptive filter for clutter removal is shown in Figure 5.2. The input signal of the filter is the sampled version of the reference signal $x_r(n)$. The output signal of the filter is subtracted from the sampled

1 Here it is assumed for simplicity that the reference signal is a clean delayed version of the transmitter signal, without noise, multipath, and target echoes. For discussion of more realistic reference channel signal models, refer to [20–23].

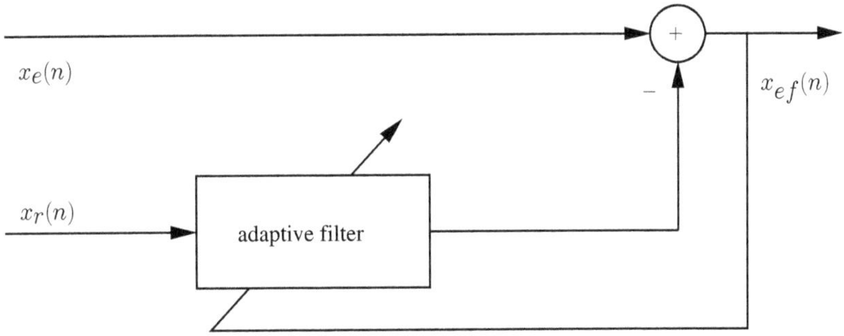

Figure 5.2 Block diagram of an adaptive filter for clutter removal.

version of the echo signal $x_e(n)$. The weights of the filter are chosen so that the residual signal $x_{ef}(n)$ is minimized in some way (depending on the filtering algorithm). This corresponds to a situation when the impulse response of the filter resembles the impulse response of the radio channel of the clutter. The principle of operation of the adaptive filter is based on the fact that the echoes without the Doppler shift are not correlated with Doppler-shifted echoes. In this way, the minimization of the residual signal $x_{ef}(n)$ should not, in principle, lead to attenuation of the target echoes with the Doppler shift. The possible effects of the target echo attenuation by the filter will be a part of the investigation.

As already mentioned, there are two broad classes of adaptive algorithms: iterative and block.[2] In iterative algorithms, the filter coefficients are recomputed with every incoming sample of the input signal. In block algorithms, the filter coefficients are calculated only once for the whole block of signal samples. The two types of filters are investigated next.

5.2 ITERATIVE ALGORITHMS

The basis for most of the adaptive filters is the theoretical Wiener filter [4]. In this filter, the error signal ($x_{ef}(n)$ in our case) is minimized in the minimum mean square error (MMSE) sense, that is the mean value of the squared absolute value of the signal is minimized $E\left[|x_{ef}(n)|^2\right]$. The solution of this problem requires knowledge

2 The term "block" in this book refers to methods in which a certain set of parameters is calculated for a block of data. These parameters do not depend on data other than the processed block. This is in contrast to some block adaptive filters, such as block least mean squares (BLMS), where coefficients are also obtained for a certain block of samples. However, they are calculated based on the current block and on coefficients corresponding to the previous block of data. Such methods will not be considered in the book.

of the true statistical parameters of the input signals. In practice, this is usually impossible. For this reason, simplifications of this theoretical filter are used.

5.2.1 NLMS Filter

One of the most popular adaptive filters is the least mean squares (LMS) [4]. The criterion used for the derivation of this filter is the minimization of the squared absolute value of the instantaneous error signal $|x_{ef}(n)|^2$. The solution of this problem leads to a relatively simple iterative formula for the calculation of the filter coefficients. The problem associated with the classical version of the LMS filter is the choice of the adaptation constant to provide filter stability and a fast convergence rate. To assure the filter stability, the adaptation constant should be smaller than a certain value connected with the eigenvalues of the input signal autocorrelation matrix [4]. Since the autocorrelation matrix of the input signal is not known a priori, it is difficult to set an appropriate value of the filter parameter. To overcome this problem, a normalization procedure can be used. A modified version of the LMS filter, called the normalized LMS (NLMS), consists of dividing the adaptation constant by the squared norm of the vector of the input signal [4]. In such a way, easier control of the choice of the adaptation constant can be achieved; the step size μ_{NLMS} is chosen from the $(0, 2)$ range.

Consider a vector of the input signal samples $\mathbf{x}_r(n) = [x_r(n), x_r(n-1), \ldots, x_r(n-M+1)]'$, where M is the order of the filter. The algorithm for the calculation of the NLMS filter is shown in Table 5.1 [4] (superscript H in the equations denotes the Hermitian transpose). In each iteration of the algorithm, the input signal $x_r(n)$ is filtered with a time-varying impulse response $\mathbf{h}(n)$. The filter's impulse response is updated in each step of the algorithm using a simple vector addition. A small positive constant a is added in the denominator of the update component of the impulse response to avoid problems if the signal is equal to zero. As can be seen, the algorithm is very simple, as it does not involve any complicated

Table 5.1
Algorithm of the NLMS Filter [4]

Initialization:
$\mathbf{h}(0) = \mathbf{0}$
Filtering:
$x_{ef}(n) = x_e(n) - \mathbf{h}^H(n)\mathbf{x}_r(n)$ $\mathbf{h}(n+1) = \mathbf{h}(n) + \frac{\mu_{\text{NLMS}}}{a+\|\mathbf{x}_r(n)\|^2} x_{ef}(n)\mathbf{x}_r^*(n)$

matrix calculations, such as matrix inversion. The computational complexity of the NLMS algorithm is $O(M)$ (M is the filter order).

The NLMS algorithm convergence performance has been tested by simulating the reference and echo signals and performing filtering. The reference signal was simulated as colored Gaussian noise, a relatively good model for an FM signal with fast music content (e.g., rock music) [24]. The echo signal was simulated as a sum of the delayed copies of the reference signal without the Doppler shift. The amplitudes of the copies decayed with the delay. The maximum simulated delay of the reference signal copy was equal to 25 samples. The order of the filter was equal to 30, so that all delayed copies of the reference signal could be removed. The signal-to-noise ratio (SNR) in the echo signal was 50 dB. The SNR was measured as the ratio of powers of the strongest replica of the reference signal to the additive white Gaussian noise.

The result of the experiment is presented in Figure 5.3. The curves show the amplitude (normalized, in decibels) of the filtered echo signal versus time. As the original curves were noisy, which makes the comparison difficult, they were smoothed with a median filter of order 1,000 samples. Different curves correspond to different values of the adaptation constant μ_{NLMS}. Each of the curves converge to the level close to -50 dB, which corresponds to the noise level (no further improvement

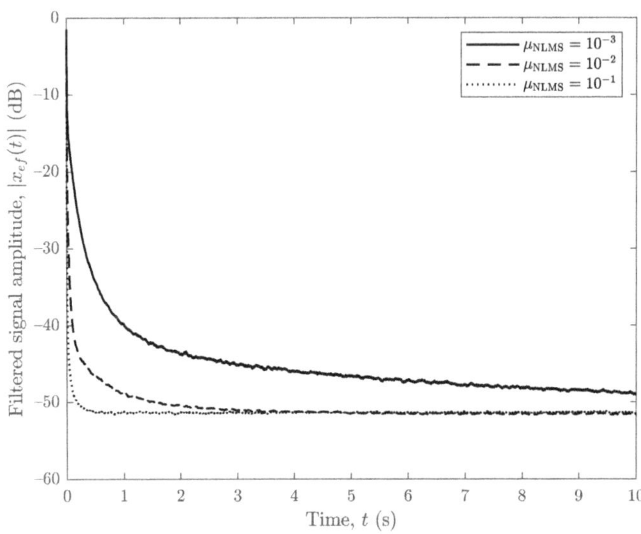

Figure 5.3 Comparison of the filtered (residual) signal of the NLMS algorithm for different values of the adaptation constant μ_{NLMS}. The curves are smoothed with a median filter of order 1,000 samples for better visualization.

is possible). As can be expected, larger values of μ_{NLMS} provide faster convergence, as the step size is larger. Nevertheless, the convergence is relatively slow in all cases, in the order of seconds. From this point of view, choosing larger values of μ_{NLMS} is advantageous. However, as will be shown in the following experiment, much lower values of the step size μ_{NLMS} are more appropriate in the passive radar application.

The idea of applying adaptive filters for clutter removal is to attenuate the delayed copies of the reference signal with zero (or with very small) bistatic velocity. Target echoes which have significant velocity should not be attenuated by the filter. From this point of view, filter selectivity in the velocity dimension is important (i.e., its ability to remove signal components with different velocity selectively).

To verify the behavior of the NLMS filter, a surveillance channel signal with target echoes has been simulated. The target echoes with the same bistatic range were distributed evenly for the $(-100, +100)$ m/s interval in the bistatic velocity. In this way the slice of the cross-ambiguity function for a fixed bistatic range (equal to one of the simulated targets) had a rectangular shape with a constant value for bistatic velocities of $(-100, +100)$ m/s. The application of the adaptive filter creates a notch across the bistatic velocity dimension; therefore, the rectangular shape of the simulated targets will be changed by the filter. In this way frequency selectivity of the filter can be investigated.The results of the experiment are shown in Figure 5.4. A rectangular shape, corresponding to the simulated targets for bistatic velocities of $(-100, +100)$ m/s, is visible. The sidelobes outside of the $(-100, +100)$ m/s interval at the level of -30 dB result from summing the sidelobes of individual target echoes. The application of the NLMS filter results in a notch in the CAF. A lower step size μ_{NLMS} provides better velocity selectivity. This property can be used for choosing the width of the filter notch, depending on the Doppler spread of the clutter. However, even for the smallest of the simulated values of the adaptation constant ($\mu_{\text{NLMS}} = 10^{-3}$), the filter removes the Doppler-shifted target echoes substantially. This is usually disadvantageous in practice, as the target echoes often approach (and cross) the zero-Doppler region. From this point of view, smaller values of μ_{NLMS} should be preferred, which is a requirement contrary to the attractiveness of fast convergence of the filter.

It is worth mentioning that the shape visible in Figure 5.4 is something different than the frequency response of the filter. The frequency response is defined as the Fourier transform of the impulse response. However, the frequency selectivity analyzed in the presented experiment is related to the behavior of a single coefficient of the impulse response in time.

The conducted experiments have shown that the NLMS filter is not very effective when applied to a typical clutter attenuation problem in passive radar.

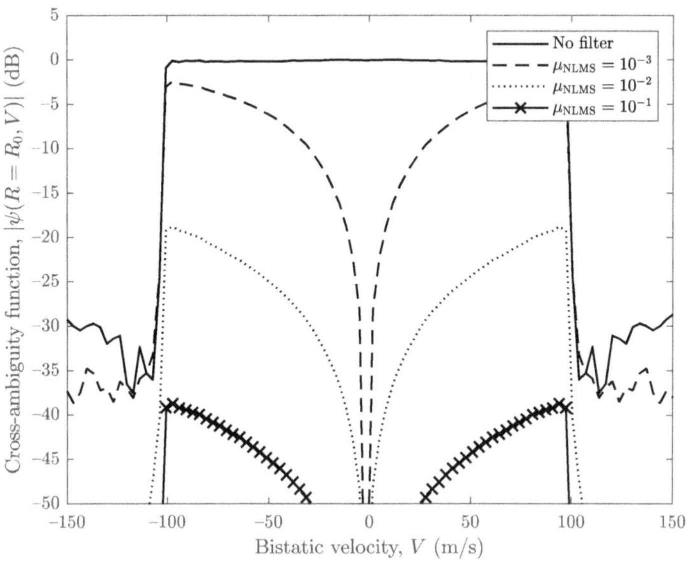

Figure 5.4 Comparison of the frequency selectivity of the NLMS filter for different values of the adaptation constant μ_{NLMS}.

Convergence of the filter is relatively slow; therefore, large values of the adaptation constant are preferred. However, smaller values of the step size have to be used to provide high velocity selectivity. The filter coefficients from one CPI can be used in the next CPI; therefore, the convergence problem is encountered only once, at the beginning of radar operation. However, when the radar scenario changes rapidly in time, fast convergence can still be an issue. Moreover, it is often convenient to treat each CPI separately, without the need to store filter coefficients. In practice, the choice of the NLMS filter parameter that yields satisfactory results is difficult, if not impossible. For this reason, other types of filters are often used. Two examples are presented next.

5.2.2 RLS Filter

One of the more complex algorithms that is often compared with the (N)LMS is the recursive least squares (RLS) filter. In this algorithm, the error defined by: $\sum_{i=1}^{n} \lambda_{\mathrm{RLS}}^{n-i} |x_{ef}(i)|^2$ is minimized. The parameter λ_{RLS} is the filter forgetting factor, which determines the adaptation capability of the filter. Values of λ_{RLS} closer to 1 correspond to a lower adaptation capability of the filter. The solution of the recursive

Table 5.2
Algorithm of the RLS Filter [4]

Initialization:
$\mathbf{P}(0) = \gamma \mathbf{I}, \quad \gamma \gg 1$ $\mathbf{h}(0) = \mathbf{0}$
Filtering:
$x_{ef}(n) = x_e(n) - \mathbf{h}^H(n)\mathbf{x}_r(n)$ $\mathbf{k}(n) = \frac{\mathbf{P}(n-1)\mathbf{x}_r(n)}{\lambda_{\text{RLS}} + \mathbf{x}_r^H(n)\mathbf{P}(n-1)\mathbf{x}_r(n)}$ $\mathbf{h}(n) = \mathbf{h}(n-1) + \mathbf{k}(n)x_{ef}^*(n)$ $\mathbf{P}(n) = \frac{1}{\lambda_{\text{RLS}}}\left[\mathbf{P}(n-1) - \mathbf{k}(n)\mathbf{x}_r^H(n)\mathbf{P}(n-1)\right]$

minimization problem leads to the equations listed in Table 5.2 [4]. Similar to the case of the NLMS filter, the input signal $x_r(n)$ is filtered with a time-varying impulse response of the filter $\mathbf{h}(n)$. The filter's impulse response is updated in each iteration using a gain vector $\mathbf{k}(n)$. The computation of this vector is relatively time-consuming, as it involves operation on the inverse correlation matrix $\mathbf{P}(n)$, which also has to be updated in each iteration.

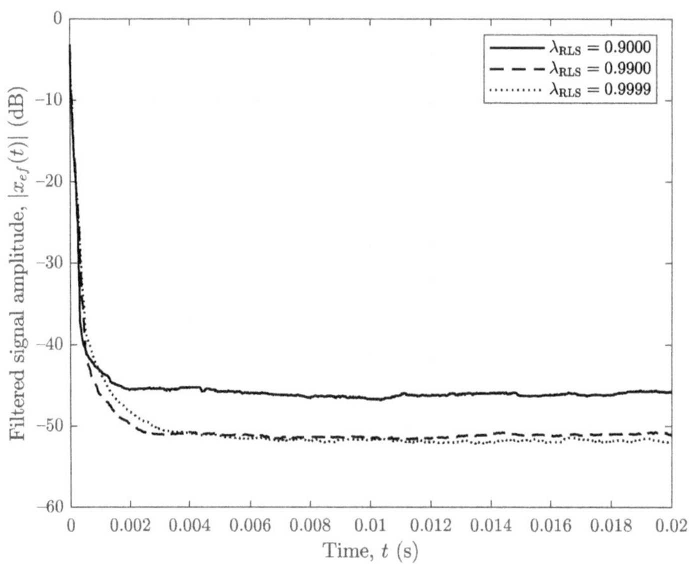

Figure 5.5 Comparison of the filtered (residual) signal of the RLS algorithm for different values of the adaptation constant λ_{RLS}.

The computational burden of the RLS filter is high, as it involves manipulations on matrices and vectors with the size of the filter order (usually tens or hundreds of taps) in each iteration. Moreover, the computational complexity grows with the square of the filter order; therefore, the filter's complexity is $O(M^2)$ (M is the filter order).

Despite its complexity, the RLS filter is used as it shows much better performance in terms of the convergence rate than the NLMS filter. This property has been tested by performing an experiment on simulated signals. The method for simulation of the signals was identical to the one used for testing the NLMS filter. In Figure 5.5 the amplitude of the signal after filtering $|x_{ef}(n)|$ is shown. Different curves correspond to different values of the filter forgetting factor λ_{RLS}. The convergence rate of the RLS filter is much faster than for the NLMS filter and very similar for different values of λ_{RLS}. In the case of the RLS filter, the steady state is achieved in a time of milliseconds, compared with the convergence time in the order of seconds for the NLMS filter. For the lowest value of the forgetting factor used in the simulations ($\lambda_{\text{RLS}} = 0.9000$), the signal level in the steady state is higher than for the remaining two cases. This results from the fact that the step size of the algorithm is

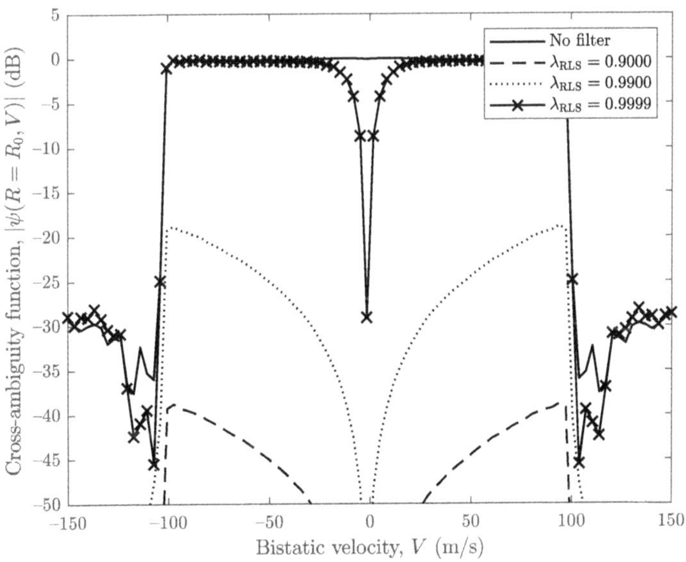

Figure 5.6 Comparison of the frequency selectivity of the RLS filter for different values of the adaptation constant λ_{RLS}.

high, due to the low value of the forgetting factor, which does not allow the filter to adapt properly to the signal.

Different values of the forgetting factor λ_{RLS} do not influence the convergence rate of the RLS filter significantly; however, they change the frequency selectivity of the filter. Figure 5.6 shows the plots of the CAF slice for simulated targets for the $(-100, +100)$ m/s velocity interval. Larger values of the forgetting factor correspond to a narrower notch of the filter. This fact, similar to the case of the NLMS filter, can be used for selecting the range of velocities, for which the echoes will be removed. The difference between the two filters is that for the RLS the parameter providing that the highest frequency selectivity also yields a very fast convergence rate.

5.2.3 LSL Filter

The RLS filter, despite its fast convergence rate, is not very attractive from the computational point of view. One of the possible alternatives is to apply an adaptive filter using the lattice structure. An example of such a filter is the least square lattice (LSL) [4]. There are two main types of LSL filters that differ in the way they compute estimation errors. In this book, a filter version with a priori estimation errors has been used. This structure has better numerical properties than the corresponding algorithm with a posteriori estimation errors. The algorithm summarizing the equations used in the LSL filter is shown in Table 5.3 [4]. The filter consists of two parts:

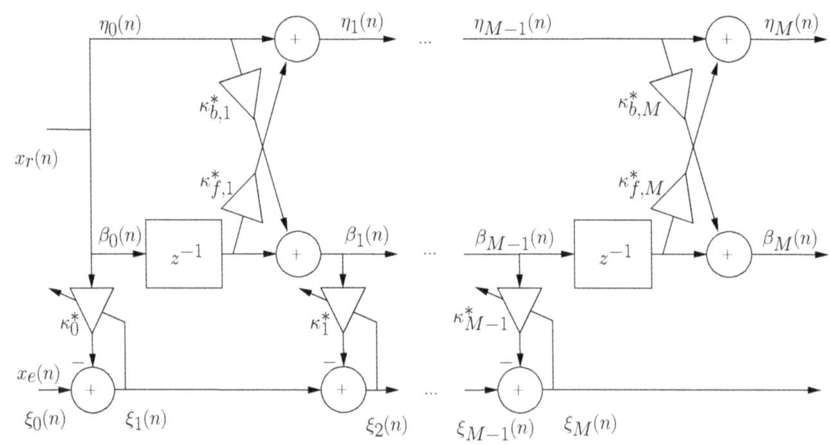

Figure 5.7 Structure of the LSL filter with a priori estimation errors for clutter removal [4].

a lattice predictor and a transversal filter (see Figure 5.7). In the lattice predictor, fed with the reference signal $x_r(n)$, forward $\eta_m(n)$ and backward $\beta_m(n)$ prediction errors are calculated. The backward prediction error signals, which are mutually orthogonal, are used in the transversal filter to suppress delayed copies of the reference signal from the echo signal $x_e(n)$. The lattice prediction operates as follows: for each new sample of the input signals, a series of iterations for consecutive lattice stages $m = 1, \ldots, M$ is performed, where M is the final order of the filter. In each of the iterations, forward $\mathcal{F}_{m-1}(n)$ and backward $\mathcal{B}_{m-1}(n-1)$ prediction-error energies (weighted forward and backward prediction-error squares) are calculated. Next, the forward $\eta_m(n)$ and backward $\beta_m(n)$ prediction errors are calculated in a single lattice stage, using the forward $\kappa_{f,m}(n)$ and backward $\kappa_{b,m}(n)$ reflection coefficients. The conversion factor $\gamma_m(n)$ is also updated recursively. In the transversal filter, fed with the echo signal $x_e(n)$, the error signal (residual) $\xi_m(n)$ is calculated, and the joint-process regression coefficient $\kappa_{m-1}(n)$ is updated recursively.

Despite relatively complicated equations describing the LSL filter, the computational burden is moderate, and it grows linearly with the filter order M; therefore, the complexity is $O(M)$.

The LSL filter performance has been tested in a similar manner to the NLMS and RLS filters. The results of the application of the LSL filter are shown in

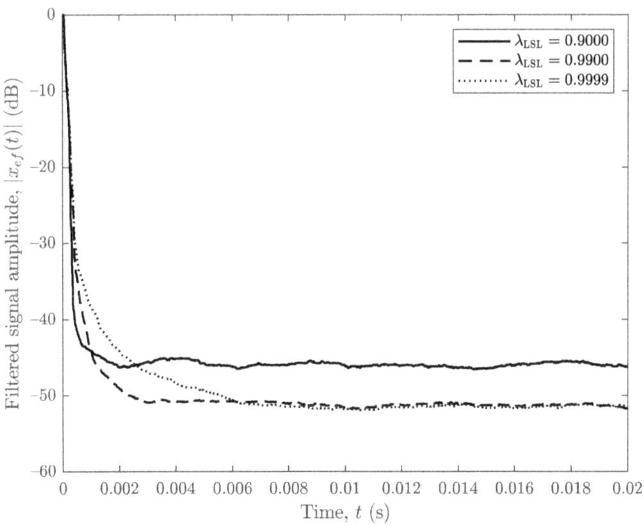

Figure 5.8 Comparison of the filtered (residual) signal of the LSL algorithm for different values of the adaptation constant λ_{LSL}.

Table 5.3
Algorithm of the LSL Filter Using a priori Estimation Errors [4]

Initialization:
For the 0th time instant:
$\mathcal{F}_{m-1}(0) = \delta$
$\mathcal{B}_{m-1}(-1) = \delta$
(δ is a small positive constant)
$\kappa_{f,m}(0) = 0$
$\kappa_{b,m}(0) = 0$
$\gamma_0(0) = 1$
$\kappa_{m-1}(0) = 0$
For the 0th filter stage:
$\eta_0(n) = x_r(n)$
$\beta_0(n) = x_r(n)$
$\xi_0(n) = x_e(n)$
Predictions:
Prediction-error energies:
$\mathcal{F}_{m-1}(n) = \lambda_{\text{LSL}}\mathcal{F}_{m-1}(n-1) + \gamma_{m-1}(n-1)\lvert\eta_{m-1}(n)\rvert^2$
$\mathcal{B}_{m-1}(n-1) = \lambda_{\text{LSL}}\mathcal{B}_{m-1}(n-2) + \gamma_{m-1}(n-1)\lvert\beta_{m-1}(n-1)\rvert^2$
Prediction errors:
$\eta_m(n) = \eta_{m-1}(n) + \kappa^*_{f,m}(n-1)\beta_{m-1}(n-1)$
$\beta_m(n) = \beta_{m-1}(n-1) + \kappa^*_{b,m}(n-1)\eta_{m-1}(n)$
Reflection coefficients:
$\kappa_{f,m}(n) = \kappa_{f,m}(n-1) - \frac{\gamma_{m-1}(n-1)\beta_{m-1}(n-1)}{\mathcal{B}_{m-1}(n-1)}\eta^*_m(n)$
$\kappa_{b,m}(n) = \kappa_{b,m}(n-1) - \frac{\gamma_{m-1}(n-1)\eta_{m-1}(n)}{\mathcal{F}_{m-1}(n)}\beta^*_m(n)$
Conversion factor:
$\gamma_m(n) = \gamma_{m-1}(n-1) - \frac{\gamma^*_{m-1}(n-1)\lvert\beta_{m-1}(n-1)\rvert^2}{\mathcal{B}_{m-1}(n-1)}$
Filtering:
$\xi_m(n) = \xi_{m-1}(n) - \kappa^*_{m-1}(n-1)\beta_{m-1}(n)$
$\kappa_{m-1}(n) = \kappa_{m-1}(n-1) + \frac{\gamma_{m-1}(n)\beta_{m-1}(n)}{\mathcal{B}_{m-1}(n)}\xi^*_m(n)$

Figure 5.8. The convergence rate is very similar to the RLS filter and also much faster than for the NLMS filter. The influence of the forgetting factor on the convergence rate is not very significant, as in the case of the RLS filter.

The tests of the frequency selectivity of the LSL filter are shown in Figure 5.9. They are very similar to those obtained for the RLS filter. Values of the adaptation constant λ_{LSL} closer to 1 provide better frequency selectivity. Similar to the RLS filter, the highest of the simulated values of λ_{LSL} (0.9999), which yields the narrowest notch, also provides a very fast convergence rate.

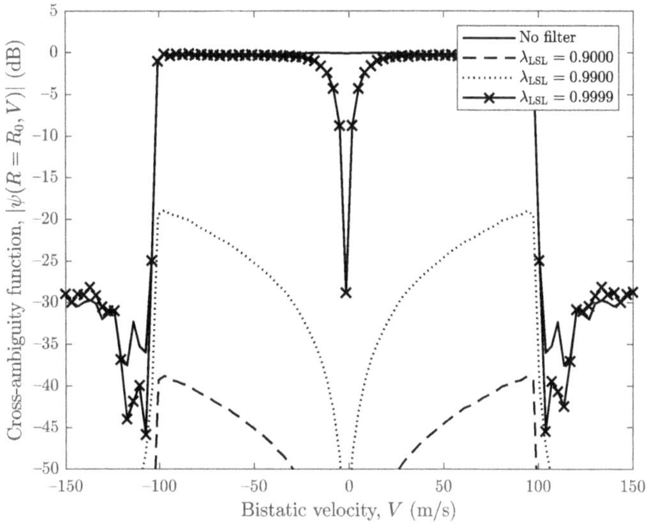

Figure 5.9 Comparison of the frequency selectivity of the LSL filter for different values of the adaptation constant λ_{LSL}.

The results presented for the three iterative algorithms (NLMS, RLS, and LSL) lead to several conclusions. The NLMS filter is simple from the computational point of view. The adaptation speed of the filter can be adjusted by the adaptation step; however, the convergence rate is relatively slow. Moreover, low values of the step size, corresponding to slow convergence, are preferable from the point of view of the frequency selectivity.

The RLS filter provides a much faster convergence rate than the NLMS, even for the filter parameter values that provide satisfactory filter frequency selectivity. A serious disadvantage of the RLS filter is its high computational complexity.

Very good results can be obtained using the LSL filter. The convergence rate and frequency selectivity are very similar to the RLS filter. At the same time, the computational complexity is relatively low, higher than that of the NLMS filter, but it grows only linearly with the filter order.

5.3 BLOCK ALGORITHMS

The NLMS, RLS, and LSL filters work in an iterative manner, that is, every time that a sample of the input signals appears, the filter coefficients are recalculated. This allows the filter to adapt to changing conditions. However, as shown in Section 5.2, the use of the iterative adaptive filters is related to the problem of convergence. Even if the convergence is fast, as in the case of the RLS and LSL filters, the transient interval has to be eliminated from further processing.

The problem of convergence can be avoided if a block algorithm is applied. Moreover, as the signal in passive radar is divided into CPIs for correlation processing, block clutter removal methods seem to be a natural solution, with each CPI processed independently from the others. When block methods are applied, one set of coefficients is calculated for the analyzed signal block. Since the coefficients are computed based on the processed block, they are matched to this block. Three types of block methods will be considered in this chapter: the LS matrix solution, block lattice filter (and its modification), and the CLEAN algorithm.

5.3.1 The LS Matrix Solution

As a basis for the effective block lattice structure, presented in Section 5.3.2, a straightforward method based on the least squares (LS) matrix solution will be presented [9, 12], sometimes referred to as the Extensive Cancelation Algorithm (ECA).

Assume that the model for clutter echoes can be written in a discrete form as:

$$x_e(n) = C_0^s x_r(n) + C_1^s x_r(n-1) + ... + C_{M-1}^s x_r(n-M-1) = \sum_{m=0}^{M-1} C_m^s x_r(n-m), \tag{5.3}$$

where C_i^s are the amplitudes of clutter returns corresponding to consecutive delays. Equation (5.3) can be rewritten in a matrix form:

$$\begin{bmatrix} x_e(n) \\ x_e(n+1) \\ \vdots \\ x_e(n+N) \end{bmatrix} = \tag{5.4}$$

$$\begin{bmatrix} x_r(n) & x_r(n-1) & \dots & x_r(n-M-1) \\ x_r(n+1) & x_r(n) & \dots & x_r(n-M) \\ \vdots & \vdots & \ddots & \vdots \\ x_r(n+N) & x_r(n+N-1) & \dots & x_r(n+N-M-1) \end{bmatrix} \begin{bmatrix} C_0^s \\ C_1^s \\ \vdots \\ C_{M-1}^s \end{bmatrix}$$

or compactly as:

$$\mathbf{X}_e = \mathbf{X}_r \mathbf{C}^s, \tag{5.5}$$

where $\mathbf{X}_e$, $\mathbf{X}_r$ and $\mathbf{C}^s$ correspond to the appropriate components of (5.4). Equation (5.5) represents an overdetermined set of equations, as typically the number of available samples N is much larger than the length M of the modeled vector of clutter amplitudes $\mathbf{C}^s$.

The block method for clutter removal based on the LS matrix solution involves finding the vector of amplitudes $\mathbf{C}^s$, and using it for the subtraction of the clutter components from the echo signal. The solution of (5.5) in the LS sense can be found by calculating the pseudo-inverse of $\mathbf{X}_r$:

$$\hat{\mathbf{C}}^s = \left(\mathbf{X}_r^H \mathbf{X}_r\right)^{-1} \mathbf{X}_r^H \mathbf{X}_e. \tag{5.6}$$

The calculated vector $\hat{\mathbf{C}}^s$ represents estimated values of the clutter amplitudes for successive delays of the reference signal. To remove the clutter from the echo signal, the following subtraction is performed:

$$\mathbf{X}_{ef} = \mathbf{X}_e - \mathbf{X}_r \hat{\mathbf{C}}^s, \tag{5.7}$$

where $\mathbf{X}_{ef}$ is the echo signal vector after the clutter removal.

The computational complexity of the presented method is $O(M^3)$ (if straightforward algorithms for matrix multiplication and inversion are used), which is very high. For this reason an alternative solution with similar performance but lower computational complexity is preferred, which will be investigated later in this chapter.

The presented method has been tested using simulated signals, with f_s = 200 kHz and f_c = 100 MHz, parameters typical for FM radio. The clutter reflections have been simulated in the same way as in the case of testing iterative filters. The noise level was −50 dB. Additionally, a target echo at bistatic range R = 30 km and bistatic velocity V = 30 m/s has been simulated. In Figure 5.10(a) the CAF of the reference and echo signals is shown. The simulated clutter reflections are visible for zero bistatic velocity; the clutter echoes are not Doppler-shifted. The clutter extends to approximately 30 km of bistatic range. The values of the CAF are normalized to the mean value of the correlation sidelobes (the mean sidelobe level corresponds to 0 dB). In the simulated scenario, the mean level corresponded to the sidelobes of the strongest peak, the simulated direct signal. The ratio of the peak to sidelobe level is approximately 50 dB, which corresponds to the BT product. The simulated target echo is not visible, as it is masked with the sidelobes of the DPI and clutter, which is a typical situation encountered in practice.

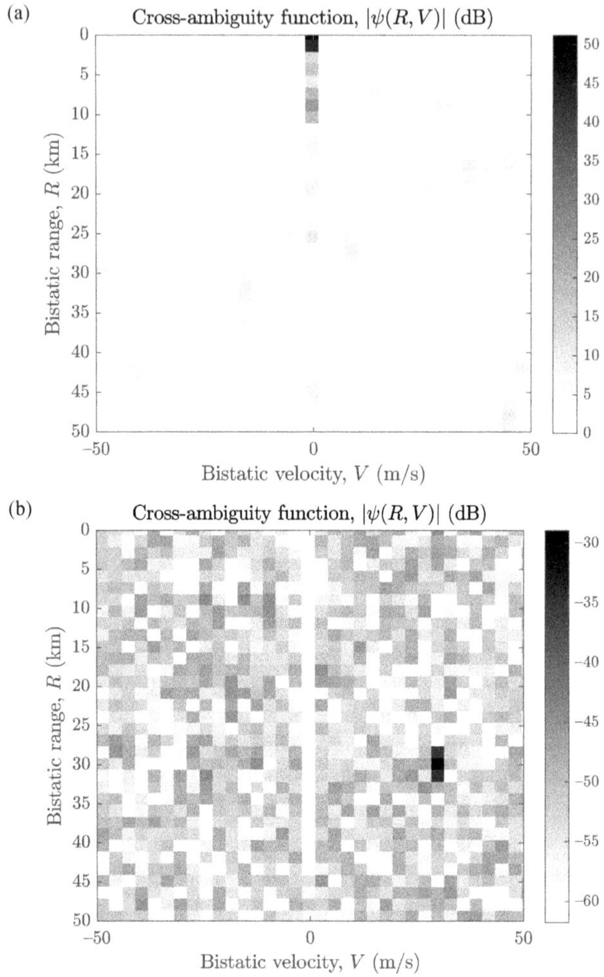

Figure 5.10 Cross-ambiguity function of the simulated reference and echo signals: (a) with clutter reflections and (b) after the application of the LS-based clutter removal (values normalized to the mean noise floor level before clutter cancelation).

The CAF calculated after the application of the clutter removal method (5.7) is presented in Figure 5.10(b). The clutter echoes have been removed, as a notch at exactly zero velocity can be seen. The notch extends up to 45 km, which corresponds to the filter order $M = 30$. The mean level of the sidelobes has been reduced from

0 dB to -50 dB, which corresponds to the simulated noise level. The simulated target echo is now visible at $R = 30$ km and $V = 30$ m/s, with the amplitude of approximately -30 dB with respect to the original random sidelobes' level. The experiment shows that desired clutter cancelation can be obtained, without the problem of filter convergence.

The presented algorithm can be extended to include not only the delayed copies of the reference signal as in (5.3), but also Doppler-shifted copies of the reference signals [12]. The rest of the algorithm remains unchanged, but this time the size of the vector $\mathbf{C}^s$ representing clutter amplitudes is $M \times F$, where M is the number of delays and F is the number of frequency bins for which the clutter is modeled.

The presented method for clutter removal is straightforward and performs well. However, its computational complexity is very high. An alternative approach is to use the block lattice structure presented next.

5.3.2 Block Lattice Filter

The structure of the block lattice filter used for clutter removal is very similar to the structure of the LSL filter, and it is shown in Figure 5.11 [4]. The main difference between the iterative LSL and block lattice filter lies in the equations of the two algorithms. The equations describing the block lattice filter operation are listed in Table 5.4 [4]. The output of each stage of the predictor is the backward prediction error $b_i(n)$ and forward prediction error $f_i(n)$, which are calculated using the reflection coefficients κ_i. As in the LSL filter, the backward prediction error signals are mutually orthogonal. This property allows the delayed reference signal copies to be removed from the echo signal independently. This operation is performed in the lower part of the filter structure. For each delay, a correlation coefficient h_i of the backward prediction error and the echo signal is calculated. It is used for the removal of the delayed copy of the reference signal from the echo signal (i.e., removal of the clutter reflections).

From the equations listed in Table 5.4, it can be seen that all required calculations are vector operations, which can be easily and efficiently implemented in hardware (e.g., in field-programmable gate array (FPGA)) or in software (e.g., in graphics processing unit (GPU)). Moreover, the computational burden grows linearly with the filter order; therefore, the complexity is $O(M)$. This is advantageous compared with the matrix operations required in the LS matrix method.

The block lattice filter has been tested on the same signal as the LS matrix method. The obtained result is shown in Figure 5.12. The cross-ambiguity function

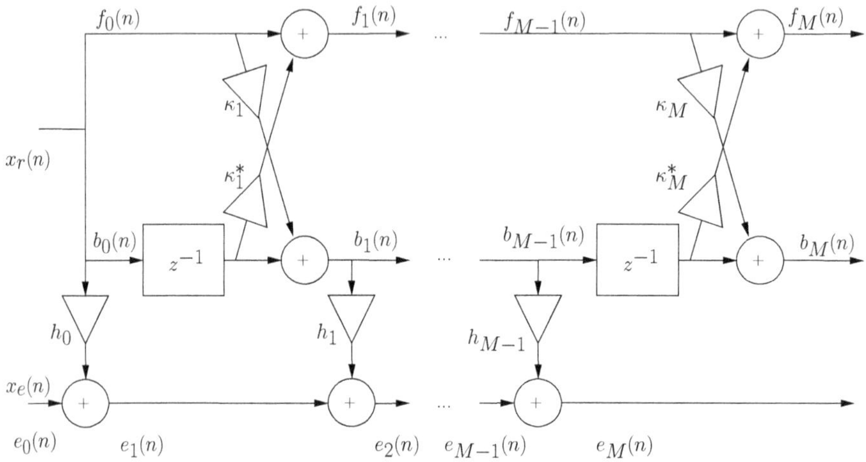

Figure 5.11 Structure of the block lattice filter.

Table 5.4
Algorithm of the Block Lattice Filter [4]

Initialization:
$b_0(n) = x_r(n)$ $f_0(n) = x_r(n)$ $e_0(n) = x_e(n)$
Predictions:
Reflection coefficient: $\kappa_{i+1} = \frac{2\sum_n b_i(n-1)f_i^*(n)}{\sum_n \lvert f_i(n)\rvert^2 + \lvert b_i(n-1)\rvert^2}$ Prediction errors: $b_{i+1}(n) = b_i(n-1) - \kappa_{i+1} f_i(n)$ $f_{i+1}(n) = f_i(n) - \kappa_{i+1}^* b_i(n-1)$
Filtering:
$h_i = \frac{\sum_n e_i(n) b_i^*(n)}{\sum_n \lvert b_i(n)\rvert^2}$ $e_{i+1}(n) = e_i(n) - h_i b_i(n)$

calculated using the signal after the application of block lattice filtering is almost identical to the one presented in Figure 5.10(b). The notch created by the lattice filter extends to approximately 45 km, which corresponds to the filter order. Similar to the case of the LS method, it is as narrow as one frequency bin. The mean sidelobe level

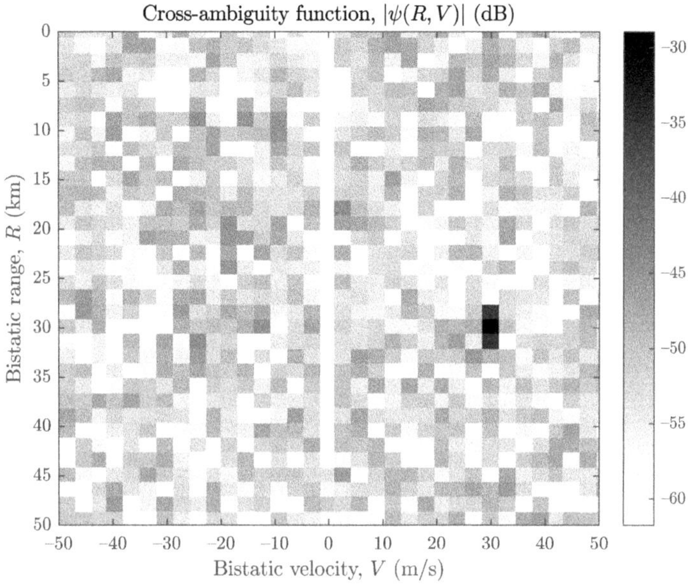

Figure 5.12 Cross-ambiguity function of the simulated reference and echo signals after the application of the block lattice clutter removal (values normalized to the mean noise floor level before clutter cancelation).

has been reduced to -50 dB, corresponding to the level of the simulated noise. The simulated target echo is visible at $R = 30$ km and $V = 30$ m/s, which was masked by the sidelobes of the DPI and clutter before the filtering.

After the presentation of the two block methods (i.e., the LS method and block lattice filter) two conclusions can be drawn. First, the block lattice filter and LS matrix methods perform very similarly. Second, the computational burden of the block lattice filter algorithm is much lower than that of the LS matrix method. These facts suggest that the lattice filter should be used in practical applications instead of the LS method. For this reason, the lattice filter will be used in further investigations.

For the iterative approaches, it has been shown that the filters reveal variable frequency selectivity, depending on the adaptation constant. This is a very important property in real-life applications, as the clutter may be spread in the Doppler domain. In the case of the block methods presented so far, it can be concluded that only the echoes from the zero-velocity cell are removed. This results from the fact that if a set of coefficients is calculated for the entire block of data corresponding to the CPI, the Doppler resolution is equal to the inverse of the CPI length, which is equivalent to one velocity resolution cell. A modification of the block lattice filter that allows the Doppler-spread clutter to be removed is investigated in the next section.

5.3.3 Modified Block Lattice Filter

The capability of the block lattice filter for removing Doppler-spread clutter can be achieved by introducing a frequency shift to the backward prediction error signals $b_i(n)$ [11]. In such an approach, the backward prediction error signal is multiplied by an exponential term, before calculating the correlation coefficient h_i:

$$b_i^f(n) = b_i(n) \cdot \exp(\mathrm{j}2\pi f n), \tag{5.8}$$

where f is the amount of the frequency shift. Consider frequency shifts $f = k \cdot \frac{1}{N}$, where N is the number of signal samples in the CPI, and $k = -N_{fb}, -N_{fb} + 1, \ldots, N_{fb} - 1, N_{fb}$. These frequency shifts correspond to the frequency bins ranging from $-N_{fb}$ to $+N_{fb}$. If the frequency-shifted backward prediction error signals $b_i(n)$ are removed from the echo signal, Doppler-spread clutter is canceled.

The frequency-shifted backward prediction errors $b_i(n)$ are no longer orthogonal, as in the original, unmodified block lattice filter. For this reason, consecutive iterations of the filter operating for different Doppler shifts are not independent of each other. This results with somewhat inferior Doppler-spread clutter removal capability. However, it can be remedied by applying the filter several times to the same data [11].

This approach has been tested for the block lattice filter in the same way as in the case of the iterative filters, that is, by simulating targets with velocities in the $(-100, +100)$ m/s interval, and applying different filters. The result of this experiment is shown in Figure 5.13. The curve denoted as "Freq. bins: 0" corresponds to a classical block lattice filter, without the frequency shift of the backward prediction error signals (as introduced in Section 5.3.2). It can be seen that a narrow notch in the cross-ambiguity function slice is created, whose width corresponds to one frequency bin. The curve marked with "Freq. bins: $(-5, +5)$" has been obtained by modulating the backward prediction error signals with a frequency shift ranging from -5 to $+5$ frequency bins (11 frequency bins in total). In this case, the width of the filter notch is wider, which allows the Doppler-spread clutter to be removed. Analogously, the curve denoted with "Freq. bins: $(-10, +10)$" has been obtained by removing clutter from frequency bins ranging from -10 to $+10$ (21 frequency bins in total). As can be seen, the capability of the block lattice filter for removing the Doppler-spread clutter can be obtained in a relatively simple way. Moreover, the velocity interval,

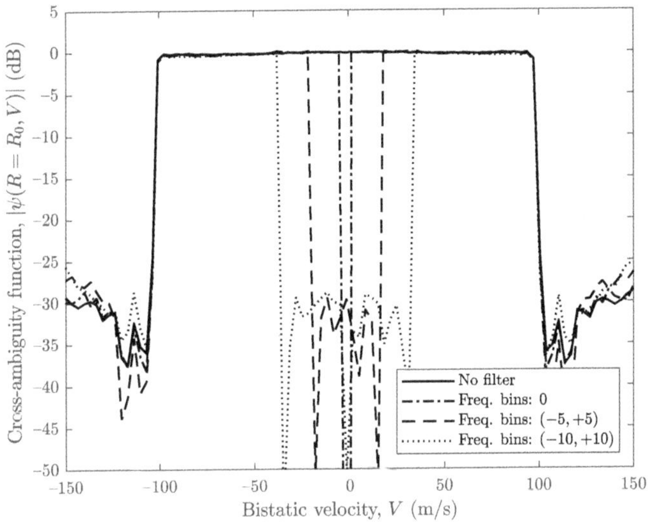

Figure 5.13 Comparison of the frequency selectivity of the modified block lattice filter for different number of the frequency bins for which the signal is removed.

for which the clutter is removed, can be set very precisely by changing the N_{fb} parameter (the number of frequency bins for which the clutter is removed).

The computational complexity of the modified block lattice filter is increased in comparison with the original block lattice filter, as additional operations for frequency-shifted signals are required. The complexity increases linearly with the number of frequency bins N_{fb} for which clutter is removed. The overall complexity is $O(M \cdot N_{fb})$ (M is the filter order).

5.3.4 The CLEAN Algorithm

One of the block methods for removing strong scatterers is the CLEAN algorithm [13, 14]. It has been applied in numerous fields, such as astronomy and radar. The idea of the algorithm when applied to radar is to remove a strong target echo, so that weaker echoes can be observed. This is very similar to the clutter removal problem analyzed in this chapter so far. The approach is, however, slightly different. In the case of adaptive filters, the aim was to remove DPI and clutter, which are located at zero velocity (or close to zero). The CLEAN algorithm can be used for the removal of arbitrary scatterers, including strong target echoes with high velocity (high Doppler frequency).

The CLEAN algorithm applied to passive radar operates in the following way: first, the cross-ambiguity function $\psi(R, V)$ is calculated. If it contains a strong peak, whose sidelobes can mask other target echoes, the position of this peak on the range-velocity plane is estimated:

$$\{R_0, V_0\} = \arg\max_{R,V} |\psi(R, V)| \tag{5.9}$$

The estimated target parameters (i.e., bistatic range R_0 and bistatic velocity V_0) are used to model a strong echo signal. This is done by delaying the reference signal $x_r(t)$ by a value proportional to R_0, and shifting it in frequency by a value proportional to V_0:

$$x_m(t) = x_r\left(t - \frac{R_0}{c}\right) \cdot \exp\left(-j2\pi\frac{V_0}{\lambda}t\right) \tag{5.10}$$

The last parameter that has to be estimated is the complex amplitude of the strong target echo. This can be done by calculating the correlation coefficient of the modeled strong echo signal $x_m(t)$ with the received echo signal $x_e(t)$:

$$A = \frac{\langle x_e(t), x_m(t)\rangle}{\langle x_m(t), x_m(t)\rangle} \tag{5.11}$$

Next, the modeled strong echo signal is subtracted from the echo signal:

$$x'_e(t) = x_e(t) - Ax_m(t) \tag{5.12}$$

In this way, the strong echo is removed from the received signal. The cross-ambiguity function is calculated again. If another strong scatterer is present, which can mask weaker echoes, the cleaning procedure is repeated. The CLEAN algorithm is summarized in Table 5.5.

The CLEAN algorithm is able to remove strong target echoes with and without Doppler shift. However, its applicability to removing clutter is limited, due to the fact that only a single scatterer is modeled in one iteration of the algorithm. Therefore, removing clutter, which is usually composed of multiple scatterers, using the CLEAN algorithm would require many iterations to provide satisfactory results. The CLEAN algorithm is suitable for removing strong target echoes that can be modeled well as a individual scatterer (e.g., the strong echo of a moving target that masks other weaker echoes).

Table 5.5
The CLEAN Algorithm

Cross-ambiguity function calculation:
$\psi(R, V)$
Estimation of strongest echo parameters:
$\{R_0, V_0\} = \underset{R,V}{\arg\max}\|\psi(R, V)\|$
Modeling of strong echo signal:
$x_m(t) = x_r\left(t - \frac{R_0}{c}\right) \cdot \exp\left(-j2\pi\frac{V_0}{\lambda}t\right)$
$A = \frac{\langle x_e(t), x_m(t)\rangle}{\langle x_m(t), x_m(t)\rangle}$
Strong echo removal:
$x'_e(t) = x_e(t) - Ax_m(t)$

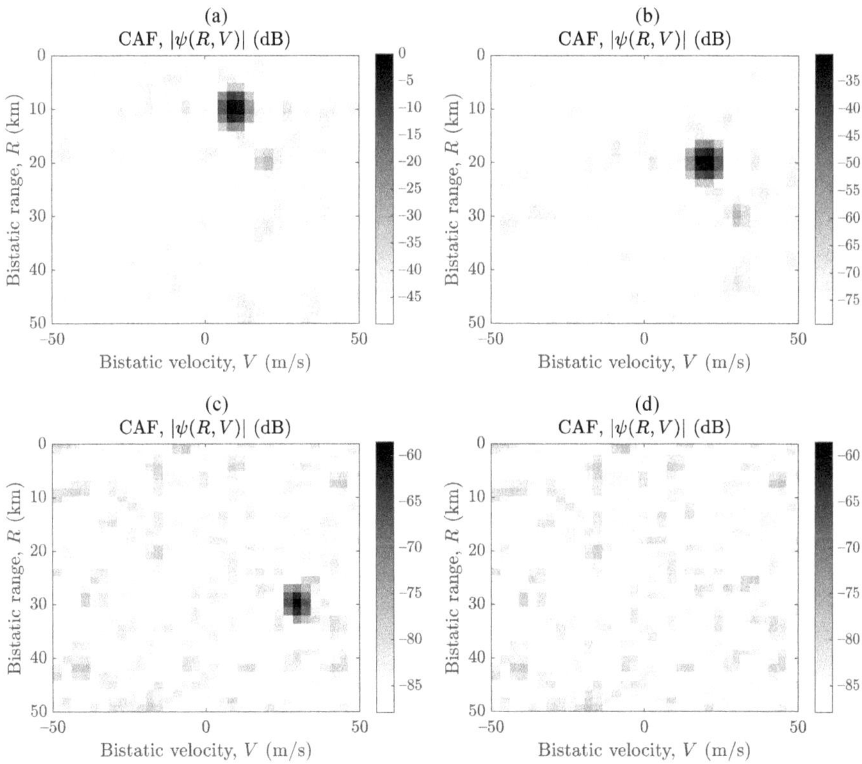

Figure 5.14 Cross-ambiguity function after (a) zeroth, (b) first, (c) second, and (d) third iteration of the CLEAN algorithm (values normalized to the strongest echo amplitude).

An example of the application of the CLEAN algorithm is shown in Figure 5.14. Figure 5.14(a) shows the initial CAF. The values of the cross-ambiguity function are normalized to the maximum value, which corresponds to the strongest echo at $R_0 = 10$ km, $V_0 = 10$ m/s. The second strongest target echo is at $R_0 = 20$ km, $V_0 = 20$ m/s, with the amplitude -30 dB with respect to the strongest echo. The sidelobe level is at -50 dB, which corresponds to the BT level. After the first iteration of the CLEAN algorithm, the strongest echo is removed, as shown in Figure 5.14(b). The sidelobe level has been reduced to -80 dB. The second iteration of the algorithm results in the CAF shown in Figure 5.14(c), with the sidelobe level at -85 dB. This corresponds to the thermal noise level. At this point, another application of the CLEAN algorithm will not improve the results anymore. If, however, it is applied one more time, the last of the simulated echoes, at $R_0 = 30$ km, $V_0 = 30$ m/s, is removed. The noise floor level is the same as in the previous case.

5.4 NUMERICAL RESULTS

Both groups of adaptive algorithms (i.e., iterative and block) have been tested on real passive radar data. First, the convergence of the iterative algorithms has been investigated using a real signal. The signal used in the experiments was acquired with the *PaRaDe* system and represents the typical characteristics of an FM-based passive radar. In Figure 5.15 the amplitude of the filtered signal versus time is plotted for three different adaptive filters: NLMS ($\mu_{\text{NLMS}} = 0.1$), RLS ($\lambda_{\text{RLS}} = 0.9999$), and LSL ($\lambda_{\text{LSL}} = 0.9999$). As can be seen, the convergence rate of the NLMS is much slower than that of the RLS and LSL. The performance of the RLS and LSL filters is almost identical. These observations confirm the results obtained for the simulated signal. As the RLS and LSL filters yield very similar results (and much better than NLMS), and the latter is more computationally effective, the LSL seems to be more practical and will be used in further experiments.

As will be shown, real clutter has some specific features that have not been reflected in the simplistic simulations presented so far. In Figure 5.16 an example of the CAF of real signals is presented. Figure 5.16(a) shows the CAF of the reference signal and raw echo signal (without any clutter filtering). The peak corresponding to the direct signal (DPI) can be seen at zero range and zero velocity. The sidelobes of the DPI (at the $BT \approx 47$ dB level below the maximum peak) dominate over other components of the signal, even the clutter is almost not visible. To demonstrate the real-life clutter characteristics, the DPI component has been removed from the signal (this corresponds to the application of the block lattice filter with order 1). The CAF

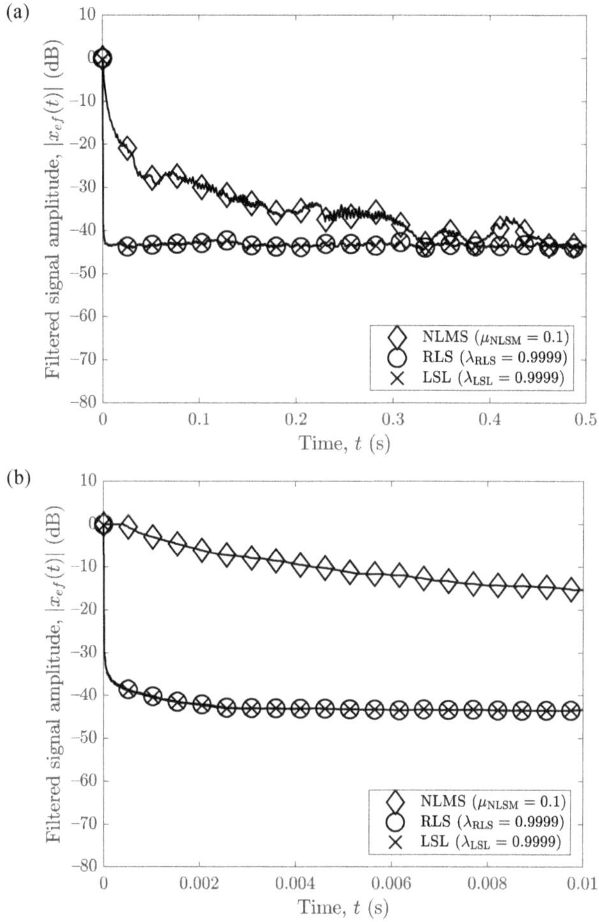

Figure 5.15 Comparison of the filtered (residual) signal of the NLMS, RLS, and LSL filters for real signals: (a) full view and (b) zoomed view (curves smoothed for better visualization).

of such a signal is shown in Figure 5.16(b). Now the clutter reflections are visible. An interesting feature of the clutter is that it is spread in the velocity direction for short ranges. For longer ranges, the clutter energy is focused in the zero-velocity resolution cell. These clutter properties can be observed for various environments (radar placed inland and on the sea shore) and also for different passive radar systems.

As a solution to the problem of different characteristics of clutter for short and long ranges, two adaptive filters with different parameters can be applied. One of

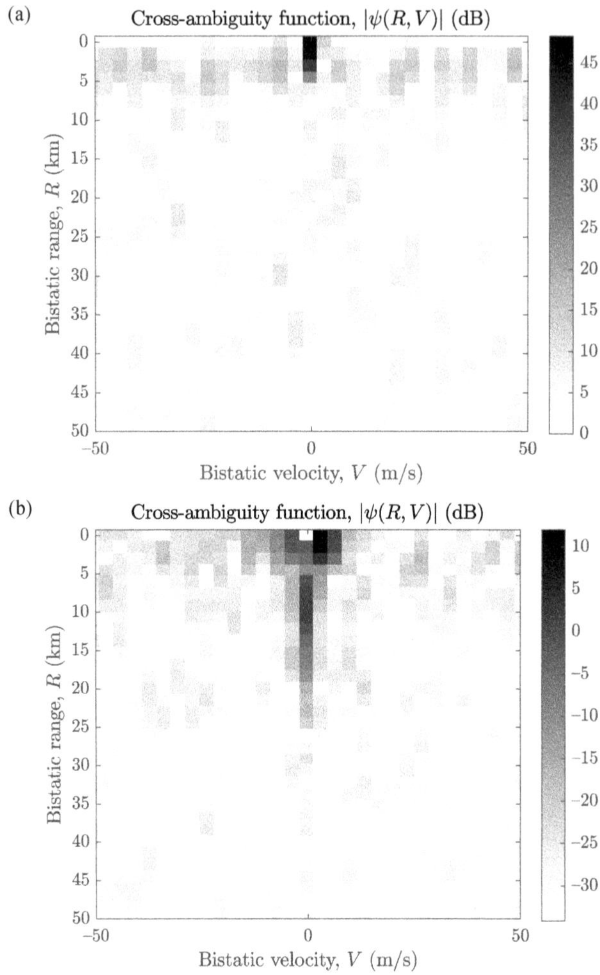

Figure 5.16 Cross-ambiguity function of the real reference signal and (a) raw echo signal and (b) echo signal with removed DPI (values normalized to the mean noise floor level before clutter cancelation).

the filters with a low order (operating only for short ranges) should be characterized by a low frequency selectivity. Another filter with a high order should have a narrow notch in the velocity dimension. An example of the application of such a technique to real data is shown in Figure 5.17. In Figure 5.17(a) the results of the application of the iterative LSL filter are shown. Two regions are visible. The first region

(a)

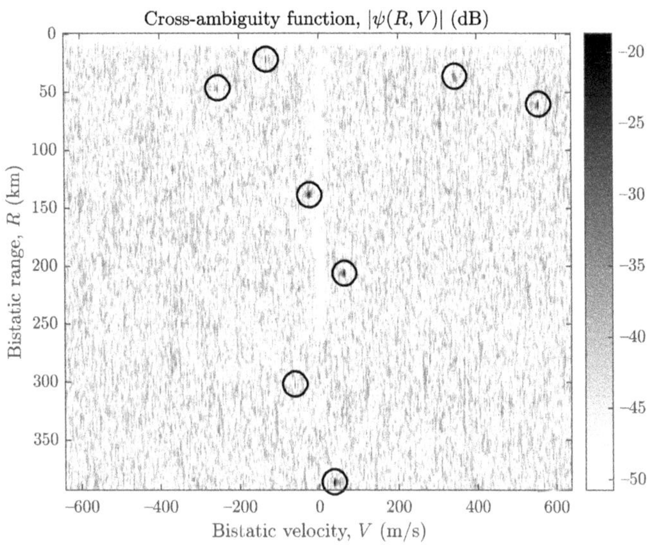

(b)

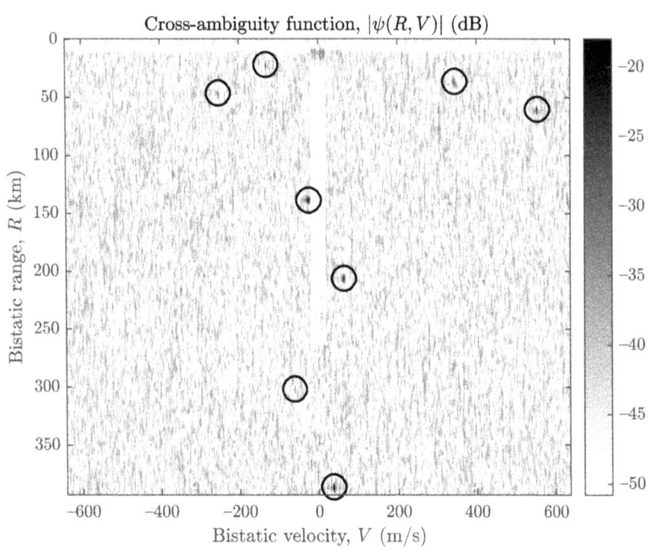

Figure 5.17 Cross-ambiguity function of the real reference and echo signals with canceled clutter using (a) the LSL filter and (b) the modified block lattice filter. Target echoes marked with circles (values normalized to the mean noise floor level before clutter cancelation).

corresponds to short ranges (up to 8 km) and covers a wide span of velocities. This corresponds to the low-order filter ($M = 5$) with a relatively small forgetting factor ($\lambda_{\text{LSL}} = 0.99$). The second region spans from 0 to 250 km and creates a narrow notch in the velocity dimension. This notch has been created using the high-order filter ($M = 170$) with the forgetting factor close to 1 ($\lambda_{\text{LSL}} = 0.9999$). The mean sidelobe level has been reduced by more than 40 dB. Numerous target echoes are visible (marked with circles), which were masked by the DPI and clutter sidelobes prior to adaptive filtering. In Figure 5.17(b), a corresponding result for the modified block lattice filter is shown. Again, two regions are visible, which were created by the low-order filter ($M = 5$) with a large value of the number of frequency bins for which the clutter is removed ($N_{fb} = 300$), and the high-order filter ($M = 170$) with a relatively narrow clutter notch ($N_{fb} = 5$). The mean sidelobe level has also reduced by approximately 40 dB. As can be seen, comparable results can be obtained with the iterative and block methods.

5.5 SUMMARY

In this chapter, one of the most crucial problems in passive radar (i.e., clutter and direct-path interference removal) has been addressed. Since the random sidelobes of the direct signal, as well as the sidelobes of the clutter, can mask the target echoes, the appropriate filtering algorithms decide the detection performance of the system.

Two broad classes of clutter filtering algorithms have been investigated: iterative and block. The adaptation constant in the iterative algorithms influences the convergence rate, which is the expected result. Moreover, the adaptation constant influences the frequency selectivity of the filter. Among the basic iterative algorithms, it has been shown that the LSL filter converges very fast and is characterized by a low computational burden.

It is worth noting that there are many more versions of adaptive filters than those analyzed in this book [4]. For example, many variations of LMS filters exist. Some of them may perform better than the investigated NLMS filter. However, the aim of this chapter is to show the properties of the most representative algorithms, and show their general behavior.

In the case of the block methods, the convergence problem does not exist. The straightforward method using the least squares solution is computationally demanding, as it involves cumbersome matrix operations. However, almost identical results can be obtained with the block lattice filter, whose computational requirements are modest. Variable frequency selectivity of the lattice filter can be obtained by shifting the backward prediction error signals in frequency.

Table 5.6 summarizes the properties of different methods for clutter removal. The listed features include convergence rate (in case of iterative algorithms), computational complexity and frequency selectivity.

Table 5.6
Comparison of Properties of Methods for Clutter Removal (M Denotes the Filter Order)

Method	*Convergence Rate*	*Computational Burden*	*Frequency Selectivity*
NLMS	Slow	Low ($O(M)$)	Low, adjustable with filter parameter
RLS	Very fast	High ($O(M^2)$)	High, adjustable with filter parameter
LSL	Very fast	Moderate ($O(M)$)	High, adjustable with filter parameter
LS matrix	n/a	Very high ($O(M^3)$)	Zero frequency bin
Block lattice	n/a	Moderate ($O(M)$)	Zero frequency bin
Modified block lattice	n/a	Moderate ($O(M \cdot N_{fb})$)	Adjustable with filter parameter N_{fb}

References

[1] M. Malanowski, *Signal Processing Algorithms and Target Localisation Methods for Continuous-Wave Passive Bistatic Radar*. No. 179 in Prace Naukowe Politechniki Warszawskiej. Elektronika, Warszawa, ul. Polna 50: Oficyna Wydawnicza Politechniki Warszawskiej, 2012.

[2] D. W. O'Hagan, C. J. Baker, and H. D. Griffiths, "Signal and Interference Analysis: Proposed Analogue Signal Suppression Techniques for PCL Radar," in *Proc. European Radar Conference – EURAD*, (Manchester, UK), pp. 296–298, September 2006.

[3] R. Saini, M. Cherniakov, and V. Lenive, "Direct Path Interference Suppression in Bistatic System: DTV Based Radar," in *Proc. International Conference on Radar*, (Adelaide, Australia), pp. 309–314, September 2003.

[4] S. Haykin, *Adaptive Filter Theory*. third ed., Upper Saddle River, NJ: Prentice Hall, 1996.

[5] R. Cardinali, F. Colone, C. Ferretti, and P. Lombardo, "Comparison of Clutter and Multipath Cancellation Techniques for Passive Radar," in *Proc. IEEE Radar Conference*, (Boston, Massachusetts, USA), pp. 469–474, April 2007.

[6] K. Wang, R. Tao, Y. Ma, and T. Shan, "Adaptive Multipath Cancellation Algorithm in Passive Radar," in *Proc. IEEE CIE International Conference on Radar*, (Shanghai, China), p. CD, October 2006.

[7] P. E. Howland, D. Maksimiuk, and G. Reitsma, "FM Radio Based Bistatic Radar," *IEE Proc. Radar, Sonar and Navigation*, vol. 152, pp. 107–115, June 2005.

[8] M. Malanowski, "Comparison of Adaptive Methods for Clutter Removal in PCL Radar," in *Proc. International Radar Symposium 2006*, (Cracow, Poland), pp. 237–240, May 2006.

[9] K. Kulpa, "Adaptive Clutter Rejection in Bi-Static CW Radar," in *Proc. International Radar Symposium 2004*, (Warsaw, Poland), pp. 61–66, May 2004.

[10] K. Kulpa, "Ground Clutter Suppression in Noise Radar," in *Proc. International Conference on Radar*, (Toulouse, France), p. CD, October 2004.

[11] K. Kulpa, "Simple Sea Clutter Canceller for Noise Radar," in *Proc. International Radar Symposium 2006*, (Cracow, Poland), pp. 299–302, May 2006.

[12] F. Colone, D. O'Hagan, P. Lombardo, and C. Baker, "A Multistage Processing Algorithm for Disturbance Removal and Target Detection in Passive Bistatic Radar," *IEEE Trans. Aerospace and Electronic Systems*, vol. 45, pp. 698 – 722, April 2009.

[13] K. S. Kulpa and Z. Czekała, "Masking Effect and Its Removal in PCL Radar," *IEE Proceedings – Radar, Sonar and Navigation*, vol. 152, pp. 174–178, June 2005.

[14] K. Kulpa, "The CLEAN Type Algorithms for Radar Signal Processing," in *2008 Microwaves, Radar and Remote Sensing Symposium*, pp. 152–157, September 2008.

[15] M. J. Paradie and J. Pernic, "Airborne VHF Environment Noise Measurements," in *Proceedings of the 1996 Tactical Communications Conference. Ensuring Joint Force Superiority in the Information Age*, pp. 389–395, April 1996.

[16] R. Dalke, R. Achatz, Y. Lo, P. Papazian, and G. Hufford, "Measurement and Analysis of Man-Made Noise in VHF and UHF Bands," in *Proceedings of 1997 Wireless Communications Conference*, pp. 229–233, August 1997.

[17] N. G. Riley and K. Docherty, "Modelling and Measurement of Man-Made Radio Noise in the VHF-UHF Bands," in *1995 Ninth International Conference on Antennas and Propagation, ICAP '95 (Conf. Publ. No. 407)*, vol. 2, pp. 313–316, April 1995.

[18] J. Zoellner, J. Robert, M. Slimani, P. Schlegel, and M. Pulsmeier, "Analysis of the Impact of Man-Made Noise on DVB-T and DVB-T2," in *IEEE International Symposium on Broadband Multimedia Systems and Broadcasting*, pp. 1–6, June 2012.

[19] E. N. Skomal, "Analysis of Airborne VHF/UHF Incidental Noise over Metropolitan Areas," *IEEE Trans. Electromagnetic Compatibility*, vol. EMC-11, pp. 76–83, May 1969.

[20] G. Cui, J. Liu, H. Li, and B. Himed, "Target Detection for Passive Radar with Noisy Reference Channel," in *2014 IEEE Radar Conference*, pp. 0144–0148, May 2014.

[21] J. Liu, H. Li, and B. Himed, "Analysis of Cross-Correlation Detector for Passive Radar Applications," in *2015 IEEE Radar Conference (RadarCon)*, pp. 0772–0776, May 2015.

[22] M. K. Bączyk, K. Kulpa, P. Samczyński, and M. Malanowski, "The Impact of Reference Channel SNR on Targets Detection by Passive Radars Using DVB-T Signals," in *2015 IEEE Radar Conference (RadarCon)*, pp. 0708–0712, May 2015.

[23] X. Zhang, H. Li, J. Liu, and B. Himed, "Joint Delay and Doppler Estimation for Passive Sensing with Direct-Path Interference," *IEEE Trans. Signal Processing*, vol. 64, pp. 630–640, February 2016.

[24] C. J. Baker, H. D. Griffiths, and I. Papoutsis, "Passive Coherent Location Radar Systems. Part 2: Waveform Properties," *IEE Proc. Radar, Sonar and Navigation*, vol. 152, pp. 160–168, June 2005.

Chapter 6

Detection and Estimation

6.1 INTRODUCTION

In this chapter, two closely related topics will be investigated, namely, target detection and parameter estimation. Target detection in passive radar is carried out on the CAF, whose dimensions are the bistatic range and bistatic velocity. The aim of detection is to decide if a target presence is declared for each of the range-velocity cells of the CAF. The decision is made by comparing the value of the CAF with a detection threshold. This is the topic of the first part of the chapter.

If the detection threshold is exceeded in a certain range-velocity cell, the position of this cell already provides a coarse estimate of the target echo parameters. However, finer estimation is also possible. The problem of the estimation of the bistatic parameters of the target is analyzed in the second part of the chapter.

6.2 TARGET DETECTION IN PASSIVE RADAR

The classical problem of target detection in radar can be formulated as follows: in each resolution cell one of two possibilities can occur:

1. Target echo is absent in the resolution cell.
2. Target echo is present in the resolution cell.

The problem of detection is making the decision whether situation 1 or 2 occurs. Situation 1 corresponds to a hypothesis H_0, where the received signal x consists of noise w only:

$$H_0 : x = w \tag{6.1}$$

Situation 2, in turn, corresponds to a hypothesis H_1, where the received signal x contains target echo s and noise w:

$$H_1 : x = s + w \tag{6.2}$$

The decision on target detection is based on thresholding of the received signal. If the threshold is exceeded by the signal, a detection is declared. As the detection is a probabilistic process due to the random nature of observed signals, an erroneous decision can be made. Two types of wrong decision can occur:

1. The target echo is present in the resolution cell, but detection is not declared. The probability of this event is *miss probability* (P_{miss}). It is related to the probability of detection P_d as: $P_d = 1 - P_{\text{miss}}$.
2. The target echo is absent in the resolution cell, but detection is declared. The probability of this event is called the *probability of false alarm* (P_{fa}).

The approach that is most often used in radar is to keep the probability of false alarm at a constant level while maximizing the probability of detection. In detection theory this approach is known as the Neyman-Pearson detector [1, 2].

The probability of false alarm P_{fa} can be calculated as:

$$P_{fa} = \int_D^{+\infty} f_0(z)\mathrm{d}z \tag{6.3}$$

where $f_0(z)$ is the probability density function (PDF) of the signal in the case of a lack of a target echo (the received signal contains noise only), and D is the detection threshold. If the PDF of noise $f_0(z)$ is known, the detection threshold can be calculated using (6.3) by setting a desired P_{fa}, calculating the integral and inverting the relationship. As can be seen, the threshold depends only on the noise characteristics, and not on the echo signal characteristics.

Once the detection threshold D is set, the probability of target detection P_d can be calculated as:

$$P_d = \int_D^{+\infty} f_1(z)\mathrm{d}z \tag{6.4}$$

where $f_1(z)$ is the PDF of the echo-and-noise signal. Function $f_1(z)$ depends on the strength of the signal with respect to the noise (SNR); thus, the probability of detection is a function of the target echo SNR.

The concept of calculating P_d and P_{fa} is shown in Figure 6.1. The horizontal axis corresponds to the signal amplitude, whereas the vertical axis shows the value of

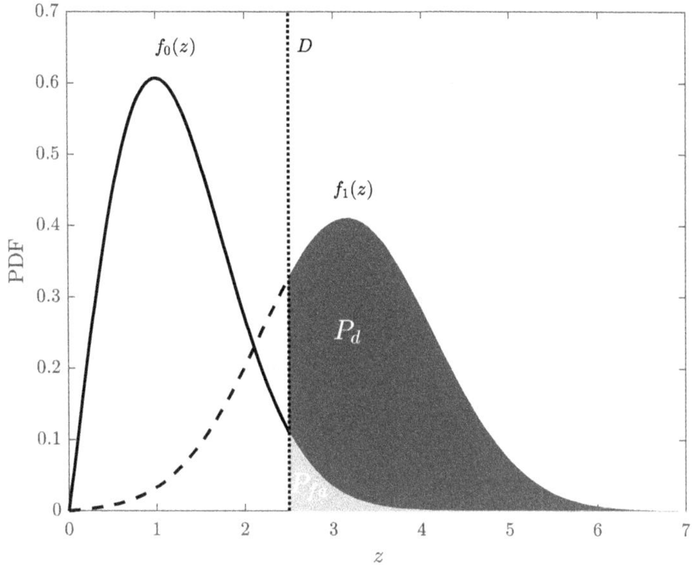

Figure 6.1 Graphic representation of calculating the probability of detection and probability of false alarm.

the PDF. The two PDF functions shown represent cases of the noise-only and noise-and-echo signals. The vertical line indicates the detection threshold. The areas under the curves represent the probabilities: P_d (for $f_1(z)$ PDF) and P_{fa} (for $f_0(z)$ PDF).

One of the most important problems in classical active monostatic radar is the estimation of the shape and parameters of the $f_0(z)$ function, which determines the detection threshold value. In the simplest and most advantageous practice cases, the baseband noise is a complex Gaussian signal (real and imaginary parts are Gaussian-distributed). In such a case, the absolute value of the noise signal is Rayleigh-distributed. Such a PDF decays very quickly, which has an important practical consequence; in order to obtain the required P_{fa}, the detection threshold is relatively low. This is not true in the case of other PDFs that are encountered in radar. For example, PDFs such as log-normal and Weibull. are characterized by heavy tails. This means that high values of the noise signal occur more frequently. This requires setting a higher detection threshold in order to obtain the required P_{fa}. This, in turn, causes P_d to decrease.

Detection in passive radar is based on the analysis of the absolute value of the CAF $|\psi(R, V)|$. Similar to most radars, target detection is declared when the value of the signal exceeds a certain threshold. Let us analyze the problem of a single target detection. Assume that the echo signal contains only a single target echo with the bistatic parameters R_0, V_0. The CAF calculated for the corresponding values of the bistatic range and velocity is:

$$\begin{aligned}\psi_1(R_0, V_0) = \int_{-T/2}^{T/2} & C'' \cdot x_r\left(t - \frac{R_0}{c}\right) \cdot \exp\left(\mathrm{j}\frac{2\pi}{\lambda}V_0 t\right) \cdot \\ & x_r^*\left(t - \frac{R_0}{c}\right) \cdot \exp\left(-\mathrm{j}\frac{2\pi}{\lambda}V_0 t\right) \mathrm{d}t = \\ & C'' \cdot \int_{-T/2}^{T/2} \left| x_r\left(t - \frac{R_0}{c}\right)\right|^2 \mathrm{d}t. \end{aligned} \tag{6.5}$$

Therefore, the correlation value is a complex constant,[1] proportional to the reference signal energy.

Considering a case when the echo signal contains noise only $x_e(t) = w(t)$, the CAF will have the following form:

$$\psi_0(R_0, V_0) = \int_{-T/2}^{T/2} w(t) \cdot x_r^*\left(t - \frac{R_0}{c}\right) \cdot \exp\left(-\mathrm{j}\frac{2\pi}{\lambda}V_0 t\right) \mathrm{d}t. \tag{6.6}$$

In such a case, the delayed and Doppler-shifted reference signal is randomized by the received noise, which results in a different noise signal. In practice, the CAF is calculated as an appropriate sum of samples. This means that the value of $\psi(R_0, V_0)$ is calculated by summing many samples (usually hundreds of thousands). Therefore, based on the central limit theorem, real and imaginary parts of $\psi(R_0, V_0)$ will tend to the Gaussian distribution. Thus, the absolute value of the cross-ambiguity function $|\psi(R_0, V_0)|$ will tend to the Rayleigh distribution. This is true also even when the distribution of $w(t)$ is non-Gaussian, as a consequence of the central limit theorem. In such a case, however, the level of the noise floor increases, which reduces the sensitivity of the system. This problem can be alleviated by the appropriate robustification of the detection procedure [3–7].

The consequence of the aforementioned properties of the cross-ambiguity function is that the problem of detection in the passive radar is a classical problem

1 It is assumed here that the target is nonfluctuating (Swerling 0 case).

of detecting a constant complex value embedded in complex Gaussian noise. The amplitude of the noise signal is Rayleigh-distributed:

$$f_0(z) = \frac{z}{\sigma^2}\exp\left(-\frac{z^2}{2\sigma^2}\right), \quad z > 0 \tag{6.7}$$

The PDF of the echo-and-noise signal has the Rice distribution:

$$f_1(z) = \frac{z}{\sigma^2}\exp\left(-\frac{z^2+|A|^2}{2\sigma^2}\right) I_0\left(\frac{z|A|}{\sigma^2}\right), \quad z > 0 \tag{6.8}$$

where $A = \psi_1(R_0, V_0)$ is the amplitude of the target echo, and $I_0(\cdot)$ is the Bessel function.

The probability of false alarm is calculated as:

$$P_{fa} = \int_D^{+\infty} \frac{z}{\sigma^2}\exp\left(-\frac{z^2}{2\sigma^2}\right) \mathrm{d}z = \exp\left(-\frac{D^2}{2\sigma^2}\right) \tag{6.9}$$

In practice, the detection threshold D is calculated from the assumed P_{fa}, which can be done by rearranging the above relation into:

$$D = \sqrt{-2\sigma^2 \ln P_{fa}} = \sigma\sqrt{-2\ln P_{fa}} \tag{6.10}$$

The detection threshold D has a linear relationship with the characteristic parameter σ of the Rayleigh distribution. This can be expressed as:

$$D = \sigma \cdot \alpha_D \tag{6.11}$$

where $\alpha_D = \sqrt{-2\ln(P_{fa})}$ is the *threshold factor*, a multiplier that has to be applied to the noise parameter σ to obtain absolute detection threshold D.

The probability of detection in such a situation can be calculated by [1, 2, 8, 9]:

$$P_d = \int_D^{+\infty} \frac{z}{\sigma^2}\exp\left(-\frac{z^2+|A|^2}{2\sigma^2}\right) I_0\left(\frac{z|A|}{\sigma^2}\right) \mathrm{d}z = Q\left(\sqrt{2\mathrm{SNR}}, \sqrt{-2\ln(P_{fa})}\right) = Q\left(\frac{|A|}{\sigma}, \alpha_D\right), \tag{6.12}$$

where $Q(\cdot)$ is the Marcum function, $\mathrm{SNR} = |A|^2/(2\sigma^2)$ is the signal-to-noise ratio at the output of the correlator. The classical curves of the probability of detection for different values of P_{fa} are plotted in Figure 6.2. It can be seen that the SNR of 15 dB provides a probability of detection very close to 1 for the typical values of the P_{fa}.

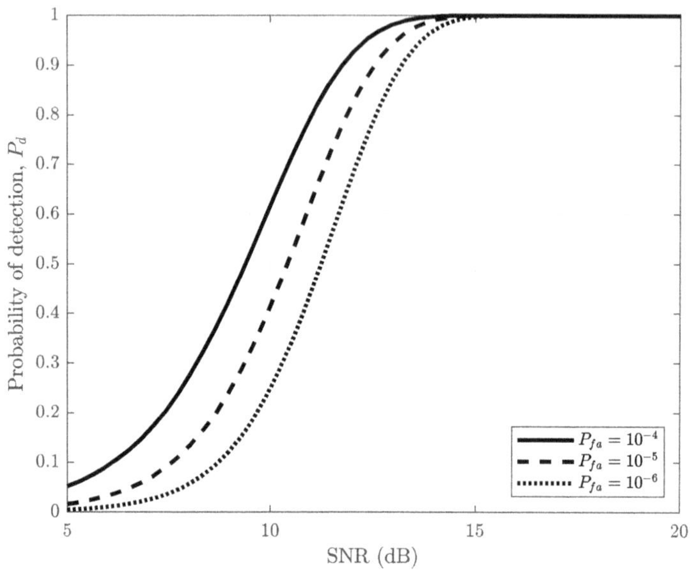

Figure 6.2 Probability of detection versus SNR after cross-ambiguity function calculation.

6.2.1 Constant False Alarm Rate (CFAR)

In practice, the noise level on the CAF is not known, and it has to be estimated. One approach is to assume that the noise level is constant over the whole CAF surface, estimate a single value of the noise, and then calculate corresponding threshold value. In reality, it often happens that the noise level distribution is not uniform, and using a single threshold value is not robust in terms of providing constant false alarm rate. An alternative approach typically used in radar is to apply a CFAR algorithm [1]. The aim of the algorithm is to adapt the detection threshold to the local characteristics of the signal to provide a constant rate of false alarms. This is done by estimating the noise level for a particular resolution cell, called the *cell under test* (CUT), based on the signal values from the neighboring resolution cells. The rationale behind this approach is the assumption that signal characteristics do not change very fast (i.e., the statistical parameters of a signal for a particular resolution cell can be deduced from nearby resolution cells). CFAR detection can be expressed as testing the following condition:

$$\text{CUT} \lesseqgtr \hat{\sigma}\alpha_D \tag{6.13}$$

where CUT is the value of the cell under test, $\hat{\sigma}$ is the noise level estimate, and α_D is the threshold factor. If the above condition is fulfilled, that is, the CUT exceeds the right side of (6.13), a detection for this resolution cell is declared. The same condition expressed on the decibel scale, which is often used in radar, is:

$$\text{CUT} \lessgtr \hat{\sigma} + \alpha_D \text{ (dB)} \tag{6.14}$$

6.2.1.1 One-Dimensional CFAR

Traditionally, in active monostatic radar, the detection was carried out on a one-dimensional signal (i.e., the range profile). The range profile is obtained by recording the echo signal after the transmission of a single radar pulse, then performing range compression by matched filtering. The samples of the signal obtained in this way represent the fast time, which can be scaled in the range from the radar.

The block diagram of a one-dimensional CFAR is presented in Figure 6.3. In a single iteration of the algorithm, the aim is to determine the detection threshold for the CUT based on the *reference cells*, or *training cells*. The number of the reference cell window size on each side of the CUT is N_w. The CUT is surrounded by the *guard cells*. The number of the guard cells is N_g on each size of the CUT. These are cells that do not take part in the estimation of the noise level. The reason for using the guard cells is that the target echo can be spread across many resolution cells (due to signal windowing or oversampling); thus, the estimation of the noise level surrounding the target echo could be distorted by the echo itself. The number of the guard cells should be chosen based on the width of the target echo.

The number of reference cells N_w has to be chosen for each radar scenario individually. If the noise is *iid* (independent and identically distributed), its characteristics do not change with range. From this point of view, the more reference cells are used, the more accurate the estimation of the noise parameter is. In practice, however, the noise is rarely *iid*. The noise characteristic changes with the range; therefore, the adaptive properties of the CFAR algorithm have to be used. This means

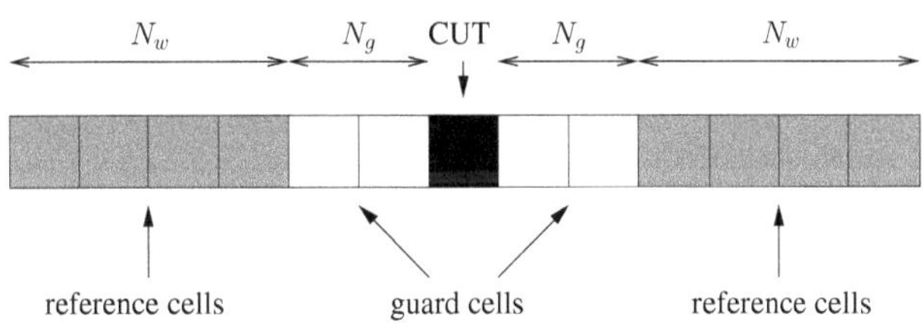

Figure 6.3 Concept of a one-dimensional CFAR.

that the reference windows cannot be too long, so that the noise level estimate can follow the trend, ignoring local perturbations. This is very similar to lowpass filtering. In fact, the calculation of the noise level estimate using CFAR can be viewed as a kind of lowpass filtering of the signal.

The noise level is estimated from the reference cells in one of many possible ways. The simplest, and still one of the most popular approaches, is *Cell Averaging* CFAR (CA-CFAR). In this case, the estimate of the noise level is obtained by simply calculating the mean value from the reference cells from both windows[2]:

$$\hat{\sigma}(n) = \frac{1}{2}\left(\frac{1}{N_w}\sum_{i=n-N_w-N_g}^{n-N_g-1} x(i) + \frac{1}{N_w}\sum_{i=n+N_g+1}^{n+N_w+N_g} x(i)\right) \tag{6.15}$$

where $x(i)$ represents the ith signal sample. Another common approach is the *Smallest-of Cell Averaging* CFAR (SOCA-CFAR). In this case, the mean values are calculated from both windows, and the smallest of them is taken as the estimate of the noise level:

$$\hat{\sigma}(n) = \min\left(\frac{1}{N_w}\sum_{i=n-N_w-N_g}^{n-N_g-1} x(i), \frac{1}{N_w}\sum_{i=n+N_g+1}^{n+N_w+N_g} x(i)\right) \tag{6.16}$$

In the *Greatest-of Cell Averaging* CFAR (GOCA-CFAR), the greatest of the mean values corresponding to the two windows is used as the noise estimate:

$$\hat{\sigma}(n) = \max\left(\frac{1}{N_w}\sum_{i=n-N_w-N_g}^{n-N_g-1} x(i), \frac{1}{N_w}\sum_{i=n+N_g+1}^{n+N_w+N_g} x(i)\right) \tag{6.17}$$

All three versions of the CFAR algorithm mentioned above share a common feature: they are based on the calculation of the arithmetic mean, which is susceptible to outliers. This means that if a value that is very different from the average level occurs in the window, the arithmetic mean will be severely influenced by this value. In the case of CFAR, this may lead to a situation where a strong target echo heavily influences the noise level estimate.

2 It is important to note that the estimates $\hat{\sigma}(n)$ obtained by the different versions of the CFAR algorithm may have different characteristics and do not have to reflect the mean level of the noise. For this reason, the threshold factor α_D has to be chosen differently for each of the CFAR algorithms. Nevertheless, the value $\hat{\sigma}(n)$ obtained by all CFAR algorithms, will be referred to as "noise level estimate."

One of the approaches to combat this effect is to use the *Order Statistic* CFAR (OS-CFAR) [10]. In this case, the values from the reference window are ordered (sorted):

$$\mathbf{x}_s = \text{sort}([x(n - N_w - N_g), \ldots, x(n - N_g - 1), \\ x(n + N_g + 1), \ldots, x(n + N_w + N_g)]) \quad (6.18)$$

Next, a particular value at index k is chosen from the sorted vector of samples:

$$\hat{\sigma}(n) = \mathbf{x}_s^k \quad (6.19)$$

The value k is in the range $1, \ldots, 2N_w$. For example, if the middle sample from the vector of sorted values is chosen, it corresponds to the median estimator of the noise level. Other values can also be chosen, resulting in a particular percentile of the distribution (e.g., median is a 50% percentile). The main idea of the OS-CFAR is that a single or a few outliers among the reference values will not influence the noise level estimation. The reason for this is that those values will be at the edges of the $\mathbf{x}_s$ vector and therefore will not influence, the median value (if the number of the outliers is smaller than the one-sided reference window length N_w).

An example of the application of different versions of the CFAR algorithm on synthetic, one-dimensional data is shown in Figure 6.4. The solid line represents the simulated signal, where a target echo is visible with the SNR of 20 dB. The echo occupies three resolution cells, which represent a widening of the target response due to the windowing. The dashed line shows the noise level estimated by the CFAR. The true noise level is at 0 dB, and the ideal CFAR algorithm should estimate this value. However, as will be shown, different versions of the CFAR algorithm yield different, sometimes substantially distorted, results. The dotted curve shows the obtained detection threshold which results from the estimated noise level summed with the threshold factor $\alpha_D = 13$ dB. In the presented example, a square-law detector was applied, which means that the squared amplitude of the complex signal ($|x|^2$) was processed by the CFAR algorithm. Other popular choices are the linear-law detector ($|x|$) and logarithmic detector ($\log_{10}|x|$). The square-law detector, although not always the best choice, allows some of the typical problems with the CFAR algorithm to be clearly shown. In all cases, $N_w = 16$ reference cells and $N_g = 3$ guard cells on each side of the CUT were used.

The first example from Figure 6.4 corresponds to the CA-CFAR. As can be seen, the detection threshold is relatively constant in the region far away from the target, oscillating around the 13 dB value. However, in the vicinity of the target, the detection threshold is substantially increased, which is an undesirable effect as

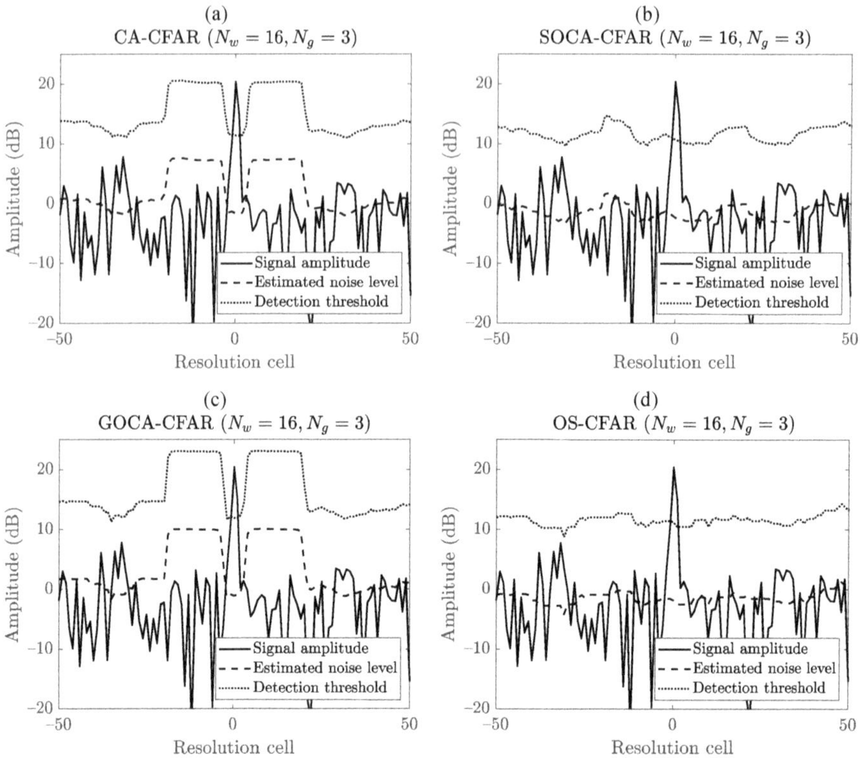

Figure 6.4 Results of the CFAR algorithm for a single target situation: (a) Cell-Averaging, (b) Smallest-of-Cell-Averaging, (c) Greatest-of-Cell-Averaging, and (d) Order-Statistic.

weaker echoes in that region could remain undetected. The reason for the increase in the detection threshold is the fact that when one of the reference windows contains a target echo, the mean value is substantially distorted by the high-value samples. A possible solution to this problem is the SOCA-CFAR algorithm, where the minimum of two mean values is taken as the noise level estimate. In this case, when one of the reference windows contains the target echo but the second window contains noise samples only, the algorithm will choose the smaller value, which represents the noise level better. As can be seen in Figure 6.4(b), in the case of a single target this approach results in a nearly constant detection threshold, which is not influenced by the presence of the target echo. An opposite strategy to SOCA-CFAR is the GOCA-CFAR, where the greater of the two mean values is chosen. This approach is used to avoid the *clutter edge* problem, introduced in one of the next examples. The detection threshold in the case of the GOCA-CFAR reveals an even larger influence of target echo than the CA-CFAR, as the greater of two mean values is selected (see Figure 6.4(c)). The fourth result (Figure 6.4(d)) corresponds to the OS-CFAR with the median value as the noise level estimator. The detection threshold is not influenced by the target echo, similar to the SOCA-CFAR case.

In Figure 6.5 a two-target situation is presented, with target echoes with an SNR equal to 20 dB and 17 dB, respectively. The echoes are separated by 10 resolution cells. In this scenario, the features of each of the four algorithms are investigated in the not uncommon situation of closely separated targets. As already indicated, in the CA-CFAR and GOCA-CFAR, the detection threshold is highly influenced by a single target echo, which could prevent other echoes in the vicinity from being detected. This is exactly what is presented in Figure 6.5(a, c).

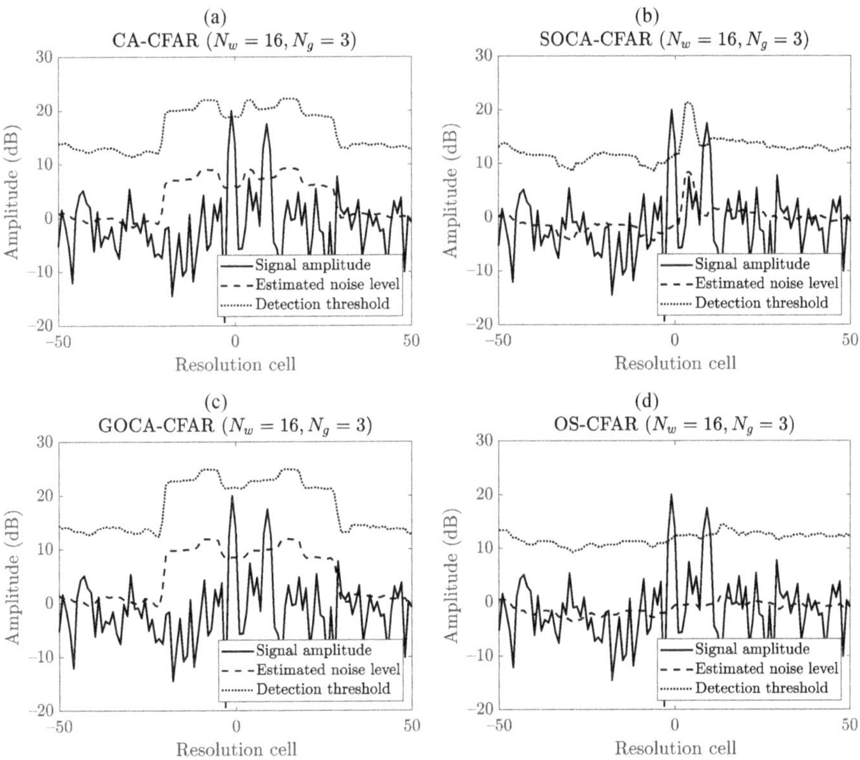

Figure 6.5 Results of the CFAR algorithm for a two-target situation: (a) Cell-Averaging, (b) Smallest-of-Cell-Averaging, (c) Greatest-of-Cell-Averaging, and (d) Order-Statistic.

The weaker target echo with an SNR of 17 dB is not detected in both cases. In addition, in the case of the GOCA-CFAR, even the stronger echo with an SNR of 20 dB was not detected. In the case of the SOCA-CFAR and OS-CFAR, both targets were detected. In the case of the SOCA-CFAR, the detection threshold is increased between the two target echoes. This results from the fact that between the two targets each of the reference windows contains a target echo; therefore, even choosing the smaller value yields a high value. Only the OS-CFAR provides a nearly constant detection threshold, which is not influenced even by two target echoes.

Let us consider the *clutter edge* case, which is one of the typical scenarios analyzed in classical monostatic radar. The clutter edge scenario happens when the signal level changes rapidly, which can correspond to a transition between two different types of the terrain (e.g., sea and land, or field and forest). This poses a potential problem for the CFAR algorithm, as false alarms can appear at the clutter edge when the detection threshold does not follow the signal change fast enough. An example of the results of different CFAR algorithms for the clutter edge situation is shown in Figure 6.6. At the resolution cell index equal to 0, the signal changes rapidly from 0 dB to 20 dB. There is no real target echo in the scene; therefore, the desired behavior of the CFAR algorithm is to adapt the detection threshold fast enough to avoid false detections, especially at the clutter edge. As can be seen, in this particular example, two algorithms performed as expected: CA-CFAR and GOCA-CFAR. The SOCA-CFAR and OS-CFAR, resulted in a threshold crossing at the edge of the clutter.

In the presented examples the same threshold factor α_D has been applied to different CFAR algorithms. This resulted in different average values of the detection threshold D. For example, in the GOCA-CFAR the detection threshold is, in general, higher than in SOCA-CFAR for the same value of α_D, because the greater of the two average values is used instead of the smaller one. This means that the value of α_D should be chosen for a particular algorithm type and reference window size in order to provide the same level of false alarms.

In general, any CFAR algorithm which estimates the noise level from the data is inferior to the case when this level is known a priori. This leads to the *CFAR losses*, which is the difference in the target echo SNR that has to be added to provide the same probability of detection P_d and the same probability of false alarm P_{fa} in the CFAR detection as in the ideal case, when the noise level is known. The CFAR losses depend on the window size, the algorithm type (e.g., CA, GOCA, SOCA), detection law (e.g., linear detector, square-law detector), and required P_d and P_{fa} [1].

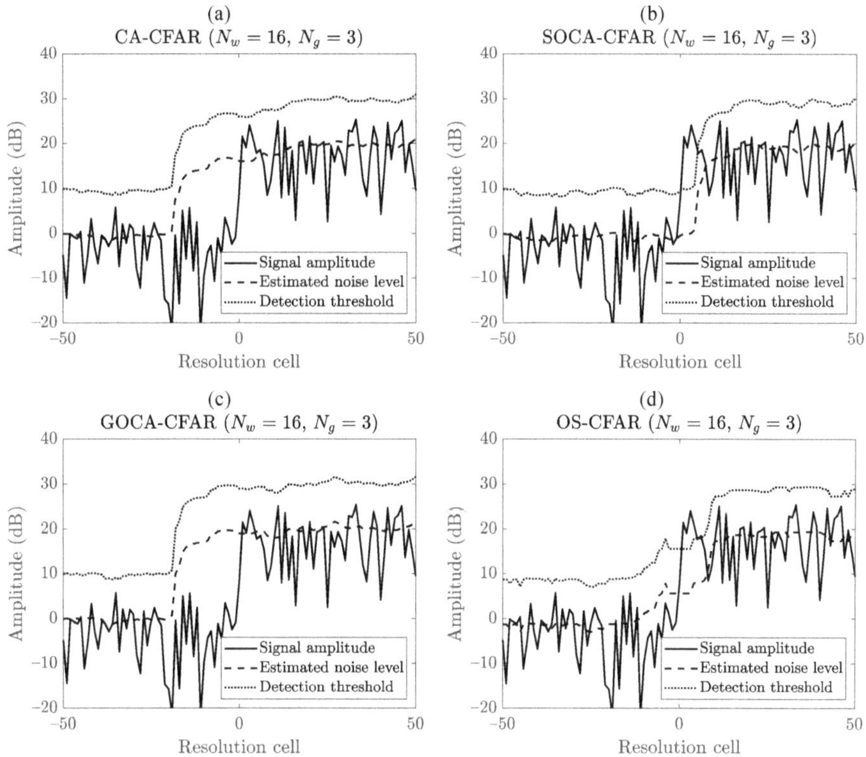

Figure 6.6 Results of the CFAR algorithm for a clutter edge situation: (a) Cell-Averaging, (b) Smallest-of-Cell-Averaging, (c) Greatest-of-Cell-Averaging, and (d) Order-Statistic.

Let us consider the application of a classical one-dimensional CFAR algorithm to the passive radar scenario. By analogy to the active monostatic radar, the CFAR can be applied to the CAF along the range dimension. This is repeated for each of the velocity resolution cells. Another possibility is to apply the CFAR in the velocity direction. As shown earlier, a typical CAF has a relatively flat level of noise (regardless of its nature: actual noise level or random correlation sidelobes) in both directions. For this reason, estimating the noise level for a particular resolution cell from the surrounding resolution cells in the velocity dimension is a reasonable assumption.

In Figure 6.7 the CAF used for the testing of different CFAR algorithms is shown. The plot at the top shows the CAF with the target echo. The target echo is embedded in the noise. The SNR after correlation is approximately 20 dB. The

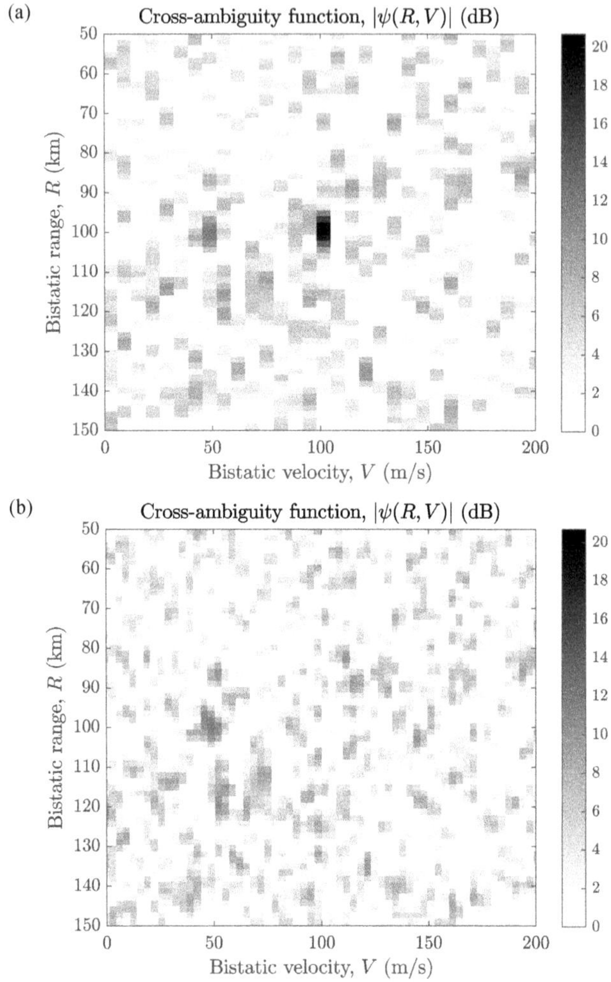

Figure 6.7 CAF with (a) target echo and noise and (b) noise only.

CAF at the bottom was calculated for the noise component of the signal (without target echo). An ideal CFAR algorithm should estimate a constant level of noise, which corresponds to the CAF shown at the bottom. The results of the application of different versions of the CFAR algorithm operating along the range dimension are shown in Figure 6.8. The CFAR variations used in the experiment were: CA-CFAR (Figure 6.8(a)), SOCA-CFAR (Figure 6.8(b)), GOCA-CFAR (Figure 6.8(c)), and

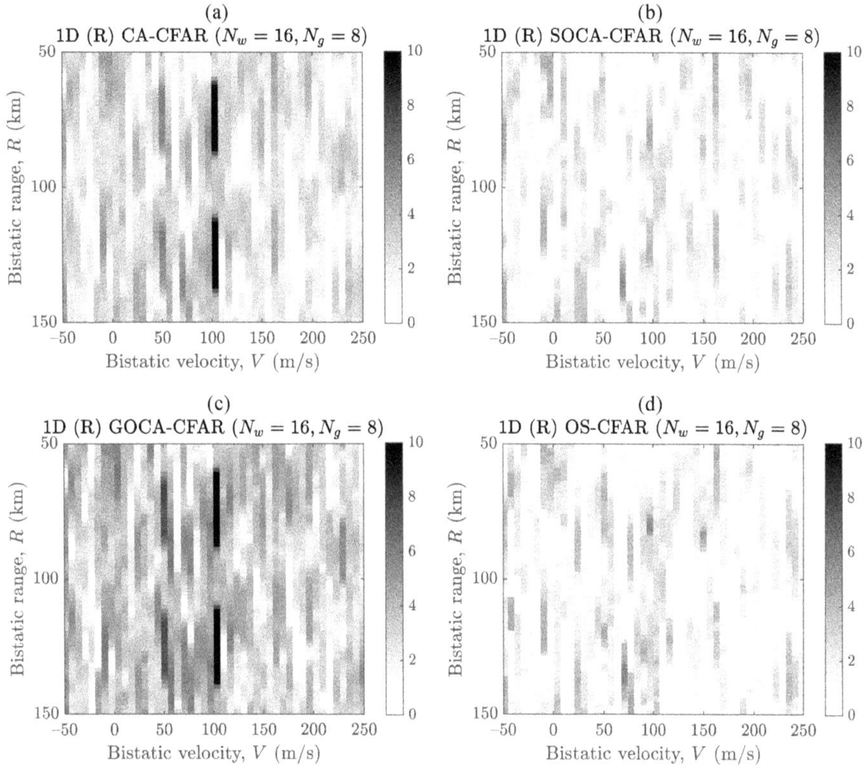

Figure 6.8 Mean level estimated by the 1D CFAR algorithm operating in the range dimension: (a) Cell-Averaging, (b) Smallest-of-Cell-Averaging, (c) Greatest-of-Cell-Averaging, and (d) Order-Statistic.

OS-CFAR (Figure 6.8(d)). Each of the four plots shows the noise level estimated by the CFAR. In each case, the number of guard cells was equal to $N_g = 8$ on each side of the CUT, and the number of reference cells was equal to $N_w = 16$ on each side of the CUT. As mentioned before, noise level estimation performed by the CFAR can be viewed as lowpass filtering (possibly nonlinear), or smoothing. This can be observed in all four images through the vertical streaks. In the case of the CA-CFAR, two maxima are visible on each side of the target echo in the range direction. This behavior is very typical for this version of the algorithm; the noise estimate is increased by the target echo before and after the target position in the same way as in the 1D case shown in Figure 6.4(a). This is a disadvantageous behavior, as it does not reflect the true noise level estimate (which should be constant), and it can lead to the missed detection of weaker target echoes due to the masking caused by the stronger echoes. A similar, yet even stronger, effect of increasing the noise level estimate can be observed for GOCA-CFAR. In the case of SOCA-CFAR and OS-CFAR, the noise level estimate is almost not influenced by the target echo, which is the desired effect.

In Figure 6.9 the results of the application of the 1D CFAR in the velocity dimension are shown for the CA-CFAR, GOCA-CFAR, SOCA-CFAR, and OS-CFAR algorithms, respectively. The same behavior is seen as in the case of the CFAR applied in the range dimension, but this time the characteristic features are visible in the velocity dimension. This is especially visible in the case of the CA and GOCO algorithms, where a high value of the estimated noise level is visible before and after the target echo in the velocity dimension.

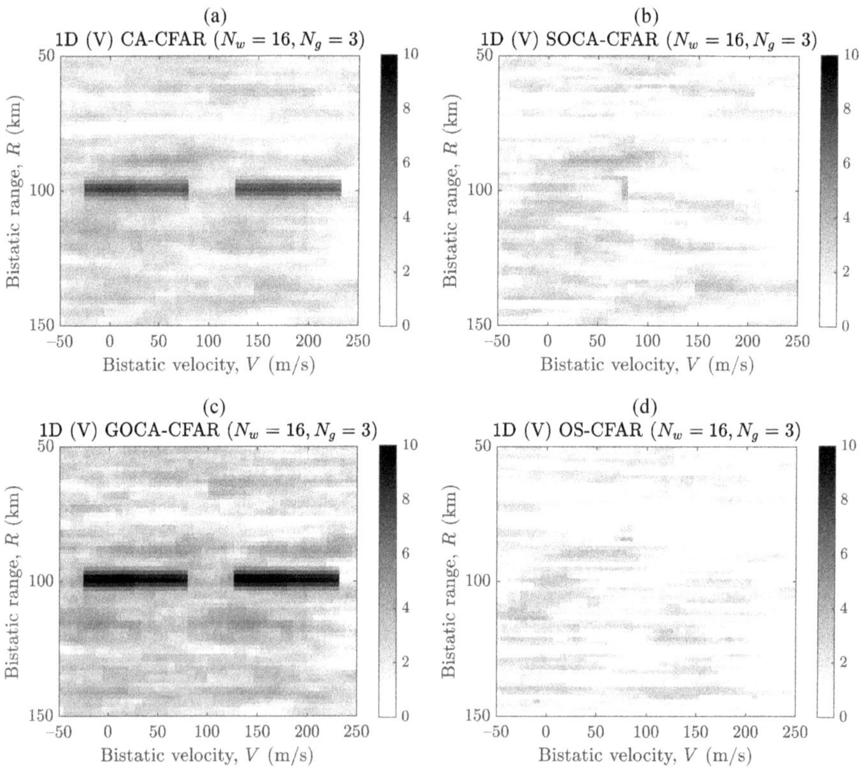

Figure 6.9 Mean level estimated by the 1D CFAR algorithm operating in the velocity dimension: (a) Cell-Averaging, (b) Smallest-of-Cell-Averaging, (c) Greatest-of-Cell-Averaging, and (d) Order-Statistic.

6.2.1.2 Two-Dimensional CFAR

The concept of 1D CFAR can be extended to two dimensions [11], which is very natural in the case of the CAF routinely obtained in passive radar. One of the possibilities of constructing the reference window is shown in Figure 6.10(a). In this case a ring is created surrounding the CUT. In the range direction the number of guard cells is N_{gr}, and the length of the reference window is N_{wr} on each side of the CUT. In the same way, the window size is defined in the velocity direction: the number of the guard cells is N_{gv} and the length of the reference window is N_{wv}. The number of resolution cells in the window is: $(2N_{wr}+1)(2N_{wv}+1)-(2N_{gr}+1)(2N_{gv}+1)$. This, in general, will be a relatively large number compared to a one-dimensional

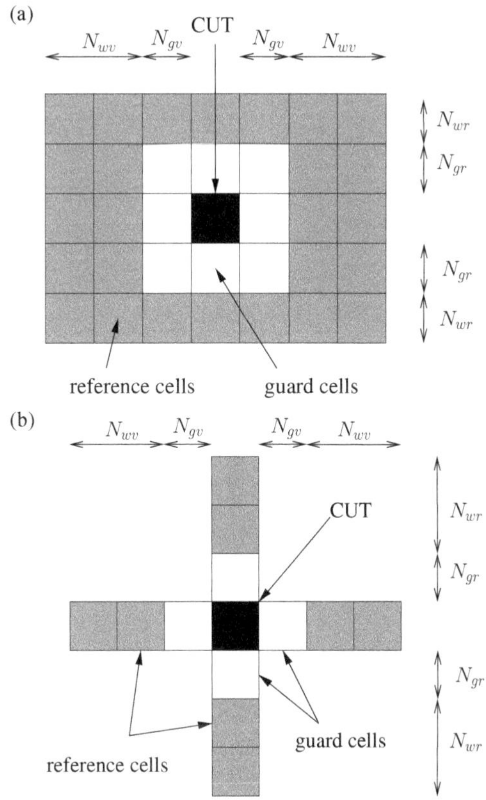

Figure 6.10 Concept of a two-dimensional CFAR with (a) ring-shaped and (b) cross-shaped reference windows.

version. In this case, the simple averaging of all values from the reference window seems to be the only straightforward approach. Dividing the window into parts, calculating the mean values of these parts, and then choosing the minimum/maximum value, or using the order statistic, are also possible, but will not be considered here.

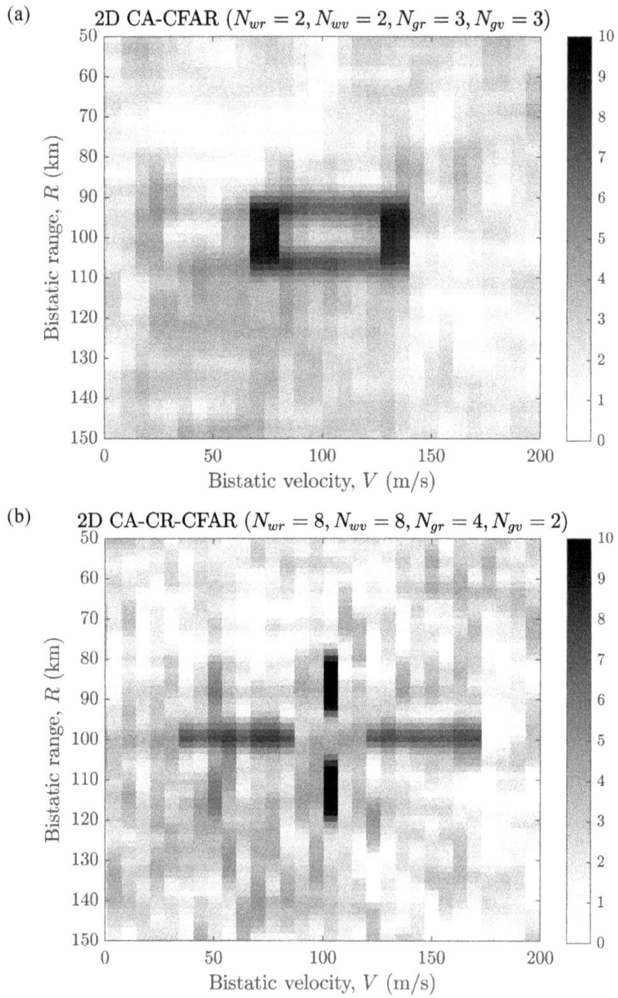

Figure 6.11 Mean level estimated by the 2D CFAR algorithm: (a) cell-averaging (ring) and (b) cell-averaging (cross).

The basic CA-CFAR will be assumed, where the simple mean is calculated from the reference window.

Another intuitive approach to constructing a 2D CFAR is shown in Figure 6.10(b). The reference window is shaped like a cross around the CUT. The total number of reference cells in this approach is $2N_{wr} + 2N_{wv}$. In this case, some combinations of the GOCA or SOCA-CFAR would be easier to apply than in the previous case, but again, the focus will be on the CA-CFAR.

The application of the two examples of a 2D CFAR to the CAF from Figure 6.7(a) is shown in Figure 6.11. In the first case (Figure 6.11(a)), the estimated noise level has a ring shape surrounding the target echo. This can be understood in the following way: the averaging realized by the CA-CFAR is in fact a 2D convolution of a ring-shaped mask with the CAF values. If the target echo has a thumbtack character, which is almost the case (see Figure 6.7(a)), then the estimated noise level will reflect the shape of the mask applied in the CFAR algorithm. A similar situation is observed in the second case (Figure 6.11(b)), where the cross-shaped mask has been applied.

6.2.2 Numerical Results

The selected CFAR algorithms described above have been applied to real passive radar data. FM-radar data were used in the experiment. The CPI was 1s, and 60s worth of data was processed. In Figure 6.12 the accumulated detections (using binary integration) over 60s are shown for the four versions of the algorithm: 1D CA-CFAR (range), 1D CA-CFAR (velocity), 2D CA-CFAR (ring), and 2D CA-CFAR (cross). In the case of the 1D CA-CFAR (range), 2D CA-CFAR (ring), and 2D CA-CFAR (cross), the first few range cells were neglected to avoid problem with no CAF values for negative ranges. The threshold factor α_D was chosen to obtain similar false alarm rate in each case. There were a few targets present in the scene, which can be observed as continuous trajectories in the plots. Apart from the target detections, a uniform distribution of false alarms is desirable. This is not the case in all the presented examples. In the first case, where 1D CA-CFAR along the range dimension was applied, a relatively uniform false alarm distribution is obtained. In the case of the same algorithm but applied to the velocity dimension, a congestion of false alarms can be observed around the zero velocity. Similar but less severe congestion is seen in the 2D versions of the algorithm. This effect, which is specific to passive radar, deserves an in-depth analysis.

As mentioned in the previous chapters, one of the main problems in passive radar is the direct path interference (DPI) and clutter filtering. Without it, the mov-

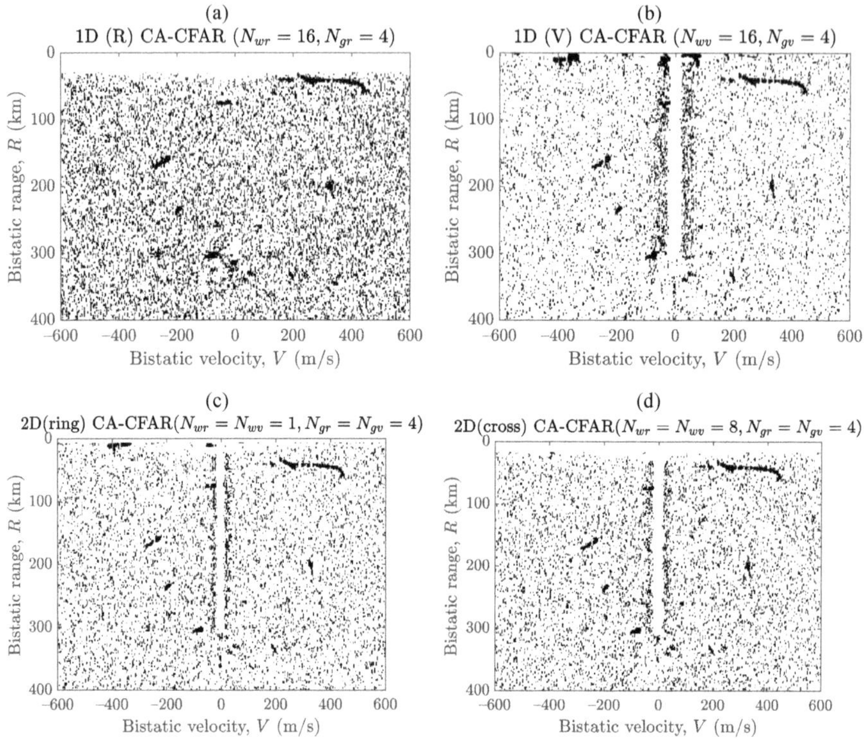

Figure 6.12 Accumulated detections from 60s with CPI of 1s for FM-based passive radar for: (a) 1D (range) CA-CFAR, (b) 1D (velocity) CA-CFAR, (c) 2D (ring) CA-CFAR, and (d) 2D (cross) CA-CFAR.

ing target echoes would be masked by the random sidelobes of the DPI and clutter echoes. The clutter is not always concentrated only on zero velocity; therefore, an applied filtering algorithm often removes the clutter from a certain span of velocities. As a result, a valley, or notch, in the CAF is created, which spans the range of zero to a maximum range, which corresponds to the order of the applied filter. This valley is a desirable effect, as the low-velocity echoes are removed, but it creates a problem for the detection algorithm. This is very similar to the clutter edge problem analyzed in the CFAR detection. In this case, however, it is not actually clutter edge , but an edge created by removing the clutter. In Figure 6.13 the averaged value of the CAF functions used in the previous example is shown, revealing the notch created by the clutter filtering. The averaging was done for ranges from 0 to 300 km, which corresponds to the order of the adaptive filter used for clutter re-

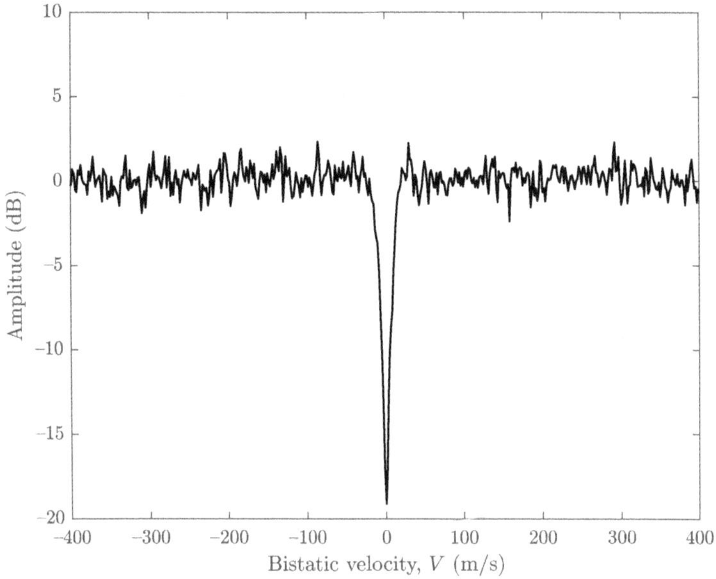

Figure 6.13 Averaged values of the CAF for ranges from 0 to 300 km with a notch at the zero velocity created by the clutter filtering.

moval. A single snapshot (one CPI) was used here. A notch at zero velocity can be clearly seen, which creates the clutter edge problem for the CFAR detector.

An example of the CAF slice for a single range, equal to 195 km, where a target echo is present, is shown in Figure 6.14. The dashed curve represents the detection threshold calculated by the CFAR. A 1D CA-CFAR operating in the velocity direction was applied here. A target echo present at the velocity of 325 m/s is correctly detected by the algorithm; however, a characteristic increase of the detection threshold can be observed on both sides of the echo. Around the zero velocity, the detection threshold is slightly lowered, as the notch in the CAF decreases the averaged value of the reference cells calculated by the CA-CFAR. This leads to a threshold crossing at the edge of the valley.

A comparison of accumulated detections for different 1D CFAR algorithms applied along the velocity direction is shown in Figure 6.15. The four plots correspond to CA-CFAR, SOCA-CFAR, GOCA-CFAR and OS-CFAR algorithms, respectively. The strongest influence of the clutter notch on the detection density can be observed for the SOCA-CFAR. This is not surprising considering the analysis of the clutter edge problem presented earlier. As the smaller of the two average values

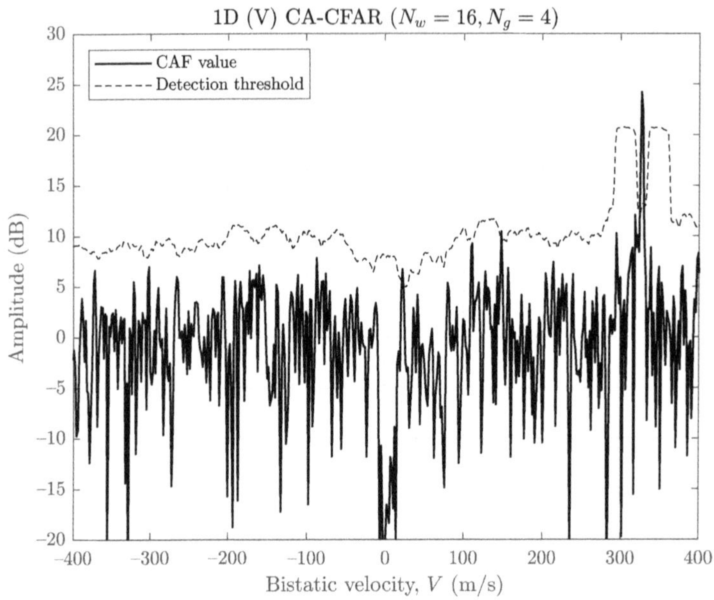

Figure 6.14 Velocity cross-section of the CAF with detection threshold calculated by the 1D CA-CFAR operating in the velocity dimension.

is used, if one of the windows is placed on the clutter notch, the noise estimate is decreased. The influence of the clutter valley can also be clearly seen for the CA and OS algorithms. The algorithm that is least influenced by the clutter filtering is the GOCA-CFAR. This is because in the vicinity of zero velocity the algorithm chooses the greater of two values, one of which is lower due to the clutter notch.

The presented analysis shows that detection in passive radar has its specific features. One of them is that the data is inherently 2D, as the CAF has the range and velocity dimensions. Another feature is the noise distribution. In passive radar the problem of heavy-tailed noise and/or clutter distributions is not so severe as in active radars, especially when the sea clutter is considered. The reason for this is that the values of the CAF are created based on the summation of thousands of samples, and even a spiky input signal results in the relatively fast decaying distribution of the CAF. A problem that is unique to passive radar is the clutter filtering, which creates a notch in the CAF. This can cause a congestion of false alarms in the vicinity of zero velocity.

It has to be noted that the parameters used in the performed analyses were chosen to show the typical problems of CFAR detection (e.g., the choice of the square-

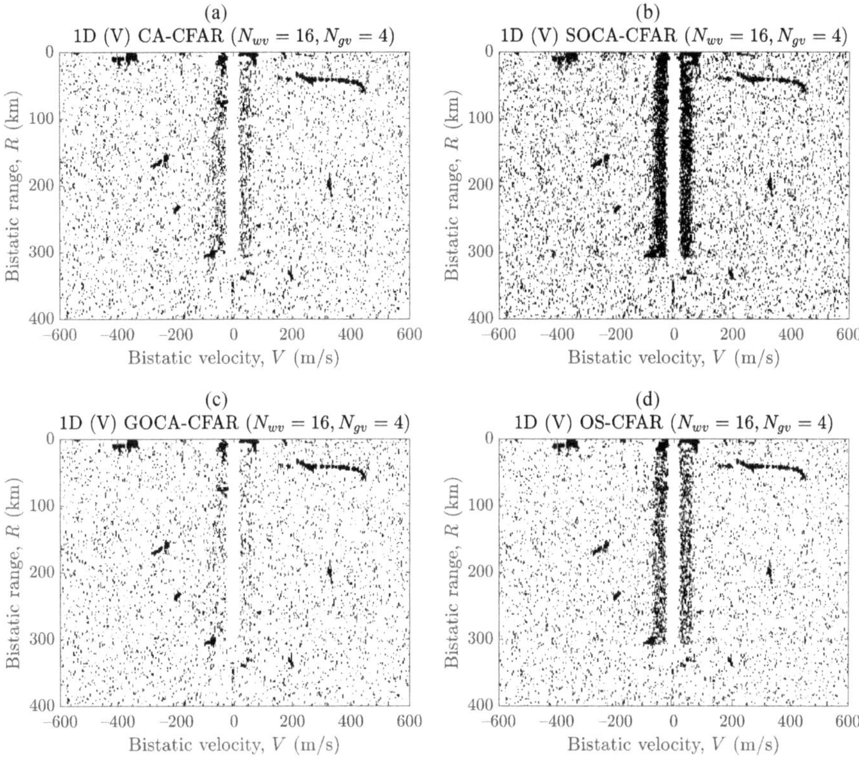

Figure 6.15 Accumulated detections from 60s with CPI of 1s for FM-based passive radar for: (a) 1D (velocity) CA-CFAR, (b) 1D (velocity) SOCA-CFAR, (c) 1D (velocity) GOCA-CFAR, and (d) 1D (velocity) OS-CFAR.

law detector). In practice, the parameters should be chosen to a specific scenario, and the aforementioned problems can be mitigated, or completely eliminated, using the right combination of standard approaches.

6.3 BISTATIC PARAMETER ESTIMATION IN PASSIVE RADAR

After the application of the CFAR algorithm, a set of detections is obtained. Usually a single target echo generates numerous detections concentrated around the true target position, as the echo is spread over several range and velocity cells (e.g., due to the applied windowing). The detections have to be combined so that a cluster is created that represents a single target. Next, the exact values of the bistatic range and bistatic velocity have to be calculated. The simplest approach is to use the estimates of parameters corresponding to the maximum value in the cluster. The accuracy of such estimation, however, is limited to the size of the resolution cell. Usually, a much better performance can be obtained by analyzing the values of the CAF function in the vicinity of the correlation peak of the echo.

One of the approaches to fine, subresolution cell parameter estimation is to fit a curve representing a single target response to the measured signal values [1, 12, 13]. In this way, the maximum of the fitted curve indicates the precise estimate of the target parameter, either in range or velocity. The shape of the curve can be chosen so that it resembles the real shape of the response of a point target. To choose such a curve, analysis of the ambiguity function can be carried out.

In practice, another approach can be used to simplify the estimation process. Instead of a curve matched to the signal characteristics, a simple parabola can be used, which approximates the values surrounding the maximum of the target echo [1]. In this way, the estimation process is reduced to the very simple calculation of the parabola coefficients based on the amplitude values of the CAF.

An example of such estimation is shown in Figure 6.16. The plot at the top shows the slice of the CAF in the bistatic range direction as well as the fitted parabola. The CAF was calculated based on a real FM signal with approximately 50-kHz bandwidth. The sampling frequency was 200 kHz and the CAF was calculated for CPI of 1s. The parabola is fitted to the three values of the CAF: the maximum one and one of each side of the maximum. As can be seen, the fitted curve reflects the shape of the target response main lobe relatively well. The plot at the bottom shows a zoom on the maximum of the CAF. In this case the estimated target range, which corresponds to the parabola maximum, is also marked. For comparison, the actual value, which is equal to $-1/3$ of the resolution cell, is shown. As can be seen, the estimated and true values are very close to each other.

The estimation accuracy depends on the SNR; the greater the SNR, the more accurate the parameter estimation. Usually, the measurement accuracy in radar can be modeled in the following way [14]:

$$\sigma_R = \frac{A_R}{\sqrt{\text{SNR}}}\Delta R, \tag{6.20}$$

$$\sigma_V = \frac{A_V}{\sqrt{\text{SNR}}}\Delta V, \tag{6.21}$$

where σ_R, σ_V are the standard deviations of the estimation error (for bistatic range

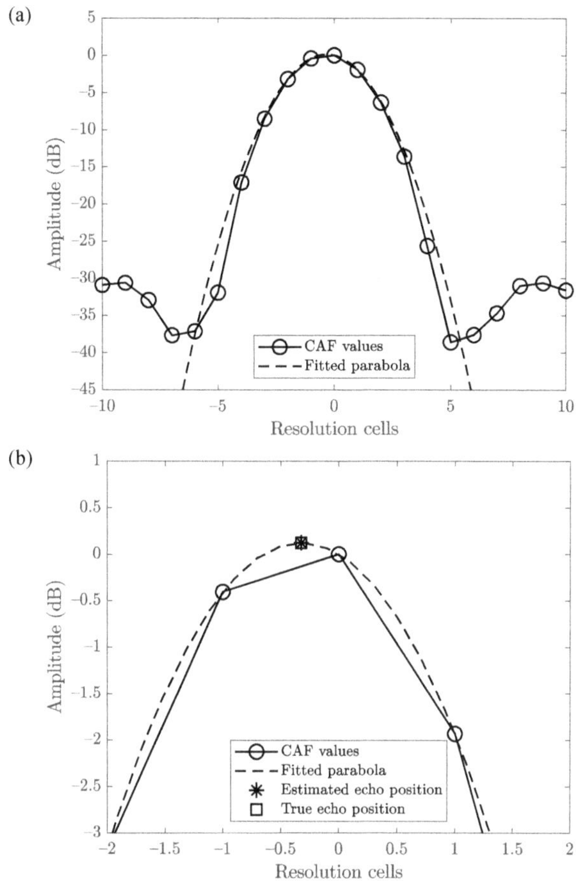

Figure 6.16 An example of fitting a parabola to the CAF values: (a) CAF values and fitted parabola and (b) zoomed view with estimated and true echo positions marked.

and bistatic velocity, respectively), ΔR, ΔV are the sizes of the resolution cell, A_R, A_V are the coefficient, which are parameters of the model, and SNR is the signal-to-noise ratio after correlation.

The accuracy of the aforementioned estimation method using a parabolic curve has been tested on a real signal. The results are shown in Figure 6.17. They were obtained for the same FM radio signal as in the last example. The sampling frequency of 200 kHz corresponds to the bistatic range resolution cell width of 1.5 km, and the CPI of 1s corresponds to the bistatic velocity resolution cell of

3 m/s (for 100-MHz carrier frequency). In Figure 6.17(a) the measured estimation accuracy of the bistatic range is shown versus the SNR. It can be seen that for SNR $= 20$ dB the accuracy is approximately 150m, which is one order of magnitude better than the size of the range resolution cell. The modeled accuracy, plotted for comparison, was obtained by choosing the coefficient A_R in (6.20) to fit the measurements. The obtained coefficient value was 0.85. Analogous results can be observed in Figure 6.17(b), where the accuracy for the bistatic velocity is plotted. Again, for SNR=20 dB, the accuracy is approximately an order of magnitude better

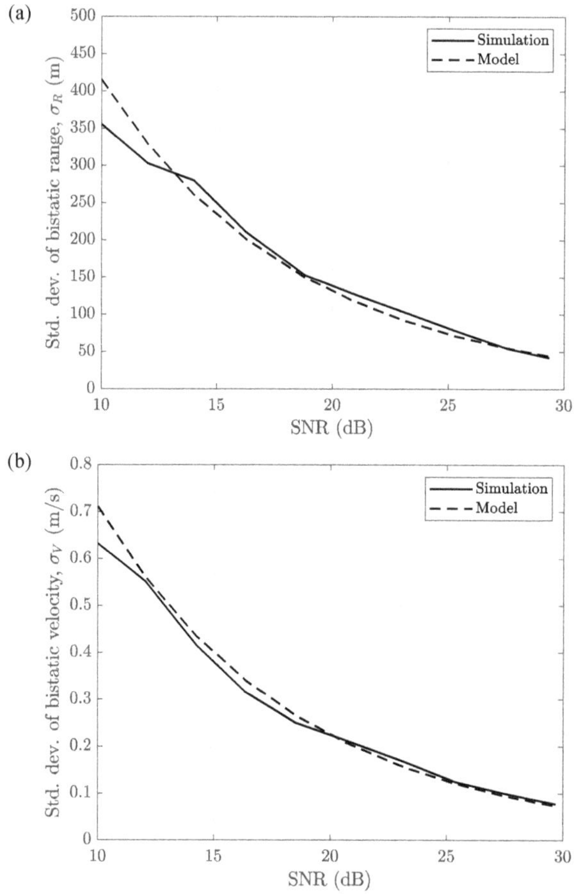

Figure 6.17 Comparison of measured and modeled bistatic range (a) and velocity (b) estimation accuracy versus SNR.

than the size of the resolution cell. In this case the obtained value of the A_V coefficient in the model (6.21) was 0.68. The coefficients A_R and A_V used to model the accuracy have to be chosen for a particular scenario, as they depend on the signal spectrum shape and applied windowing.

The proposed procedure for fine estimation of the bistatic range and bistatic velocity is applicable when the target can be treated as a point scatterer. This is true, for example, when the bistatic range is estimated for an aircraft in FM-based passive radar, where the range resolution is much larger than the size of the target. In some cases, however, such as DVB-T-based radar observing a large ship, the target echo can spread over numerous range cells due to the relation of the target size to the resolution cell size. In such a case, the proposed estimation method may provide results different than expected.

After the detection and estimation process, radar plots are created. A plot corresponds to a single target detected in a single CPI. The plot can contain numerous measurements and parameters, such as the bistatic range and velocity, whose estimation was considered in this chapter, but also echo amplitude, SNR, azimuth and elevation angles, and others.

6.4 SUMMARY

In the chapter, target detection and bistatic parameter estimation were introduced. Those processing stages in passive radar have some specific features, which are different than in the case of classical active radar. First of all, the data in passive radar are by nature two-dimensional, but the dimensions are range and velocity, whereas in active radar they are range and azimuth. A one-dimensional CFAR can be applied in passive radar along the range or velocity dimensions. Alternatively, two-dimensional CFAR can be applied. Second, due to the application of the clutter filtering, a notch in the cross-ambiguity is created, which can cause congestion of the false alarms. Third, the probability density function of the noise in the cross-ambiguity function usually has a fast-decaying distribution (such as the Rayleigh distribution), whereas heavy-tailed distributions are often encountered in active radar.

In the case of parameter estimation, a straightforward approach is to use the index of the bistatic range and bistatic velocity corresponding to the maximum of the cross-ambiguity function. The accuracy of this approach is, however, limited by the quantization of the range and velocity dimensions. A better accuracy can be obtained (e.g., by fitting a parabolic curve to the amplitudes of the target echo, as shown in this chapter). This is especially important in low-resolution cases, such as bistatic range of FM-based passive radar, where substantial increase in the accuracy can be obtained, compared with the resolution.

References

[1] M. A. Richards, *Fundamentals of Radar Signal Processing*. New York: McGraw-Hill, 2005.

[2] R. D. Hippenstiel, *Detection Theory*. Boca Raton, FL: CRC Press, 2002.

[3] K. Kulpa, Z. Gajo, and M. Malanowski, "Robustification of Noise Radar Detection," *IET Proc. Radar, Sonar and Navigation*, vol. 2, pp. 284–293, August 2008.

[4] K. Kulpa, J. Misiurewicz, Z. Gajo, and M. Malanowski, "A Simple Robust Detection of Weak Target in Noise Radars," in *Proc. European Radar Conference – EURAD*, (Munich, Germany), pp. 1554–1557, October 2007.

[5] M. Malanowski and K. Kulpa, "Robust Detection in Continuous-Wave Noise Radar – Experimental Results," in *Proc. International Radar Symposium 2010*, (Vilnius, Lithuania), pp. 347–350, June 2010.

[6] M. Malanowski and K. Kulpa, "Target Detection in Continuous-Wave Noise Radar in the Presence of Impulsive Noise," *Acta Physica Polonica A*, vol. 119, pp. 333–338, April 2011.

[7] Z. Gajo and M. Malanowski, "Noise Radar Target Detection in the Presence of Alpha-Stable Noise," in *Proc. SPIE – Photonics Applications in Astronomy, Communications, Industry, and High-Energy Physics Experiments IV*, (Wilga, Poland), pp. 693734–1–693734–8, May 2007. vol. 6937.

[8] M. I. Skolnik, *Radar Handbook*. second ed., Boston, MA: McGraw-Hill, 1990.

[9] P. Z. Peebles, *Radar Principles*. New York: John Wiley & Sons, 1998.

[10] H. Rohling, "Radar CFAR Thresholding in Clutter and Multiple Target Situations," *IEEE Trans. Aerospace and Electronic Systems*, vol. AES-19, pp. 608–621, July 1983.

[11] M. Kronauge and H. Rohling, "Fast Two-Dimensional CFAR Procedure," *IEEE Trans. Aerospace and Electronic Systems*, vol. 49, pp. 1817–1823, July 2013.

[12] M. Nalecz and K. Kulpa, "Range and Azimuth Estimation Using Raw Data in DSP-Based Radar System," in *Proc. International Conference on Microwaves and Radar – MIKON*, (Cracow, Poland), pp. 871–875, May 1998.

[13] M. Malanowski and K. Kulpa, "Experimental Analysis of Passive Radar Accuracy," in *Proc. International Radar Symposium 2009*, (Hamburg, Germany), pp. 331–335, September 2009.

[14] M. I. Skolnik, *Introduction to Radar Systems*. third ed., Boston, MA: McGraw-Hill, 2003.

Chapter 7

Bistatic Tracking

7.1 INTRODUCTION

In this chapter, target tracking in bistatic coordinates (i.e., bistatic range and bistatic velocity) is analyzed [1]. In general, the aim of the target tracking algorithm is to associate the measurements from consecutive time instants and to perform parameter estimation based on past and current measurements. In the case of the tracker considered in this chapter, state estimation is performed in the bistatic coordinates, which do not indicate the Cartesian position of the target. This will be obtained in a further processing step, addressed in Chapter 8.

Despite the fact that raw bistatic measurements can be used directly for target localization in Cartesian coordinates, an advantageous approach is to apply tracking in the bistatic coordinates. Because the tracking algorithm uses past and present measurements, target state estimation accuracy is increased in comparison with raw measurements. The increased accuracy of the estimation leads to better performance of the localization algorithm in the Cartesian coordinates domain.

Additional advantage of using the bistatic tracking as an intermediate stage between the detection and the Cartesian tracking is substantial reduction of false alarms. This is obtained by initializing only tracks, which correspond to a history of plots that represent a plausible target trajectory.

Most of the signals processed in passive radar are continuous in nature. As has been extensively discussed in Chapter 4, the CPI (coherent processing interval) in passive radar depends on the integration interval that is best suited to a particular transmission type (e.g., FM-based radars have integration intervals of approximately 1s). The consecutive blocks are processed one after another, and a set of bistatic measurements is available for each of the blocks. The blocks can overlap in time;

however, the most convenient approach is to use nonoverlapping blocks. Therefore, if the block length is equal to T, the separation between the measurements from consecutive blocks (track refresh time) is also T. This approach will be assumed in further analysis.

First, the tracking algorithm, based on the Kalman filter, is introduced. Next, the influence of various parameters on the tracking accuracy is investigated. An example of target tracking and its accuracy is shown in real data at the end of the chapter.

7.2 KALMAN FILTERING

Consider a nearly constant acceleration (NCA) model of target motion in bistatic coordinates [2–4]:

$$\mathbf{x}(k+1) = \mathbf{F}\mathbf{x}(k) + \mathbf{u}(k), \tag{7.1}$$

where $\mathbf{x}(k)$ is the state vector at the discrete time instant k, $\mathbf{F}$ is the transition matrix, and $\mathbf{u}(k)$ is the zero-mean, Gaussian process noise. Assume that the state vector consists of the bistatic range, velocity and acceleration $\mathbf{x}(k) = [R(k), V(k), A(k)]'$, that is, the consecutive derivatives of a polynomial motion model in bistatic coordinates. Then (7.1) can be rewritten in an explicit form as:

$$\begin{bmatrix} R(k+1) \\ V(k+1) \\ A(k+1) \end{bmatrix} = \begin{bmatrix} 1 & T & T^2/2 \\ 0 & 1 & T \\ 0 & 0 & 1 \end{bmatrix} \begin{bmatrix} R(k) \\ V(k) \\ A(k) \end{bmatrix} + \mathbf{u}(k), \tag{7.2}$$

where $\mathbf{u}(k)$ is the discrete process noise vector with the covariance matrix defined as:

$$E\left[\mathbf{u}(k)\mathbf{u}'(j)\right] = \delta(k,j)\mathbf{Q}, \tag{7.3}$$

where $\delta(k,j)$ is the Kronecker delta function defined as [4]:

$$\delta(k,j) = \begin{cases} 1 & k = j \\ 0 & k \neq j \end{cases}, \tag{7.4}$$

and

$$\mathbf{Q} = q \begin{bmatrix} T^5/20 & T^4/8 & T^3/6 \\ T^4/8 & T^3/3 & T^2/2 \\ T^3/6 & T^2/2 & T \end{bmatrix}, \tag{7.5}$$

and q is the power spectral density of the process noise. The above formulation means that the process noise is uncorrelated in time, and the covariance matrix of $\mathbf{u}(k)$, representing dependencies of vector elements for a particular time instant, is equal to $\mathbf{Q}$.

At the bistatic tracking stage, the true motion parameters of the target in the Cartesian coordinates, such as position or velocity vector, are not known yet. For this reason, the motion model applied in the bistatic coordinates is very generic, and does not reflect the physical motion parameters of the target directly; that is, the bistatic acceleration is different than the physical acceleration of the target. In practice, the nearly constant acceleration model provides good correspondence with the actual target parameters only in a short timescale. However, the use of the process noise in the model, which adds a random component to the deterministic model, allows the tracker to adapt to the changing target parameters.

The measurement carried out by passive radar can be modeled as:

$$\mathbf{z}(k) = \mathbf{H}\mathbf{x}(k) + \mathbf{w}(k), \tag{7.6}$$

where $\mathbf{z}(k)$ is the measurement vector, $\mathbf{H}$ is the observation model matrix, and $\mathbf{w}(k)$ is the vector of the measurement errors, which are assumed to be zero-mean Gaussian-distributed.

In a standard processing procedure used in passive radar, bistatic range and bistatic velocity measurements are available. It is worth noting that both measurements in passive radar are unambiguous, which is the result of the continuous-wave operation and noise-like properties of most of the utilized signals. Moreover, the accuracy of velocity measurement is relatively high due to the long CPI. It is therefore natural to use both measurements directly in the tracking algorithm. This is in contrast with classical pulsed radar, where unambiguous range and velocity measurement requires using special techniques, such as staggering of the pulse repetition interval [5, 6]. In addition, the velocity accuracy measurement is often low, which results from short observation time.

Therefore, the measurement vector in passive radar typically consists of the bistatic range and velocity, that is:

$$\mathbf{z}(k) = \begin{bmatrix} \hat{R}(k) \\ \hat{V}(k) \end{bmatrix}, \tag{7.7}$$

the observation model matrix is:

$$\mathbf{H} = \begin{bmatrix} 1 & 0 & 0 \\ 0 & 1 & 0 \end{bmatrix}, \tag{7.8}$$

and the covariance matrix of the measurement error vector is:

$$\mathbf{R} = E\left[\mathbf{w}(k)\mathbf{w}'(k)\right] = \begin{bmatrix} \sigma_R^2 & 0 \\ 0 & \sigma_V^2 \end{bmatrix}. \tag{7.9}$$

Because both the state transition equation (7.1) and the measurement model (7.6) are linear, standard Kalman filtering equations can be applied [3]. Kalman filtering consists of two stages: prediction and update. The prediction of the state vector estimate $\hat{\mathbf{x}}(k|k)$ is performed according to the assumed model (7.1) without the unknown process noise component:

$$\hat{\mathbf{x}}(k+1|k) = \mathbf{F}\hat{\mathbf{x}}(k|k). \tag{7.10}$$

The prediction causes an increase in the estimation uncertainty expressed by the associated state covariance matrix $\mathbf{P}(k|k)$:

$$\mathbf{P}(k+1|k) = \mathbf{F}\mathbf{P}(k|k)\mathbf{F}' + \mathbf{Q}. \tag{7.11}$$

When a new measurement is available, it is used to update the state estimate. The state covariance matrix is also updated. First, the innovation vector is calculated, which is the difference between the current measurement and the prediction from the previous time instant:

$$\mathbf{v}(k+1) = \mathbf{z}(k+1) - \mathbf{H}\hat{\mathbf{x}}(k+1|k), \tag{7.12}$$

where $\hat{\mathbf{z}}(k+1) = \mathbf{H}\hat{\mathbf{x}}(k+1|k)$ is the predicted measurement. The covariance of the innovation can be calculated by:

$$\mathbf{S}(k+1) = E\left[\mathbf{v}(k+1)\mathbf{v}'(k+1)\right] = \mathbf{H}\mathbf{P}(k+1|k)\mathbf{H}' + \mathbf{R}. \tag{7.13}$$

The optimal Kalman gain is computed as follows:

$$\mathbf{K}(k+1) = \mathbf{P}(k+1|k)\mathbf{H}'\mathbf{S}^{-1}(k+1). \tag{7.14}$$

The calculated innovation and the Kalman gain are used to update the state vector estimate:

$$\hat{\mathbf{x}}(k+1|k+1) = \hat{\mathbf{x}}(k+1|k) + \mathbf{K}(k+1)\mathbf{v}(k+1). \tag{7.15}$$

The new measurement decreases the uncertainty of the estimation, which corresponds to the following transformation of the state covariance matrix:

$$\mathbf{P}(k+1|k+1) = \mathbf{P}(k+1|k) - \mathbf{K}(k+1)\mathbf{H}\mathbf{P}(k+1|k). \tag{7.16}$$

7.2.1 Filter Initialization

The state vector and its covariance in the Kalman filter have to be initialized at the beginning of the tracking process. One of the possibilities is to initialize the state vector in the following way:

$$\hat{\mathbf{x}}(0|0) = \begin{bmatrix} \hat{R}(0) \\ \hat{V}(0) \\ 0 \end{bmatrix}. \tag{7.17}$$

This means that the first measurements of the bistatic range and bistatic velocity are directly assigned to the corresponding state vector elements. The bistatic acceleration in the state vector is set to 0, as the acceleration measurement is not available if the classical correlation processing is used. The uncertainty of the estimates of the bistatic range, velocity, and acceleration are reflected by the appropriate covariance matrix:

$$\mathbf{P}(0|0) = \begin{bmatrix} \sigma_R^2 & 0 & 0 \\ 0 & \sigma_V^2 & 0 \\ 0 & 0 & \sigma_A^2 \end{bmatrix}. \tag{7.18}$$

The variances of the bistatic range σ_R^2 and bistatic velocity σ_V^2 correspond to the measurement accuracy, as initially no history is available. Those values can be calculated based on the signal-to-noise ratio of the target echo, as shown in Section 6.3 (see (6.20) and (6.21)). The variance of the bistatic acceleration σ_A^2, in turn, corresponds to the assumed uncertainty of the acceleration, as its estimate is not available. Assuming that the bistatic acceleration is uniformly distributed in the $\langle -A_{\max}, A_{\max} \rangle$ range, the variance of the bistatic acceleration can be calculated from:

$$\sigma_A^2 = \frac{A_{\max}^2}{3}. \tag{7.19}$$

It has to be noted again that the bistatic parameters do not directly represent the physical parameters of the target. One can resort to choosing the value of $A_{\max}$ heuristically based on the typical behavior of the targets of interest.

7.3 THE TRACKING ALGORITHM

The tracking algorithms used in radars often employ the Kalman filter, such as the one already presented. Additionally, they have to be equipped with a logic that handles the track initialization, track maintenance, and track termination procedures [4].

In this book it will be assumed that the track initialization and termination are based on the M/N logic [4, 7–9]. In such a case, each unassociated measurement (plot) forms a tentative track. An association gate is set up for this track, which indicates a region of predicted measurement position in the next observation. The gate is defined with the use of the innovation vector $\mathbf{v}(k)$ and the innovation covariance matrix $\mathbf{S}(k)$ as:

$$\mathbf{v}'(k)\mathbf{S}^{-1}(k)\mathbf{v}(k) \leq \gamma, \tag{7.20}$$

where γ is a parameter chosen so that the gate probability P_g (the probability that the measurement, if the target was detected, will fall within the gate) has a desired value (e.g., 0.99). Inequality (7.20) defines an ellipse on the bistatic range-velocity plane, centered around the predicted target position.

If the measurement in the next observation falls within the association gate, it is assigned to the tentative track. If the measurement is assigned to a track in M out of N consecutive observations, the tentative track is upgraded to the confirmed track status. By choosing appropriate values of M and N, a substantial resistance to false track initiation can be obtained, which is one of the main reasons for using the bistatic tracking stage. An analogous procedure is applied in the case of track termination; if in M out of N consecutive observations a plot is not assigned to a track, the track is terminated.

During the process of tracking it may happen that multiple plots fall within the association gate of a track. In such a case, the nearest neighbor (NN) approach is assumed to be used [4]. In this approach the plot closest (according to the defined distance criterion) to the predicted track position is assigned to this track. Other approaches can also be applied, such as strongest neighbor (SN), probabilistic data association (PDA), or joint PDAF (JPDA) [4], but they will not be considered here.

The concept of target tracking in the bistatic coordinates is shown in Figure 7.1. The track is initialized with the first measurement, $\mathbf{z}(0)$. The track is predicted to the next position $\hat{\mathbf{z}}(1) = \mathbf{H}\hat{\mathbf{x}}(1|0)$. Around this position, the association gate is set up, according to (7.20), where the next measurement is expected. As the track was initialized with zero acceleration, the predicted track position has the same velocity. The next measurement, $\mathbf{z}(1)$, appears in the gate, and it is used to update the track. This causes the next association gate, constructed around $\hat{\mathbf{z}}(2)$ to be smaller, as the track covariance is reduced thanks to the track update. In the third iteration, there is no measurement present in the association gate set up around $\hat{\mathbf{z}}(3)$. This increases the covariance of the track, since there is only a prediction stage and no update stage. For this reason, the next association gate, centered at $\hat{\mathbf{z}}(4)$, is larger.

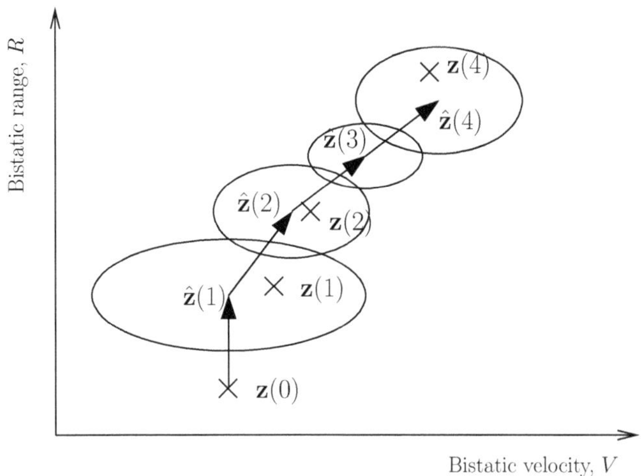

Figure 7.1 The idea of target tracking in the bistatic coordinates.

Target tracking in the bistatic coordinates is relatively simple, and there are few parameters that have to be set. The main one is the intensity of the process noise q (see (7.5)). This parameter must be chosen based on the expected behavior of the observed targets. In general, if the targets are not maneuvering too much, lower values of q can be used. If targets perform fast maneuvers, larger values of q should be applied. Larger value of q causes the association gate to grow faster in time, especially in the bistatic velocity dimension. This allows the tracker to accommodate the changes of the bistatic velocity of the target, but at the same time, this can lead to multiple false detections falling into the association gate, which can lead to track loss or tracking false targets.

7.4 NUMERICAL RESULTS

In this section selected results of computer simulations for FM radio parameters will be presented. Next, the experimental results obtained with real signals recorded with the *PaRaDe* system will be shown. A similar analysis can be found in [10], where numerical simulations have been performed for parameters corresponding to different transmitter types: FM radio, DAB radio, and DVB-T television. In [11], in turn, experimental results of bistatic tracking accuracy for FM radio have been presented.

7.4.1 Simulations

The tracking performance is influenced by numerous parameters of the signal used, such as the bandwidth and the carrier frequency. The radar designer can also influence the tracking performance to a certain extent by choosing the processing parameters.

One of the basic parameters that can be changed in passive radar is the length of the CPI (integration time) T. Shorter CPI provides faster track updates, which can be advantageous from the point of view of tracking maneuvering targets. However, short CPI yields lower SNR, which leads to a reduced probability of detection, as well as reduced accuracy of parameter estimation. Longer integration time results in higher SNR and better parameter estimation accuracy. However, there are some limitations to the integration time that can be applied depending mainly on the target maneuverability (see Chapter 4).

The other processing parameter that can be changed is the detection threshold. Lower detection thresholds lead to a higher probability of detection, which increases the tracking performance. However, lower thresholds cause larger numbers of false alarms, which can reduce the accuracy of the tracking algorithm (due to erroneous associations) and lead to the initialization of false tracks. In the analysis it was assumed that the relationship between the probability of false alarm P_{fa} and detection threshold D is:

$$P_{fa} = \exp\left(-\frac{D^2}{2\sigma^2}\right) \tag{7.21}$$

which corresponds to a Rayleigh-distributed noise with known parameters (compare Chapter 6).

In this section the influence of various parameters on the tracking performance defined as the accuracy is investigated. The analysis is based on computer simulations of tracking a single target in bistatic coordinates for the parameters typical for FM-based passive radar (B=50 kHz, f_c=100 MHz, T=1s). In the experiment, the simulated target state vector was evolving with time according to (7.1). The signal-to-noise ratio of the target echo at the output of the correlator was set to SNR=13 dB. The target was detected with the probability defined by (6.12), where the threshold was calculated to provide $P_{fa} = 10^{-6}$. The gate threshold γ was set to 16, which corresponds to the gate probability $P_g = 0.9997$. The power spectral density of the process noise was set to $q = 1\ \text{m}^2/\text{s}^5$. Now consider the tracking error of the bistatic range:

$$\delta_{Rt} = \text{std}\left\{\mathbf{x}_1(k) - \hat{\mathbf{x}}_1(k|k)\right\}, \tag{7.22}$$

where $\mathbf{x}_1(k|k)$ denotes the first element of the state vector (bistatic range). The tracking error of the bistatic velocity can be defined in an analogous way:

$$\delta_{Vt} = \text{std}\left\{\mathbf{x}_2(k) - \hat{\mathbf{x}}_2(k|k)\right\}. \tag{7.23}$$

An example of the tracking error of the bistatic range and bistatic velocity versus time is shown in Figure 7.2(a, b), respectively. In the experiment, theoretical averaging from (7.22) and (7.23) was approximated by averaging the results of

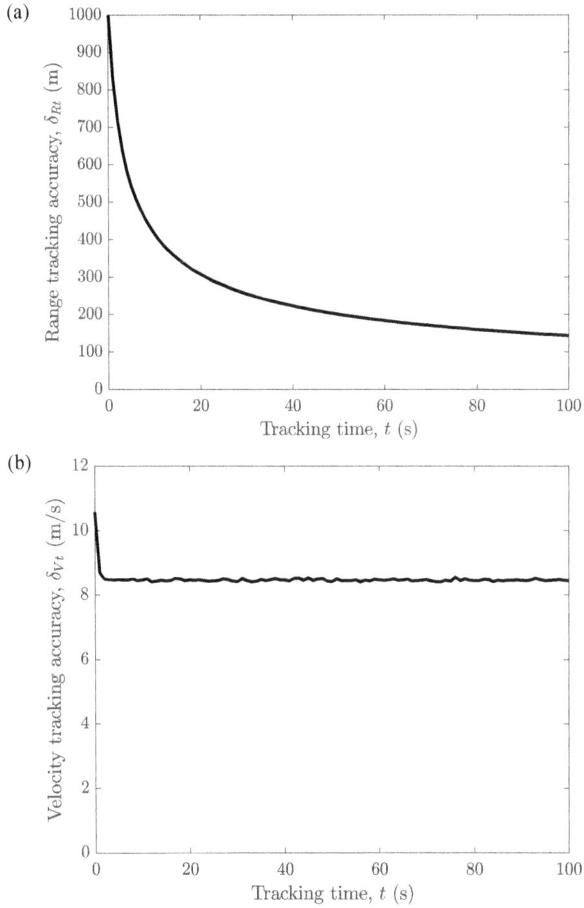

Figure 7.2 (a) Accuracy of tracking of the bistatic range and (b) bistatic velocity versus tracking time for typical FM-based radar parameters.

different runs of the simulation. The plots reveal that bistatic range tracking accuracy converges relatively slowly. This is related to the poor range resolution of an FM-based radar. As the error of the measurement is large, the tracking algorithm increasingly relies on the history of measurements. The process is slow. However, the steady state of the bistatic velocity tracking error is reached almost immediately. This results from the good velocity resolution, which limits the influence of the measurement history on the tracking. Therefore, the tracking filter relies mostly on the current measurement, not on the measurement history, as in the case of the bistatic range. It can be seen that the curve of bistatic velocity accuracy is more irregular than for bistatic range accuracy. This results from the fact that bistatic acceleration changes (caused by the process noise) directly influence the bistatic velocity, which leads to irregularities. The bistatic range is less influenced by the changes of the bistatic acceleration, as the acceleration increments are integrated and thus smoothed.

The aforementioned results can be confirmed by observing averaged values of the Kalman gain. The Kalman gain corresponding to the bistatic range can be defined as: $K_R = \mathbf{K}_{11}(k)$, where $\mathbf{K}_{11}(k)$ denotes the matrix element at position 1,1. The corresponding velocity gain can be defined as $K_V = \mathbf{K}_{22}(k)$. The range and velocity gains versus time are plotted in Figure 7.3. The range Kalman gain slowly decreases to 0, which corresponds to the higher influence of the measurements history. The velocity gain quickly reaches a steady value, relatively close to 1, which means that the current measurement is mainly used when estimating the state vector.

The results of the simulations show that in the case of the FM radio signals the incorporation of the measurement history mainly influences the estimation accuracy of the bistatic range. The availability of the measurement history does not improve the accuracy of the velocity estimation significantly. At the same time, the raw measurements are characterized by low range accuracy (due to the low signal bandwidth), and high velocity accuracy (due to the long integration time). For this reason, any potential increase of accuracy in bistatic range measurement provided by tracking is of great importance. Especially as the bistatic range accuracy is crucial when localizing targets in Cartesian space.

In many cases, the steady-state tracking accuracy is used as the quality factor. It is reasonable if the steady-state is reached soon after the track is established. This is not the case for the bistatic range in an FM-based radar, as shown in Figure 7.2(a). For this reason, the tracking accuracy after a fixed time, T_{track} after tracking has started, will be used. The tracking algorithm provides the current state estimate, but the prediction for the future can also be calculated. In many applications, for example, related to collision avoidance, the prediction error is more important than

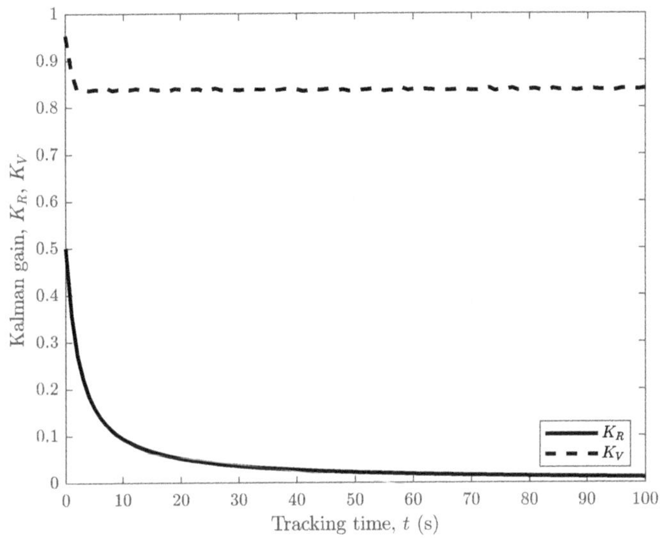

Figure 7.3 Kalman gain of the bistatic range and bistatic velocity versus tracking time for typical FM-based radar parameters.

the error of current estimate. For this reason, the investigated accuracy will be the accuracy for prediction for time T_{pred}. Taking the aforementioned considerations into account, define an error vector written symbolically as:

$$\mathbf{\Delta} = \mathbf{x}(T_{\text{track}} + T_{\text{pred}}) - \hat{\mathbf{x}}(T_{\text{track}} + T_{\text{pred}}|T_{\text{track}}), \tag{7.24}$$

where $\mathbf{x}(T_{\text{track}} + T_{\text{pred}})$ denotes the true state vector after $T_{\text{track}} + T_{\text{pred}}$ seconds and $\hat{\mathbf{x}}(T_{\text{track}} + T_{\text{pred}}|T_{\text{track}})$ is the state estimate predicted for $T_{\text{track}} + T_{\text{pred}}$ from T_{track}. In practice, the appropriate time instants are rounded to the multiples of integration time T. In addition, when integration time T is applied, the state estimate corresponds to the middle of the signal block; therefore, $T/2$ has to be added to the prediction time. This will be especially important for longer T.

Now, based on (7.24), the tracking error of the bistatic range can be defined as:

$$\delta_R = \text{std}\{\mathbf{\Delta}_1\}, \tag{7.25}$$

where $\mathbf{\Delta}_1$ is the first element of the error vector $\mathbf{\Delta}$. The tracking error of the bistatic velocity is defined in a similar way as:

$$\delta_V = \text{std}\{\mathbf{\Delta}_2\}. \tag{7.26}$$

First, the influence of the integration time on the tracking accuracy of bistatic range δ_R and bistatic velocity δ_V is investigated. Figure 7.4 shows the plots of the tracking accuracies versus the integration time. Different curves correspond to different values of the power spectral density of the process noise, representing target maneuvers. The SNR_{in} (before correlation) was set to -37 dB, which for $T = 1$s

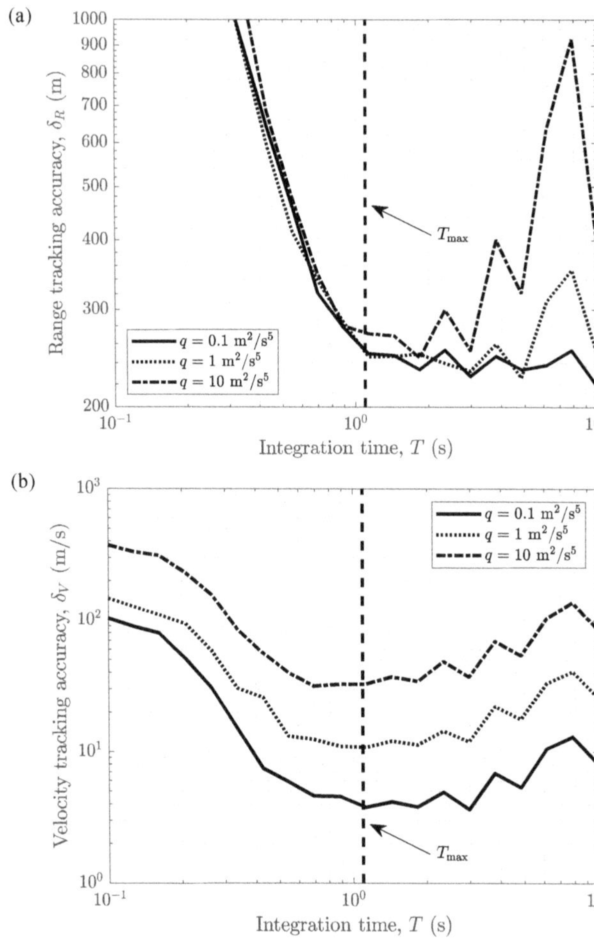

Figure 7.4 (a) Bistatic range and (b) bistatic velocity tracking accuracy versus integration time for different values of the power spectral density of the process noise. (©2009 IEEE Reprinted, with permission, from [10].)

and assumed bandwidth $B = 50$ kHz results in SNR $= 13$ dB (after correlation). For short integration time the accuracy is very low. This is caused by low integration gain, resulting in low SNR. The low values of SNR lead to a small probability of detection, as well as low accuracy of parameter estimation (compare (6.20) and (6.21)). Higher values of SNR are obtained for longer integration times. The increase in the probability of detection and accuracy of parameter estimation is, however, outweighed by the long refresh interval, which has a negative influence on the tracking accuracy. The maneuvers of the target, represented by the power spectral density of the process noise q, have a substantial influence on the velocity tracking error (see Figure 7.4(b)). The influence of q on the range tracking accuracy is smaller, especially for shorter integration times (see Figure 7.4(a)). In the figures the integration time $T_{\max}$ mark indicates the maximum time that can be used when no extended correlation processing is applied. The maximum integration time is calculated as time that does not lead to significant losses in SNR for given bistatic acceleration (compare (4.38) with $C_{mv} = 2.0$) [10, 12]:

$$T_{\max} = \sqrt{\frac{2\lambda}{A}}, \tag{7.27}$$

which is equal approximately to 1.1s for the assumed bistatic acceleration A=5 m/s^2. It can be seen that a typical value of the integration time used in FM-based radar of 1s provides results close to optimal for a wide range of the parameters.

Let us now examine the tracking accuracy of the bistatic range versus the two processing parameters: the integration time T and the probability of false alarm P_{fa}, for different values of the input signal-to-noise ratio SNR_{in}. In the simulations the process noise spectral density was constant and equal to $q = 1$ m^2/s^5. In Figure 7.5 (a) the plots of the accuracy versus the integration time are shown. The weak target echoes (i.e., with lower SNR_{in}) require higher integration time to obtain the same accuracy as for stronger target echoes. In Figure 7.5(b) the accuracy curves versus the probability of false alarm P_{fa} are shown. The value of P_{fa} determines the detection threshold, which, in turn, influences the number of false alarms and the probability of detection. For low P_{fa} the detection threshold is high, which reduces the probability of detection. High P_{fa} value leads to a high probability of detection; however, false alarms disturb the tracking accuracy, as the false alarms are used to update the track; a drawback of the NN approach, which has been used in simulations [13]. For stronger target echoes, the influence of P_{fa} is negligible, as the target is detected almost independently of the threshold.

The presented results have shown that the radar designer can influence the tracking performance in terms of accuracy by adjusting the two main processing pa-

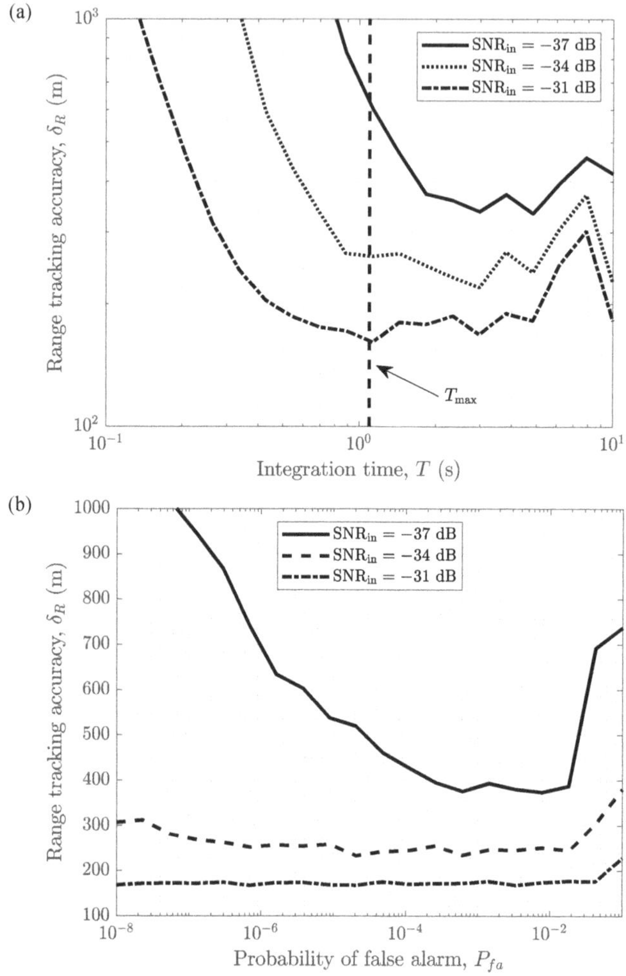

Figure 7.5 Bistatic range tracking accuracy versus (a) integration time and (b) probability of false alarm (b) for different values of the input signal-to-noise ratio. (©2009 IEEE Reprinted, with permission, from [10].)

rameters: the integration time T and probability of false alarm P_{fa}. The typical value of the integration time for FM radio equal to 1s seems to be a good choice. It has been shown that relatively high values of P_{fa} are acceptable from the point of view of the tracking accuracy. However, other aspects of the tracking, such as track formation, also have to be considered. In [14] a simple analysis of the influence of the integration time on track formation can be found. A joint optimization of the processing parameters from the point of view of track formation was presented in [15].

7.4.2 Real-Data Results

Experimental tracking accuracy has been investigated by analyzing the real data obtained from the *PaRaDe* system [11]. A single target, which was tracked for over 100s, was selected, and the results were compared with the reference data received by a commercial ADS-B (automatic dependent surveillance–broadcast) receiver, which provides the geographical position of the aircraft, as measured with the onboard navigation system. This position was converted to the bistatic coordinates using the position of the transmitter and the receiver. The result of this operation was compared with the tracked target parameters. It has been assumed that the accuracy of ADS-B data is much better than the radar measurements, and the ADS-B was treated as the ground truth.

Figure 7.6(a) shows the comparison of the raw measurements (bistatic plots), tracked range, and reference ADS-B data. The horizontal dashed lines indicate the borders of the bistatic range resolution cells. It can be seen that the raw measurements are spread around the true values of the bistatic range. The measured standard deviation of the raw measurements is approximately 250m. This is almost one order of magnitude better than the size of the range resolution cell (approximately 1,500m). The increased measurement accuracy in comparison with the resolution cell size results from the application of a fine estimation method using a parabolic curve (see Section 6.3). Tracking, however, can increase the accuracy even further. The tracked bistatic range values are marked with * when the measurement was available, and with ∘ when the target was not detected in a particular observation. It can be seen that the tracked value closely follows the ADS-B data. In Figure 7.6(b) the error Δ_R between the tracked value and the ADS-B value versus time is plotted. The dashed and dash-dotted lines show the theoretical accuracy at the level of 3-σ and 1-σ, respectively (which correspond to $\mathbf{P}_{1,1}(k|k)$, the covariance matrix element at position $1, 1$). As the tracking algorithm used the NN for data association, the theoretical results have been obtained using the methods presented in [13], where the performance of the NN in the mean sense was analyzed. The error has a tendency to decrease, and it is consistent with the theoretical values. After the initial error value in the order of 300m, the tracking converges so that the error is approximately 100m.

In the next example, shown in Figure 7.7, the bistatic data collected with the *PaRaDe* system in another scenario over 75s with the CPI of 1s are analyzed. In the

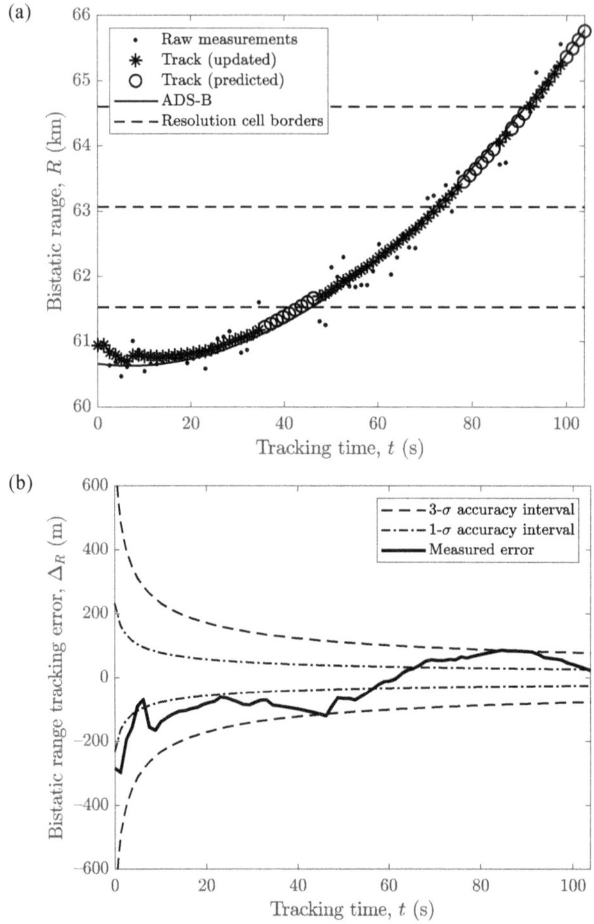

Figure 7.6 (a) Comparison of raw measurements of the bistatic range, tracked value, and ADS-B data. (b) Error of the tracked bistatic range versus tracking time. Theoretical 1-σ and 3-σ intervals plotted for comparison.

first image, Figure 7.7(a), all detected radar plots are shown. As can be seen, the plots are distributed relatively uniformly. Real target echoes can be distinguished by plot congestions which form a trajectory. The second image, Figure 7.7(b), shows the bistatic tracks formed from the plots from the first image. The results indicate the effectiveness of the tracking algorithm, as the false plots were eliminated, and only the true targets are tracked.

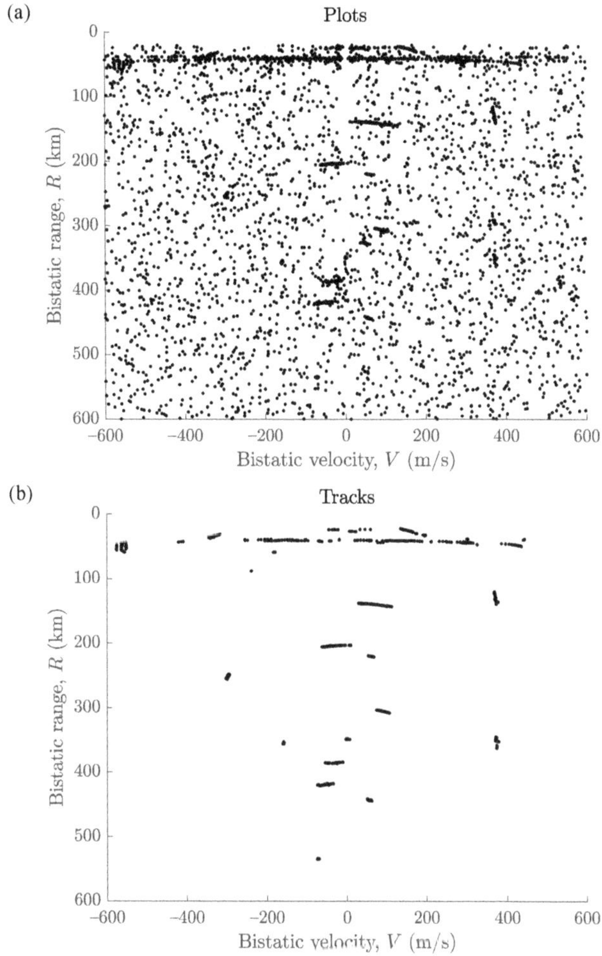

Figure 7.7 Bistatic data from 75s time interval with CPI of 1s for FM-based passive radar: (a) plots and (b) tracks.

7.5 SUMMARY

In this chapter, target tracking in bistatic coordinates has been presented. This kind of tracking is not required, as the bistatic measurements can be used directly in the localization and Cartesian tracking algorithms; however, bistatic tracking is advantageous and is often applied in practice. Bistatic tracking allows the measurements to be filtered, that is, at the output of the tracker the measurements are target-originated with a high probability, and the false alarms are eliminated. This may significantly reduce the burden connected with the subsequent processing steps.

Another advantage of bistatic tracking is the increase in the accuracy of bistatic measurements. Due to this feature, target localization can be performed more precisely. Tracking also provides an estimate of the target position if it has not been detected in an observation, as the predicted state vector is calculated. This allows localization to be performed even in the case when not all detections are available simultaneously.

The simulations show that by adjusting the main processing parameters, the CPI length and the detection threshold, the tracking performance can be optimized. The real data analysis proved that the bistatic tracking can significantly increase estimation accuracy, as well as substantially reduce the false alarm rate.

References

[1] M. Malanowski, *Signal Processing Algorithms and Target Localisation Methods for Continuous-Wave Passive Bistatic Radar*. No. 179 in Prace Naukowe Politechniki Warszawskiej. Elektronika, Warszawa, ul. Polna 50: Oficyna Wydawnicza Politechniki Warszawskiej, 2012.

[2] R. E. Kalman, "A New Approach to Linear Filtering and Prediction Problems," *Trans. the ASME Journal of Basic Engineering*, vol. 82, pp. 35–45, March 1960.

[3] Y. Bar-Shalom and X. R. Li, *Estimation and Tracking: Principles, Techniques, and Software*. Norwood, MA: Artech House, 1993.

[4] Y. Bar-Shalom and X. R. Li, *Multitarget-Multisensor Tracking: Principles and Techniques*. Storrs, CT: Yaakov Bar-Shalom, 1995.

[5] M. I. Skolnik, *Introduction to Radar Systems*. third ed., Boston, MA: McGraw-Hill, 2003.

[6] M. A. Richards, *Fundamentals of Radar Signal Processing*. New York: McGraw-Hill, 2005.

[7] Y. Bar-Shalom, K. C. Chang, and H. M. Shertukde, "Performance Evaluation of a Cascaded Logic for Track Formation in Clutter," *IEEE Trans. Aerospace and Electronic Systems*, vol. 25, pp. 873–878, November 1989.

[8] F. R. Castella, "Sliding Window Detection Probabilities," *IEEE Trans. Aerospace and Electronic Systems*, vol. 12, pp. 815–819, November 1976.

[9] J. Dezert and T. Kirubarajan, "Track Formation in Clutter Using A Bi-Band Imaging Sensor," in *Proc. 3rd International Conference on Information Fusion – FUSION*, (Paris, France), pp. TUD2/24–TUD2/32, July 2000. vol. 1.

[10] M. Malanowski and K. Kulpa, "Analysis of Bistatic Tracking Accuracy in Passive Radar," in *Proc. RadarCon 2009*, (Pasadena, California, USA), p. CD, May 2009.

[11] M. Malanowski and K. Kulpa, "Experimental Analysis of Passive Radar Accuracy," in *Proc. International Radar Symposium 2009*, (Hamburg, Germany), pp. 331–335, September 2009.

[12] M. Malanowski and K. Kulpa, "Analysis of Integration Gain in Passive Radar," in *Proc. International Conference on Radar*, (Adelaide, Australia), p. CD, September 2008.

[13] X. R. Li and Y. Bar-Shalom, "Tracking in Clutter with Nearest Neighbor Filters: Analysis and Performance," *IEEE Trans. Aerospace and Electronic Systems*, vol. 32, pp. 995–1010, July 1996.

[14] M. Malanowski, "Influence of Integration Time on Tracking Performance in PCL Radar," in *Proc. SPIE – Photonics Applications in Astronomy, Communications, Industry, and High-Energy Physics Experiments IV*, (Wilga, Poland), pp. 69372Z–1–69372Z–8, May 2007. vol. 6937.

[15] M. Malanowski and K. Kulpa, "Optimization of Confirmation Time of Bistatic Tracks in Passive Radar," *IEEE Trans. Aerospace and Electronic Systems*, vol. 47, pp. 1060–1072, April 2011.

Chapter 8

Target Localization

8.1 INTRODUCTION

In the classical surveillance radar with a rotating antenna, the targets are localized in the Cartesian space by combining the range and azimuth measurements. The range estimate is obtained by measuring the time delay between the transmission and the reception of the pulse. A constant range in a monostatic radar corresponds to a sphere, with the radar in its center. The azimuth measurement is obtained by using an antenna with a narrow beam and pointing it in a particular direction. When an echo is received for this antenna position, it is assumed that the target azimuth corresponds to the current antenna direction. The combination of range and azimuth angles provides a position in a 2D Cartesian space. If the elevation measurement is also available, the 3D target position can be calculated.

In the case of the passive radar, target localization is more complicated. As discussed in Chapter 2, a constant bistatic range corresponds to an ellipsoid with the foci in the transmitter and receiver positions. It is worth mentioning that if the transmitter and receiver were in the same place, the ellipsoid would simplify to a sphere, which is the case for the monostatic radar. The angle measurement is not always available in passive radar. Even if it is available, its accuracy is rather coarse. The reason for this is that passive radar typically operates at much lower frequencies than active radars: VHF/UHF versus L/S/C/X-bands. As the frequencies used in passive radar are low, it would be impractical to build antennas with the size much larger than the wavelength; thus, the angle measurement accuracy is limited. However, the approach of combining the range and angle measurement is still valid for the passive radar, which will be discussed in this chapter. An alternative solution of the localization problem is to use the fact that passive radar usually operates in a multistatic configuration. The most common configuration is a single receiver and multiple transmitters. For each of the transmitter-receiver pairs for which a target is detected, a bistatic ellipsoid can be formed. Each ellipsoid contains the target position, therefore, by intersecting different ellipsoids, the location of the target can be found. In the case of a single receiver, all ellipsoids have a common focus. A more general case, with multiple receivers, can also be used in practice, but will not be considered in this book.

In this chapter, the problem of localizing a target in the Cartesian coordinates is addressed, taking into consideration the two aforementioned approaches. One is finding the Cartesian coordinates by intersecting bistatic ellipsoids [1]. Two algorithms originally proposed for TDOA (time-difference-of-arrival) systems are adapted to the passive radar case. A comparison of these two algorithms is presented. The theoretical accuracy of localization is analyzed and compared with the results of computer simulations. An example of target localization using real data is shown. The other approach is using bistatic ellipse and direction-of-arrival measurement. The accuracy analysis is presented for this approach. The real measurements of a drone localization are presented.

8.2 MULTISTATIC TARGET LOCALIZATION

The concept of multistatic target localization in passive radar is shown in Figure 8.1. In the scenario three transmitters (△) and one receiver (○) are assumed. Based on each of the three transmitter-receiver pairs, a bistatic ellipsoid can be formed. The foci of each of the ellipsoids are at the corresponding transmitter positions and the receiver position, common for all ellipsoids. In most of the situations the transmitters and the receiver are located close to the ground level; therefore, an almost symmetrical scenario with respect to the ground plane is observed. In consequence, all three ellipsoids intersect at two points: one corresponding to the true target position (□), and one for negative altitude. The correct choice can be made by comparing the altitude of the two estimated positions.

In this chapter, the algorithms for multistatic target localization will be introduced first. As the problem of target localization using bistatic measurements is closely related to TDOA localization, a similar approach can be used [2]. The accuracy of target localization is investigated through theoretical calculations and by performing computer simulations. The algorithms are also verified with real data.

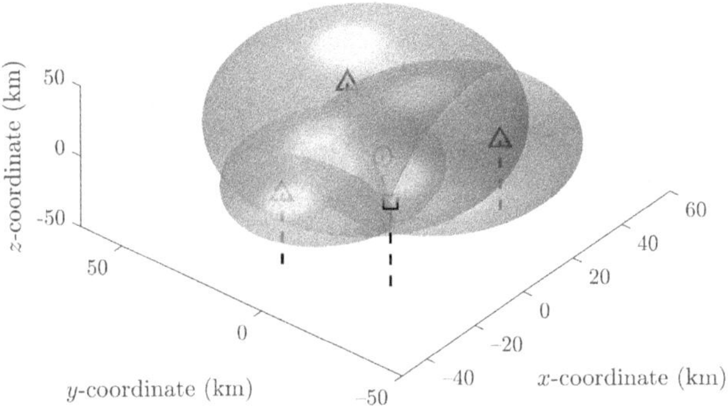

Figure 8.1 A concept of target localization based on calculating the intersection points of bistatic ellipsoids (△ – transmitters, ○ – receiver, □ – target).

8.2.1 Algorithms for Target Localization

Let us consider the geometry of the passive radar scenario shown in Figure 8.2. In this case, one receiver and multiple transmitters are considered. This is equivalent to a situation with one transmitter and multiple receivers. For the sake of simplicity, it is assumed that the receiver is at the origin of the coordinate system: $[0, 0, 0]'$. In the figure three transmitters are marked, but in general $N_{\text{Tx}} \geq 3$ can be used. The ith transmitter position is: $\mathbf{x}_{ti} = [x_{ti}, y_{ti}, z_{ti}]'$, where $i = 1, \ldots, N_{\text{Tx}}$. The target position is $\mathbf{x} = [x, y, z]'$.

The range between the target and the ith transmitter can be calculated as:

$$R_{ti} = \sqrt{(x_{ti} - x)^2 + (y_{ti} - y)^2 + (z_{ti} - z)^2} = \|\mathbf{x}_{ti} - \mathbf{x}\|, \tag{8.1}$$

where $\|\mathbf{x}\|$ denotes the Euclidean norm: $\sqrt{\mathbf{x}'\mathbf{x}}$. The baseline (the transmitter-receiver range) for the ith transmitter is:

$$R_{bi} = \sqrt{x_{ti}^2 + y_{ti}^2 + z_{ti}^2} = \|\mathbf{x}_{ti}\|. \tag{8.2}$$

A very important parameter to the algorithms analyzed in the following part of the chapter is the distance between the target and the receiver:

$$R_t = \sqrt{x^2 + y^2 + z^2} = \|\mathbf{x}\|, \tag{8.3}$$

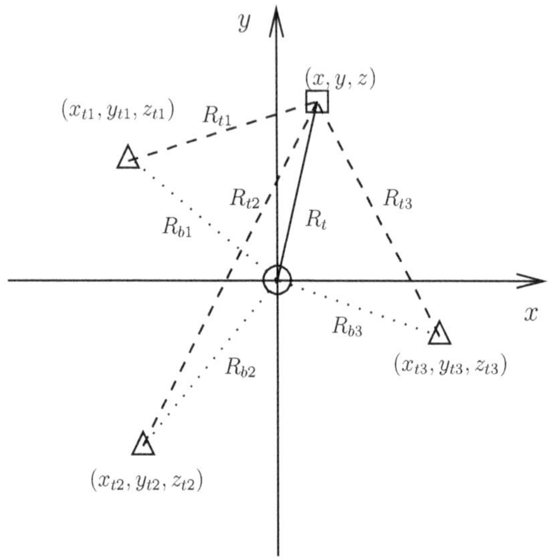

Figure 8.2 Geometry of target localization in passive radar.

It is a single value, common for all transmitters.

The value that is measured in passive radar is the bistatic range. As introduced in Chapter 2, the bistatic range is the difference between the indirect path and the direct path (i.e., the transmitter-target-receiver range minus the baseline). In the problem of target localization, it is easier to deal with the sum of transmitter-target and target-receiver ranges. This can be easily calculated by adding the baseline R_{bi}, which is known from the position of the transmitter and the receiver, to the measured bistatic range R_i:

$$R_{si} = R_i + R_{bi} = (R_{ti} + R_t - R_{bi}) + R_{bi} = R_{ti} + R_t = \sqrt{(x_{ti} - x)^2 + (y_{ti} - y)^2 + (z_{ti} - z)^2} + \sqrt{x^2 + y^2 + z^2}. \tag{8.4}$$

For further derivation, vector and matrix notation will be introduced. First, let us define the vector of the range sums R_{si}:

$$\mathbf{r} = \begin{bmatrix} R_{s1} \\ R_{s2} \\ \vdots \\ R_{sN_{\text{Tx}}} \end{bmatrix}. \tag{8.5}$$

Assume that the target position is an independent variable $\tilde{\mathbf{x}} = [\tilde{x}, \tilde{y}, \tilde{z}]'$. By substituting this value into (8.4), a vector of range sums can be calculated $\tilde{\mathbf{r}}(\tilde{\mathbf{x}}) = [\tilde{R}_{s1}, \tilde{R}_{s2}, \ldots, \tilde{R}_{sN_{\mathrm{Tx}}}]'$. The aim of target localization can be formulated as the minimization of the Euclidean norm of the error vector of the measured range sums $\mathbf{r}$ and the range sums $\tilde{\mathbf{r}}(\tilde{\mathbf{x}})$ corresponding to the hypothetical target position $\tilde{\mathbf{x}}$:

$$\hat{\mathbf{x}} = \arg\min_{\tilde{\mathbf{x}}} \|\mathbf{r} - \tilde{\mathbf{r}}(\tilde{\mathbf{x}})\|. \tag{8.6}$$

The minimization problem formulated above is nonlinear, as the range sum (8.4) is nonlinear with respect to the target position $\tilde{\mathbf{x}}$. The problem could be solved by numerical optimization, however, this would not be very efficient, as numerous iterations are usually required to find the solution. Moreover, the solution would be sensitive to the choice of the starting point, as the function (8.6) may have local minima. Let us consider an example illustrating this problem. The positions of the transmitters used in the simulation are listed in Table 8.1. The receiver was placed in the origin of the coordinate system $[0, 0, 0]'$ km. The target position was $\mathbf{x} = [11, 44, 5]'$ km. Figure 8.3 shows the norm of the error vector $\|\mathbf{r} - \tilde{\mathbf{r}}(\tilde{\mathbf{x}})\|$ calculated for the altitude of 5 km (the same as the simulated target). The asterisk symbols denote local minima in the error vector norm values. As can be seen, one of the minima corresponds to the true target position (marked with a square). However, two minima also exist in positions different than the correct one.

The presented example shows that it is desirable to find a closed-form solution to the problem, which would additionally provide a unique target position estimate. As will be shown, the first requirement can be satisfied with the proposed solution. The optimization problem (8.6) will now be rearranged into a different, closely related problem, which has a closed-form solution. Let us rearrange (8.4) in the following way:

$$R_{si} - \sqrt{x^2 + y^2 + z^2} = \sqrt{(x_{ti} - x_t)^2 + (y_{ti} - y_t)^2 + (z_{ti} - z_t)^2}. \tag{8.7}$$

Table 8.1
Relative Positions of the Transmitters Used for Comparison of the Algorithms

Transmitter	*x-offset (km)*	*y-offset (km)*	*z-offset (km)*
Tx1	20.00	0.00	0.00
Tx2	−30.00	5.00	0.15
Tx3	−10.00	−15.00	0.10
Tx4	10.00	−25.00	0.05

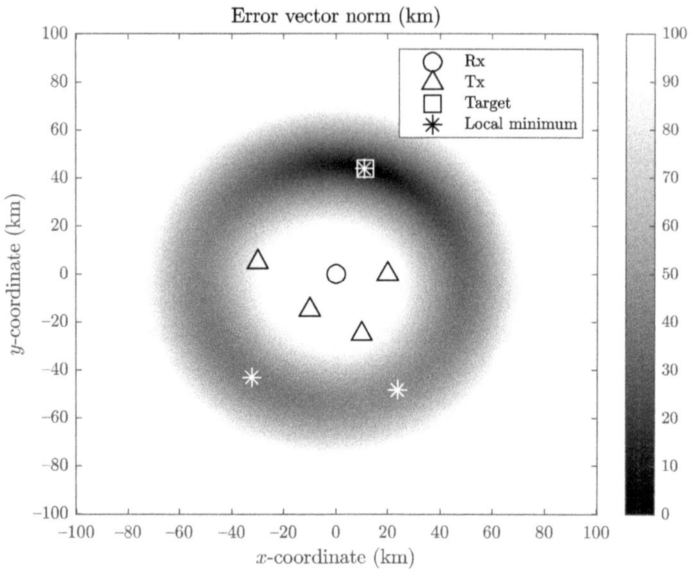

Figure 8.3 Value of the error vector norm for a multistatic scenario. Local minima present not only in the true target position.

Now both sides of the equation will be squared, which results in:

$$x_{ti}x + y_{ti}y + z_{ti}z - R_{si}\sqrt{x^2 + y^2 + z^2} = \frac{1}{2}\left(x_{ti}^2 + y_{ti}^2 + z_{ti}^2 - R_{si}^2\right). \quad (8.8)$$

In order to write the equations for multiple transmitters in a compact way, a matrix containing transmitter positions is created:

$$\mathbf{S} = \begin{bmatrix} x_{t1} & y_{t1} & z_{t1} \\ x_{t2} & y_{t2} & z_{t2} \\ \vdots & \vdots & \vdots \\ x_{tN_{\text{Tx}}} & y_{tN_{\text{Tx}}} & z_{tN_{\text{Tx}}} \end{bmatrix}, \quad (8.9)$$

In addition, the following vector is defined:

$$\mathbf{z} = \frac{1}{2}\begin{bmatrix} x_{t1}^2 + y_{t1}^2 + z_{t1}^2 - R_{s1}^2 \\ x_{t2}^2 + y_{t2}^2 + z_{t2}^2 - R_{s2}^2 \\ \vdots \\ x_{tN_{\text{Tx}}}^2 + y_{tN_{\text{Tx}}}^2 + z_{tN_{\text{Tx}}}^2 - R_{sN_{\text{Tx}}}^2 \end{bmatrix}. \quad (8.10)$$

Using the introduced matrix $\mathbf{S}$ and vector $\mathbf{z}$, (8.8) can be rewritten as:

$$\mathbf{Sx} = \mathbf{z} + \mathbf{r}R_t. \tag{8.11}$$

In the above equation $\mathbf{S}$, $\mathbf{z}$, and $\mathbf{r}$ are known, and $\mathbf{x}$ and R_t are unknown. Let us assume that R_t can be calculated somehow, then the least square solution of (8.11) can be found as:

$$\hat{\mathbf{x}} = (\mathbf{S}'\mathbf{S})^{-1}\mathbf{S}'\mathbf{z} + (\mathbf{S}'\mathbf{S})^{-1}\mathbf{S}'\mathbf{r}R_t. \tag{8.12}$$

It is worth mentioning that solution (8.12) can be found only when $(\mathbf{S}'\mathbf{S})^{-1}$ exists. This requires matrix $\mathbf{S}$ to have full rank. In some cases, for example when all transmitters and the receiver are at the same height, the $\mathbf{S}$ matrix is rank deficient, which means that $(\mathbf{S}'\mathbf{S})^{-1}$ cannot be calculated. Such cases are unlikely in practice, and will not be considered here.

As already mentioned, target localization in passive radar is closely related to the TDOA localization. For this reason, the algorithm derivation in this chapter closely followed analogous methods for the TDOA. In literature, two solutions to the problem of finding R_t in (8.12) can be found: spherical-interpolation (SI) [3, 4] and spherical-intersection (SX) [5, 6]. In the SI method, the value of R_t is calculated from a certain quotient. In the SX method, R_t is calculated as a solution of a second-order equation. The two algorithms were compared in [7] for the TDOA case. The results showed that the SI performs much better than the SX method. As will be shown in the following part of this chapter, the situation is opposite in passive radar, where SX outperforms the SI algorithm.

8.2.1.1 Spherical-Interpolation Method (SI)

By introducing a matrix $\mathbf{S}^* = (\mathbf{S}'\mathbf{S})^{-1}\mathbf{S}'$, (8.12) can be rewritten as:

$$\hat{\mathbf{x}} = \mathbf{S}^*(\mathbf{z} + \mathbf{r}R_t), \tag{8.13}$$

Now let us calculate the difference between the right and left sides of (8.11) while substituting (8.13):

$$\epsilon = \mathbf{z} + \mathbf{r}R_t - \mathbf{SS}^*(\mathbf{z} + \mathbf{r}R_t) = (\mathbf{I} - \mathbf{SS}^*)(\mathbf{z} + \mathbf{r}R_t). \tag{8.14}$$

As can be seen, the unknown $\hat{\mathbf{x}}$ was eliminated from the equation, leaving only R_t. The value of ϵ is the error vector, whose norm should be minimized, which allows R_t to be found. This is done by introducing the following matrix:

$$\mathbf{T} = \mathbf{I} - \mathbf{SS}^*. \tag{8.15}$$

This matrix has some special features. It is symmetrical; therefore, $\mathbf{T} = \mathbf{T}'$, and idempotent; therefore, $\mathbf{T}^2 = \mathbf{T}$. These properties, allow R_t to be calculated by minimizing the norm of the error vector $||\epsilon||$ defined in (8.14):

$$\hat{R}_t = -\frac{\mathbf{r}'\mathbf{T}\mathbf{z}}{\mathbf{r}'\mathbf{T}\mathbf{r}}. \tag{8.16}$$

In this method, the condition that the distance to the target should be consistent with the target position (i.e., $\hat{R}_t = ||\hat{\mathbf{x}}||$) was not enforced. For this reason, $\hat{R}_t$ calculated from (8.16) and $||\hat{\mathbf{x}}||$ can be different. As will be shown in the experiments, the correct calculation of $\hat{R}_t$ has a crucial significance for the accuracy of target position estimation.

The final result of the SI method is obtained by calculating the $\hat{R}_t$ from (8.16) and substituting it into (8.12).

The SI method can be used only in a situation where more than 3 transmitter-receiver pairs are available (in the case of 3 pairs, the residual vector ϵ is zero).

8.2.1.2 Spherical-Intersection Method (SX)

The second approach to calculating the receiver-target range R_t is based on the spherical-intersection SX method [5, 6]. This method, unlike the SI, can be used with three transmitter-receiver pairs. This method is derived as follows: two auxiliary vectors are introduced:

$$\mathbf{a} = (\mathbf{S}'\mathbf{S})^{-1}\mathbf{S}'\mathbf{z}, \tag{8.17}$$

and

$$\mathbf{b} = (\mathbf{S}'\mathbf{S})^{-1}\mathbf{S}'\mathbf{r}. \tag{8.18}$$

Using these vectors, (8.12) can be rewritten as

$$\mathbf{x} = \mathbf{a} + \mathbf{b}R_t. \tag{8.19}$$

Now, by substituting $\mathbf{x}'\mathbf{x} = ||\mathbf{x}||^2$, inserting the above equation into (8.3), and squaring both sides, $\mathbf{x}$ is eliminated and a quadratic equation with respect to R_t is obtained:

$$(\mathbf{b}'\mathbf{b} - 1)R_t^2 + 2\mathbf{a}'\mathbf{b}R_t + \mathbf{a}'\mathbf{a} = 0. \tag{8.20}$$

The solution of this equation can be easily found as:

$$\hat{R}_t = \frac{-2\mathbf{a}'\mathbf{b} \pm \sqrt{4(\mathbf{a}'\mathbf{b})^2 - 4(\mathbf{b}'\mathbf{b} - 1)\mathbf{a}'\mathbf{a}}}{2(\mathbf{b}'\mathbf{b} - 1)}. \tag{8.21}$$

After $\hat{R}_t$ is found, it is substituted into (8.12), which allows the position estimate $\hat{\mathbf{x}}$ to be calculated.

If the discriminant of the quadratic equation (8.20) is positive, two solutions for $\hat{R}_t$ exist. A decision has to be made which solution to use. Usually in passive radar, the transmitters and the receiver are placed close to the ground plane; therefore, the geometry is almost symmetrical with respect to this plane. The two values of $\hat{R}_t$ obtained from the quadratic equation often correspond to the true target position, and its (almost perfect) mirror reflection with respect to the x-y plane. To decide which value of $\hat{R}_t$ to choose, they can be substituted into (8.12), and the altitude of the estimates can be compared. In the passive radar scenario choosing the estimate with the higher altitude is usually the right choice, as the flying targets are of interest.

The two algorithms derived for the passive radar differ from their TDOA counterparts in sign in the right side of (8.11) [7]. The reason for this is that here the sums of ranges instead of differences are used.

8.2.2 Accuracy Analysis of Target Localization

An important feature of every estimation procedure is the ability to calculate its accuracy. Usually, the accuracy is expressed with the covariance matrix. This is often connected with a silent assumption that the error distribution is Gaussian, which does not have to be true. In the considered case, the covariance matrix of the input data of the estimation procedure (i.e., the bistatic data) is known. The problem is how to convert the accuracy of the input data to the accuracy of the obtained target estimate, given the nonlinear relationship between the bistatic and Cartesian coordinates. The approach that is often used in such a situation is linearization of the nonlinear function, and assuming that for small errors, a linear relationship is accurate enough. A similar approach will be used here.

Let us denote the variance of the bistatic range error corresponding to the ith transmitter as σ_{Ri}^2. The covariance matrix $\mathbf{R}$ of the measurement error of the bistatic parameters is given by:

$$\mathbf{R} = \operatorname{cov}(\mathbf{r}) = \begin{bmatrix} \sigma_{R1}^2 & & & 0 \\ & \sigma_{R2}^2 & & \\ & & \ddots & \\ 0 & & & \sigma_{RN_{\mathrm{Tx}}}^2 \end{bmatrix}. \tag{8.22}$$

Depending on which type of bistatic measurements are used for target localization, appropriate accuracy measure should be applied. In the case when raw bistatic plots are used, the bistatic range accuracy can be calculated according to (6.20). If the bistatic tracks are the input to the localization procedure, their bistatic accuracy is taken from the covariance matrix of the Kalman filter – value $\mathbf{P}_{1,1}(k+1|k+1)$ calculated using (7.16) (where $\mathbf{P}_{1,1}(k+1|k+1)$ denotes the first element from the matrix diagonal). In general, the accuracy of bistatic track is better than the accuracy of bistatic plot. For this reason, using the tracks instead of unprocessed plots is advisable to provide better localization accuracy.

The nonlinear relationship between the bistatic measurements and Cartesian position estimate can be approximated with the first-order Taylor series expansion. In this case, the Cartesian covariance matrix can be approximated as:

$$\mathbf{P} = \mathrm{cov}(\hat{\mathbf{x}}) \approx \left(\frac{\partial \mathbf{x}}{\partial \mathbf{r}}\right) \mathbf{R} \left(\frac{\partial \mathbf{x}}{\partial \mathbf{r}}\right)', \tag{8.23}$$

where $(\partial \mathbf{x}/\partial \mathbf{r})$ is the Jacobian. In order to calculate the Jacobian, analysis similar to [3] is carried out. First, (8.11) is differentiated with respect to $\mathbf{r}$:

$$\mathbf{S}\frac{\partial \mathbf{x}}{\partial \mathbf{r}} = \frac{\partial \mathbf{z}}{\partial \mathbf{r}} + \frac{\partial (\mathbf{r}R_t)}{\partial \mathbf{r}}. \tag{8.24}$$

The left side is closely related to the searched Jacobian. The two components on the right side require further analysis. The first component can be calculated as:

$$\frac{\partial \mathbf{z}}{\partial \mathbf{r}} = -\begin{bmatrix} R_{s1} & & & 0 \\ & R_{s2} & & \\ & & \ddots & \\ 0 & & & R_{sN_{\mathrm{Tx}}} \end{bmatrix} = -\mathbf{\Lambda}. \tag{8.25}$$

The derivative in the second component is:

$$\frac{\partial (\mathbf{r}R_t)}{\partial \mathbf{r}} = \mathbf{I}R_t + \mathbf{r}\frac{\partial R_t}{\partial \mathbf{r}}, \tag{8.26}$$

where $\mathbf{I}$ is the identity matrix. To calculate $\partial R_t/\partial \mathbf{r}$, the chain rule is used:

$$\frac{\partial R_t}{\partial \mathbf{r}} = \frac{\partial R_t}{\partial \mathbf{x}}\frac{\partial \mathbf{x}}{\partial \mathbf{r}} = \frac{\mathbf{x}'}{\|\mathbf{x}\|}\frac{\partial \mathbf{x}}{\partial \mathbf{r}}. \tag{8.27}$$

Substituting (8.25) and (8.26) into (8.24) yields:

$$\mathbf{S}\frac{\partial \mathbf{x}}{\partial \mathbf{r}} = -\mathbf{\Lambda} + \mathbf{I}R_t + \mathbf{r}\frac{\mathbf{x}'}{\|\mathbf{x}\|}\frac{\partial \mathbf{x}}{\partial \mathbf{r}}. \tag{8.28}$$

By rearranging (8.28), the Jacobian matrix can be expressed as:

$$\frac{\partial \mathbf{x}}{\partial \mathbf{r}} = \left(\Delta' \Delta\right)^{-1} \Delta' \left(\mathbf{I} R_t - \mathbf{\Lambda}\right), \tag{8.29}$$

where

$$\Delta = \mathbf{S} - \mathbf{r}\frac{\mathbf{x}'}{\|\mathbf{x}\|}. \tag{8.30}$$

Now, with the Jacobian calculated using (8.29), the covariance matrix in the Cartesian coordinates can be calculated with (8.23).

8.2.3 Numerical Results

8.2.3.1 Simulation Results

In order to compare the performance of the SI and SX localization algorithms, numerical simulations with a single receiver and four transmitters were carried out. The use of four transmitters allowed the SI method to be applied. The positions of the transmitters used in the simulation are listed in Table 8.1. The receiver was placed in the origin of the coordinate system $[0, 0, 0]'$ km. The target position was $\mathbf{x} = [5, 5, 5]'$ km. The measurements of the bistatic range were distorted by an additive zero-mean Gaussian error with the standard deviation σ_{Ri}=10m, for $i = 1, \ldots, 4$. This might correspond to a DVB-T-based passive radar, for example.

In the first experiment the influence of the correct estimation of the target-receiver range R_t is investigated. As can be seen from (8.12), the calculated target position $\hat{\mathbf{x}}$ is linear in R_t. In Figure 8.4 the x, y, and z components of the target position $\hat{\mathbf{x}}$ are plotted versus the target-receiver range R_t (solid line). Note the values on the horizontal axis. In the case of the x and y coordinates, the steepness is very low (all values are very close to 5 km). In the case of the z coordinate, a slight change of the R_t value corresponds to significant changes of the calculated altitude of the target. The true target position is marked with $\square$. This position may not necessarily be on the curves corresponding to $\hat{\mathbf{x}}$ due to the measurement errors. The positions marked with $\triangle$ and $\triangledown$ correspond to the two solutions of the SX method. As can be seen, the two solutions correspond to positions in the x and y coordinates that are very close to each other. This is also the case for the result of the SI method, marked with $\diamondsuit$. The situation is entirely different for the z coordinate. The two solutions of the SX method correspond to the calculated value of the target equal to $+5$ km and -5 km. This results from the mentioned effect of the nearly mirror reflection of the ellipsoid intersection with respect to the ground plane. In the case of the SI method,

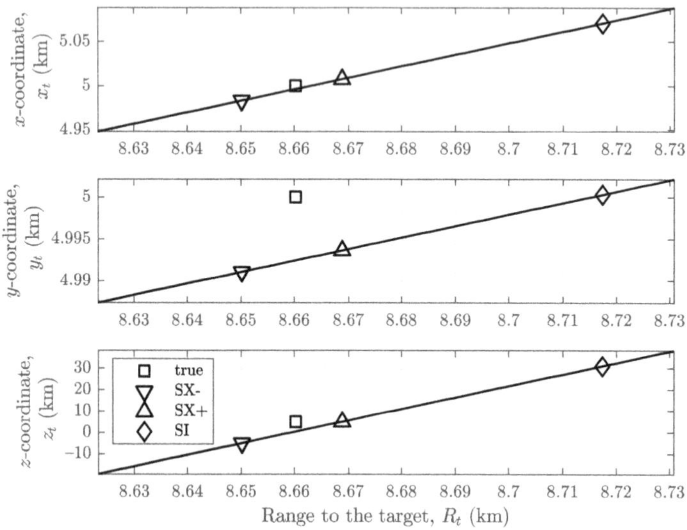

Figure 8.4 Target position versus estimated range to the target R_t.

the calculated target altitude is +30 km, which is very far from the correct value of +5 km.

The example presented above indicates that the precise estimation of the target-receiver range R_t has a fundamental influence on the estimation accuracy. Moreover, the SI method revealed an inferior performance in comparison to the SX method in a single run of the experiment. The two methods will now be compared with a statistical experiment. The same scenario was used as in the last experiment, and 1,000 Monte Carlo simulations were performed, with random errors of the bistatic measurements. The calculated values of the standard deviation in the x, y and z coordinates are shown for the SX and SI methods in Table 8.2. The results suggest that, in the analyzed scenario, the SX method performs much better than the SI method. This result is different than for the TDOA system [7]. The different performance of the two methods can be understood when taking into consideration the differences in the TDOA and passive radar scenarios. The algorithms derived for the two systems are very similar, but are not the same. In the case of the TDOA the time differences are measured, whereas in passive radar time (range) differences are measured. Because the SX method provides better results for the passive radar scenario, this approach will be investigated in the following part of the chapter.

The SX algorithm has been tested further by performing a series of simulations in a three-transmitter scenario. The scenario was based on the localization of the

Table 8.2
Comparison of the Estimation Accuracy

Algorithm	σ_x *(m)*	σ_y *(m)*	σ_z *(m)*
Spherical interpolation (SI)	63.16	13.54	25748.91
Spherical intersection (SX)	6.36	12.28	18.96

transmitters and receiver in a real measurement campaign, which is described at the end of this section. The relative positions of the three transmitters located in Poland: Pruszkow, Raszyn, and Warsaw, are listed in Table 8.3.

First, the distribution of the target position estimate is investigated for a fixed target position. The true target is located at $\hat{\mathbf{x}} = [20, 20, 10]'$ km. The bistatic measurements were disturbed by an additive zero-mean Gaussian random variable with the standard deviation $\sigma_R = 300$m; this value is typical for an FM-based passive radar. Figure 8.5(a) shows the geometry scenario with the receiver, transmitters and target positions. The bistatic ellipses are also marked. In Figure 8.5(b) a zoom of the target area is presented. The dots show target position estimates for different realizations of the bistatic measurement errors. The error ellipse is shown with a dashed line. It is calculated from the covariance matrix (8.23) in the following way:

$$\Delta\mathbf{p}'\mathbf{P}^{-1}\Delta\mathbf{p} = d \tag{8.31}$$

where $\Delta\mathbf{p} = [\Delta x, \Delta y, \Delta z]'$ is a position difference vector. The quadratic form $\Delta\mathbf{p}'\mathbf{P}^{-1}\Delta\mathbf{p}$ is a scalar d, which is a random variable with χ^2 distribution with three degrees of freedom. The quadratic form can be written in an expanded way as:

$$[\Delta x, \Delta y, \Delta z] \begin{bmatrix} \sigma_x^2 & \sigma_{xy} & \sigma_{xz} \\ \sigma_{yx} & \sigma_y^2 & \sigma_{yz} \\ \sigma_{zx} & \sigma_{zy} & \sigma_z^2 \end{bmatrix}^{-1} \begin{bmatrix} \Delta x \\ \Delta y \\ \Delta z \end{bmatrix} = d \tag{8.32}$$

The error ellipse is defined by thresholding the value d by γ_e, so that the probability of exceeding the threshold has a predetermined value P_e:

$$P_e = \Pr\{d \leq \gamma_e\} \tag{8.33}$$

Table 8.3
Relative Positions of the Transmitters Used in the Measurement Campaign

Transmitter	*x-offset (km)*	*y-offset (km)*	*z-offset (km)*
Tx1 (Pruszkow)	9.63	−2.66	0.20
Tx2 (Raszyn)	19.00	−14.53	0.26
Tx3 (Warsaw)	27.40	3.15	0.22

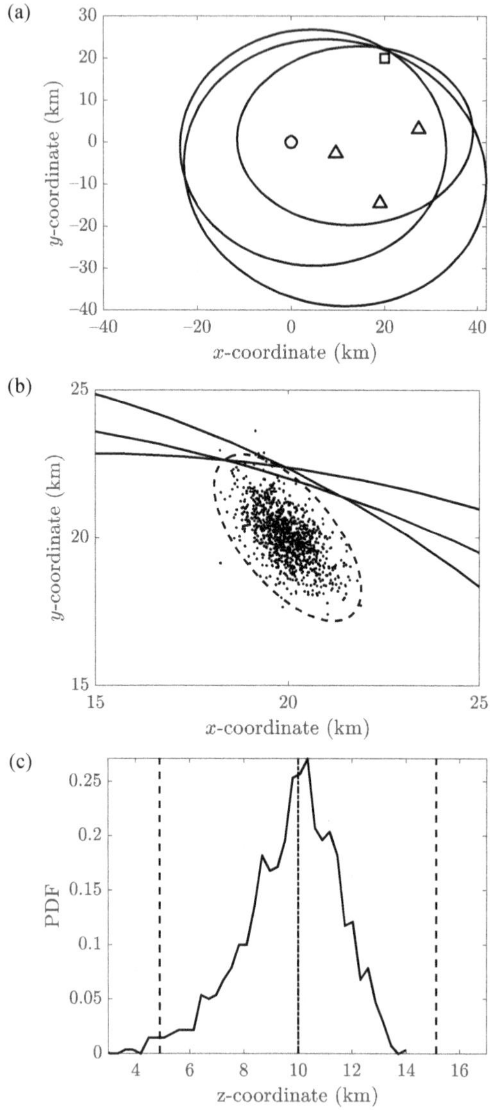

Figure 8.5 (a) Localization scene geometry. (b) Distribution of localized target positions (dots) and error ellipse (dashed line) in the xy dimension. (c) Estimated probability density function of localized target positions (solid line) and the borders of the error interval (dashed lines) in the z dimension.

For example, the threshold γ_e for $P_e = 0.99$ is equal to 11.34.

As can be observed, the error ellipse corresponds well to the random target positions. It is worth noting that the bistatic ellipses (shown with solid lines) do not intersect at the target position. This results from the fact that those ellipses are created as a cross-section of the bistatic ellipsoid at the ground level. As the target is located 10 km above the ground level, the bistatic ellipses do not correspond exactly to the target position. In Figure 8.5(c) the vertical cross-section (along the z-axis) is shown. The solid curve is the estimated PDF (probability density function) of target altitude. The vertical line at 10 km shows the true target altitude. The dashed vertical lines indicate the borders of the error ellipse in the z-coordinate. It can be seen that the distribution is close to Gaussian and the maximum value of the distribution is close to the true value.

In the next experiment a comparison of the theoretical and experimental accuracy is carried out on an area of 50×50 km. The target altitude was 10 km in all cases, which is typical for airliners. The standard deviation of the bistatic range measurement was $\sigma_R = 300$m, which again can be considered typical value for an FM-based passive radar. The theoretical accuracy was calculated using (8.23). The experimental accuracy was measured by calculating the standard deviation of the calculated target position and the true position, in the x, y, and z coordinates. The experiment was repeated 1,000 times (each time random bistatic errors were generated) for every position on the grid. The contour plots of the theoretical and experimental accuracy are shown in Figure 8.6. The three plots correspond to the standard deviation in x, y, and z coordinates, respectively. The experimental accuracy is marked with a solid line, whereas the theoretical accuracy is marked with a dashed line. As can be seen, in the case of the x and y coordinates, very good agreement is obtained. In the case of the altitude dimension, the theoretical and measured accuracy slightly diverge; the experimental accuracy is better than that predicted by theory. The reason for this is that from the two solutions of the SX algorithm, the one corresponding to the higher altitude is chosen as the final solution. This operation is not reflected in the derivation of the accuracy (8.23); hence, a slight discrepancy occurs.

Another way of representing the localization accuracy is by showing the error ellipses for different positions of the target. An example of such visualization is shown in Figure 8.7. On each point of a 10×10 km grid, an error ellipse was calculated and visualized. Again, similar to the results shown in Figure 8.6, high dependency on the geometry can be seen. However, a general tendency can be clearly observed that the accuracy decreases with the distance from the receiver. Moreover, different orientations of ellipse can be seen.

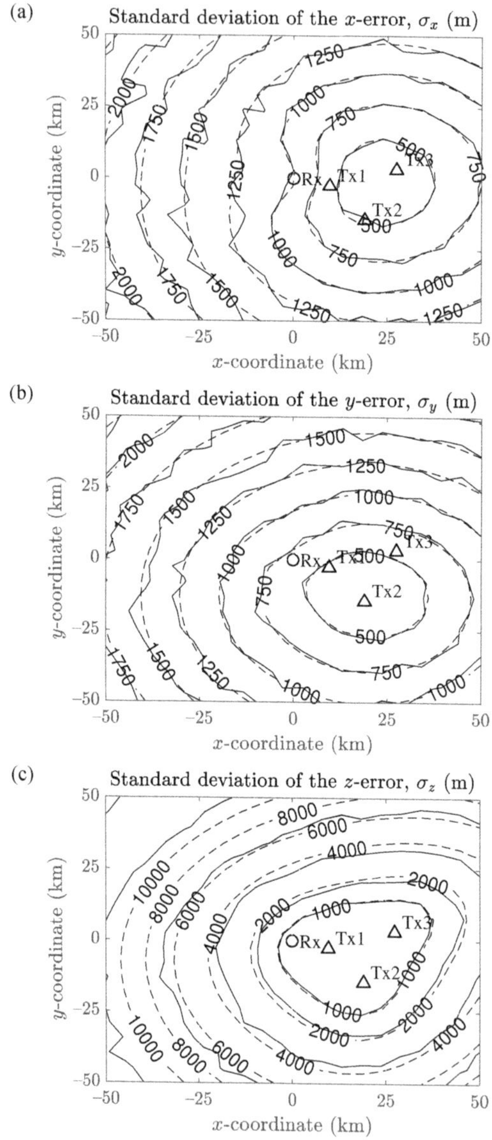

Figure 8.6 Simulated (solid line) and theoretical (dashed line) standard deviation of the position error using ellipsoid intersection in the (a) x, (b) y, and (c) z direction (the values of the contour lines, marked only for the simulated data, are the same also for the theoretical data). (©2012 IEEE Reprinted, with permission, from [2].)

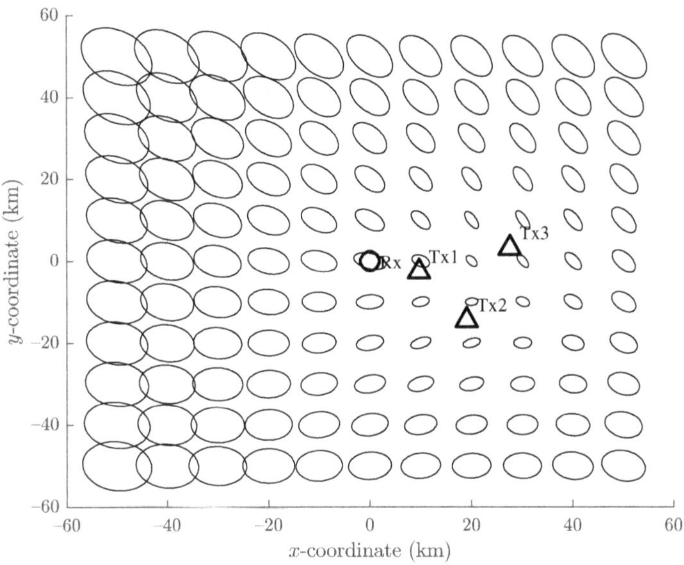

Figure 8.7 Error ellipses for different positions of the target for localization using ellipsoid intersection.

Table 8.4
Standard Deviation of the Bistatic Range Errors

Transmitter	σ_{Ri} *(m)*
Tx1 (Pruszkow)	120
Tx2 (Raszyn)	567
Tx3 (Warsaw)	266

8.2.3.2 Real-Data Results

The SX algorithm has been tested on real measurement data acquired with the *PaRaDe* system [8, 9]. In the experiment, a single target was selected at the position $[8.3, 8.4, 2.3]'$ km. As in the case of the investigation of the accuracy of the bistatic tracking in Chapter 7, the data from an ADS-B receiver have been used as a reference. This permitted the calculation of the estimation accuracy in Cartesian coordinates, as well as in bistatic coordinates (after the conversion of the target position to bistatic coordinates).

First, the positions obtained from the ADS-B in Cartesian coordinates were converted into bistatic data. These results were compared with the measurements from the passive radar. The standard deviations of the bistatic range measurements corresponding to the three transmitters are listed in Table 8.4. The bandwidth of the FM-radio signal is in the order of tens of kilohertz, which provides a range resolution of a few kilometers. However, the achievable accuracy of range measurement can be much better, especially for high signal-to-noise ratios, as discussed in Chapter 6. In the presented example, the precise range measurement was carried out by fitting a parabolic curve to the target echo [10]. The range estimate corresponded to the maximum of the fitted parabola. The achievable range accuracy is in the order of hundreds of meters, which is an order of magnitude better than the size of the range resolution cell [11].

Calculations of the standard deviation of the position error for the scenario are presented in Table 8.5. It was assumed that the position of the target does not change significantly, so that the accuracy remains fairly constant throughout the experiment. The estimated accuracy was calculated as the standard deviation of the difference between the position from the ADS-B receiver and from the radar in the appropriate direction (x, y, or z). The theoretical values of the standard deviation corresponded to the values from the diagonal of the **P** matrix calculated according to (8.23). The values of σ_{Ri} needed for the calculation of the **P** matrix were taken from Table 8.4. As the experiments show, the theoretical results are confirmed by the measurements.

Table 8.5
Standard Deviation of the Position Estimation Errors

Standard Deviation	*Estimated*	*Theoretical*
σ_x (m)	558	777
σ_y (m)	1248	1388
σ_z (m)	3826	5612

8.2.4 The Ghost Target Phenomenon

8.2.4.1 Problem Formulation

Target localization based on the intersection of the bistatic ellipsoids is associated with a problem known as the *ghost target* phenomenon [12–17]. This problem is related to the fact that when multiple bistatic ellipsoids are intersected, there is no a priori information about the source of the measurements (i.e., which target generated the measurement); therefore, all combinations of ellipsoids have to be tested. This may lead to random intersections of ellipsoids which originated from different targets. As a result, false targets are obtained, which are called ghost targets.

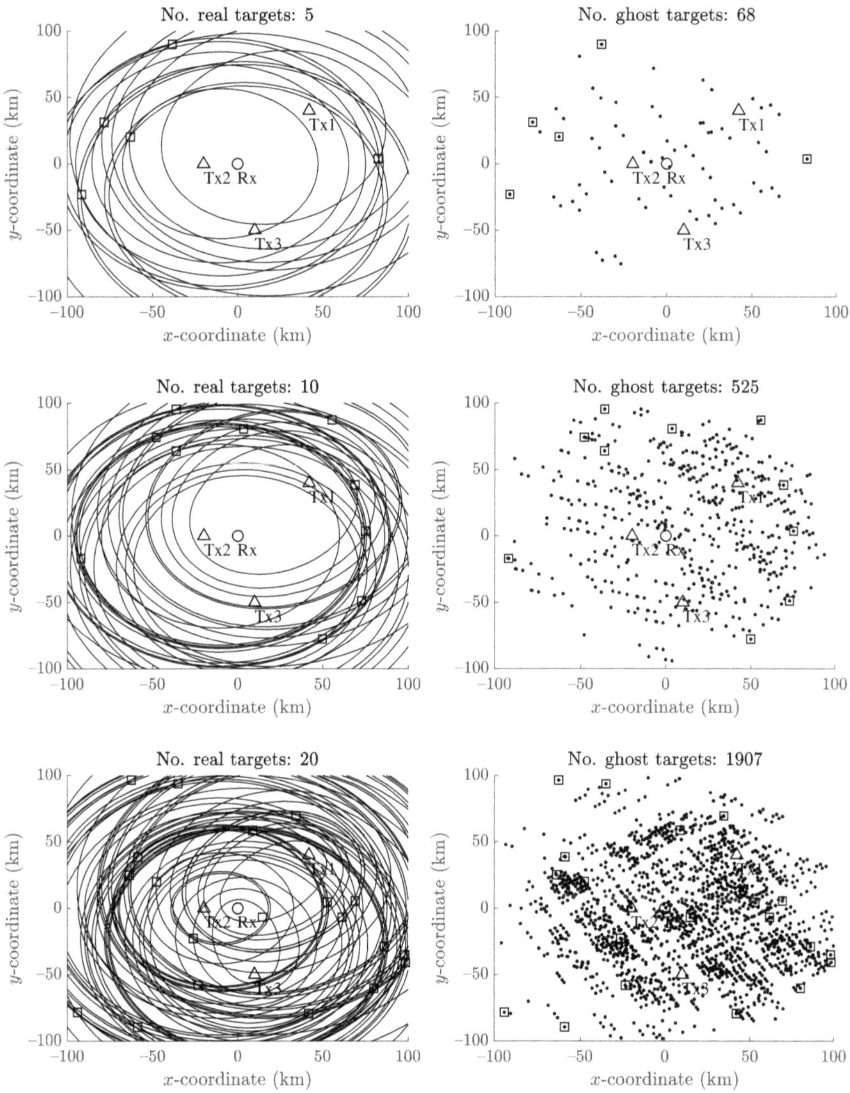

Figure 8.8 Left: Simulated target positions (squares) and corresponding bistatic ellipses (solid line). Right: localized target positions (dots) and simulated target positions (squares).

Let us investigate the ghost target problem with an example. Assume that there is a single receiver and three transmitters, as shown in Figure 8.8. In the first case, five random targets were generated. The x and y coordinates were distributed uniformly in the $(-100, +100)$ km interval, whereas the z coordinate was distributed uniformly in the $(0, 10)$ km interval. The ideal bistatic ranges (without error) were used to calculate the intersection of the ellipsoids. As no information on the source of measurement is available in the localization algorithm, all combinations of the bistatic measurements were tested. In this case it was $5^3 = 125$ tests. In some of the cases, the bistatic measurements that did not originate from the same target resulted in the intersection of the ellipsoids. The results of the experiment are shown in Figure 8.8 (top row). In the plot on the left side, the positions of the simulated targets are marked with squares. The bistatic ellipses, which are formed by the intersection of the bistatic ellipsoids with the x-y plane, are shown in a solid line. As can be seen, the bistatic ellipses do not necessarily intersect in the target positions, as they are plotted on the ground level, whereas the targets have nonzero altitude. The plot on the right side shows the result of the localization algorithm. The dots indicate positions of the localized targets. For comparison, the real target positions are also marked with a square. It can be seen that the localization algorithm found all true target positions correctly, but a number of ghost targets also appeared. In this particular scenario, 68 ghost targets were obtained. The problem with ghost targets is more severe when more measurements are available. For 10 true targets, 525 ghost targets were obtained (shown in the middle row of Figure 8.8). For 20 true targets, in turn, 1,907 ghost targets were created by the localization algorithm (shown in the bottom row of Figure 8.8).

8.2.4.2 Mitigation Strategies

The ghost target phenomenon is one of the most difficult problems to solve in passive radar. It is also a problem that does not have its counterpart in the classical monostatic active radar. There are several strategies that can be applied to mitigate this problem, some of which are:

- Use of the azimuth measurement;
- Rejection of targets with unrealistic parameters;
- Elimination of association conflicts;
- Initialization of tentative Cartesian tracks and verification over time.

The first of the approaches is the use of the direction of arrival measurement. As mentioned before, the angle measurement in passive radar is not very accurate due to the low frequencies used. Nevertheless, even when a coarse measurement is available, it can substantially reduce the number of false ellipsoid intersections. When an intersection is obtained from a set of bistatic measurements, the azimuth angle can be calculated for the estimated position of a potential target. Then this angle can be compared against the measured direction of arrival associated with all the bistatic data. If an inconsistency is encountered (i.e., the azimuth angle ϕ corresponding to the intersection is not contained in the azimuth sector $\Delta\phi$ for each of the bistatic measurements), the intersection is treated as a ghost target and rejected.

The results of the application of this approach are shown in Figure 8.9. The vertical axis corresponds to the number of true targets simulated randomly in the same scenario as in the previous example. The horizontal axis shows the average number of ghost targets obtained from the localization process. The curve plotted with a solid line shows the initial situation, when no angle information is available. This corresponds to the azimuth sector of 360° ($\Delta\phi = 360°$). As can be seen, the number of ghost targets grows rapidly, reaching several thousand for 20 true targets. Three other curves correspond to the azimuth sector of 90°, 45°, and 30°, respec-

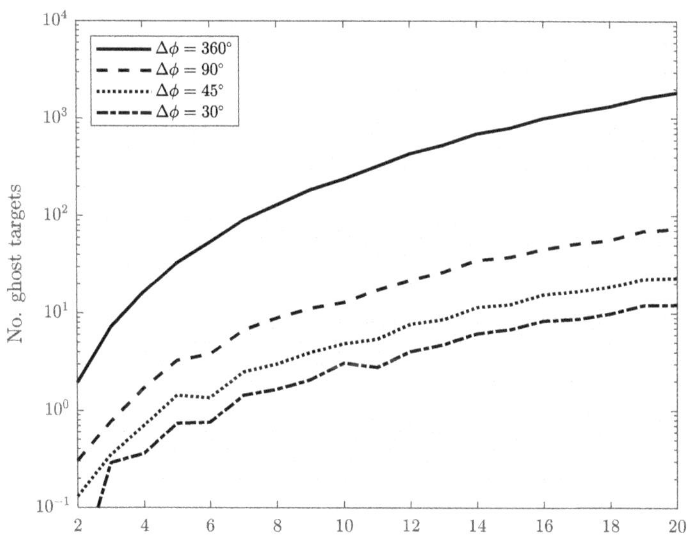

Figure 8.9 Number of ghost targets versus the number of true targets for different values of the azimuth sector.

tively. As can be seen, even a very coarse azimuth measurement (e.g. 90°-sector) provides a substantial reduction in the number of false intersections. For the 30°-sector, the reduction is more than one order of magnitude, reaching tens instead of thousands of ghost targets for 20 true targets. Note that the number of ghost targets indicated in the plot can be less than 1, as a result of averaging of different simulation results.

Additional reduction can be achieved by rejecting target localizations with unrealistic parameters, especially the altitude of the target. Figure 8.10 shows the 3D view of the scene with targets localized based on ellipsoid intersection. In this case, 20 true targets were simulated, and no azimuth measurement was taken into account. As can be seen, numerous ghost targets are present, most of which have an altitude of tens of kilometers. In many applications such high targets are of no interest or not realistic. For this reason, it can be safely assumed that targets localized above a certain altitude are ghost targets and can be rejected. However, care should be taken during this operation with regards to localization accuracy. As shown before, the altitude estimation accuracy obtained by the intersection of bistatic ellipsoids is relatively coarse. Therefore, an altitude of, for example, 20 km, might not necessarily correspond to a ghost target, even when the maximum expected altitude is 10 km. It might be the result of estimation error due to the bistatic range inaccuracy and geometrical relationships.

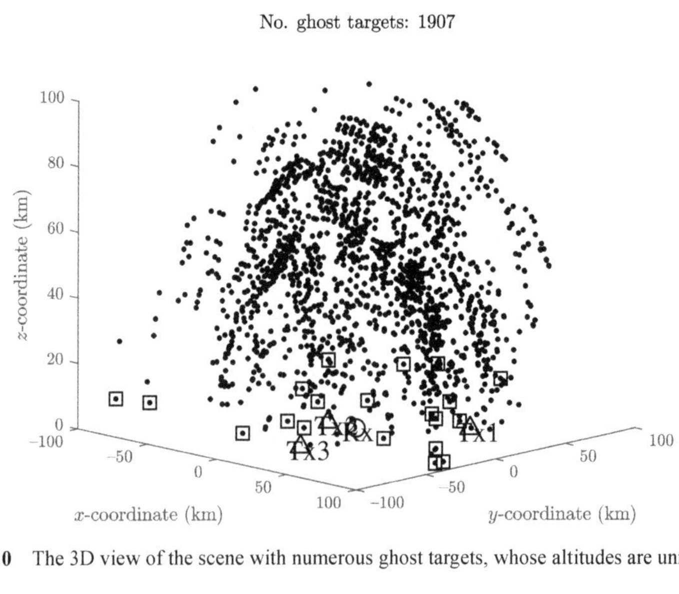

Figure 8.10 The 3D view of the scene with numerous ghost targets, whose altitudes are unrealistic.

The third of the approaches to the reduction of the ghost target phenomenon consists in eliminating the conflicts in measurement associations. A single bistatic measurement should be assigned to only one target localization. In practice, the most of ghost targets exist because a single bistatic measurements is used numerous times, producing false ellipsoid intersections. Therefore, once the bistatic measurement is used, it should be discarded from further analysis. The problem is, however, how to decide to which target localization a bistatic measurement should be assigned. A possible strategy is to start from a target localization that is the most reliable, for example, where multiple ellipsoids intersect. Then all bistatic measurements that are assigned to this localization are eliminated from other target localizations.

The fourth strategy to mitigate the ghost target problem is to initialize Cartesian tracks[1] only tentatively and then monitor their behavior and eliminate those tracks, which diverge from the assumed motion model. If the consecutive measurements form a feasible trajectory for a certain period of time, the track is confirmed. In practice, the ghost targets behave realistically only on a short timescale. In a longer period of time, however, their motion diverges from typical motion model used in the tracker. The disadvantage of this approach is increased computational complexity associated with maintaining multiple false Cartesian tracks. In addition, a delay is introduced caused by the use of the tentative track state.

The ghost target problem can even be exacerbated in a single-frequency-network (SFN) scenario [18–25]. SFN is an approach used in certain transmission standards, such as DVB-T or DVB-T2, which consist in transmitting the same content from different transmitters. The reception of multiple copies of the same signal by the television receivers is not a problem. They deal with numerous transmitters in the same way as they deal with the multipath effect. For the passive radar, however, the SFN poses an additional problem. In an SFN scenario, a single target can generate as many detections as there are transmitters in the network. In this way not only is the target that produced a given bistatic measurement ambiguous, but also the transmitter is not known.

8.2.5 Number of Combinations

Let us analyze the number of all combinations that have to be considered to test ellipsoid intersection. Assume that there are N_{targ} targets present in the surveilled airspace, and each target generates a measurement for one of the N_{pair} transmitter-receiver pairs. Next assume that a set of measurements from N_{inter} pairs is tested for ellipsoid intersection. The minimum number is 3 (in this case the SX method can

1 Cartesian tracking is the topic of Chapter 9.

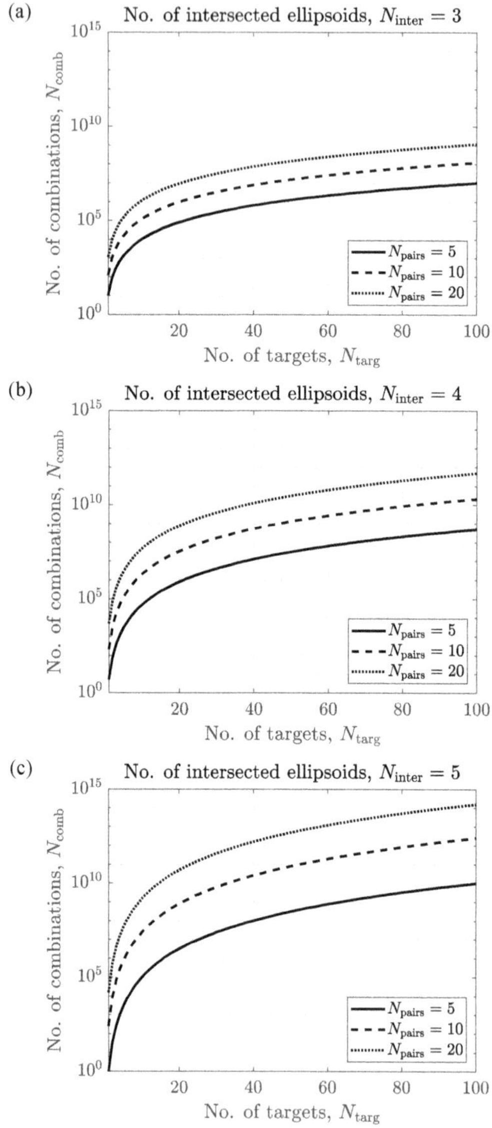

Figure 8.11 Number of combinations to be tested versus number of targets N_{targ} for different number of transmitter-receiver pairs N_{pair} and number of intersected ellipsoids N_{inter}.

be applied), but more measurements can be used (then SX and SI approaches can be used). Consider an example with $N_{\text{targ}} = 10$ targets and $N_{\text{inter}} = 3$ intersected ellipsoids. If $N_{\text{pair}} = 3$ pairs are available, this means that each 10 measurements from the first pair has to be tested with each 10 measurements from the second pair, and 10 measurements from the third pair. This means that $10^3 = 1,000$ tests have to be performed. If more pairs are available, $N_{\text{pair}} > 3$, each combination of N_{inter} out of N_{pair} has to be considered. This can be calculated as: $\binom{N_{\text{pair}}}{N_{\text{inter}}}$. The total number of combinations to be tested N_{comb} can therefore be calculated as:

$$N_{\text{comb}} = (N_{\text{targ}})^{N_{\text{inter}}} \binom{N_{\text{pair}}}{N_{\text{inter}}} \tag{8.34}$$

The calculated number of combinations for different sets of parameters is shown in Figure 8.11. The vertical axis corresponds to the number of targets, which is assumed to be the number of measurements for each transmitter-receiver pair (in practice, thc number of measurements will be different for each pair, as targets will not usually be detected by each pair). The three curves on each plot correspond to a different number of the transmitter-receiver pairs, equal to 5, 10, and 20, respectively. The three plots correspond to a different number of simultaneously intersected ellipsoid N_{inter}. This number was equal to 3, 4, and 4, respectively. The results show that the number of required tests grows rapidly with the number of targets and number of intersected ellipsoids (note the vertical logarithmic scale with each tick denoting one order of magnitude).

The above analysis shows that the number of necessary tests can saturate even a very powerful computing system. Different approaches can be used to alleviate the problem. One of the approaches is to divide the surveilled area into sectors. The bistatic measurements can be assigned to these sectors, and the hypothesis testing is performed for individual sectors. In this way, not all bistatic measurements have to be considered in the intersection testing. Another approach is to solve the problem iteratively, by testing only some of the hypotheses in one time instant and another batch of hypotheses in another time instant.

8.3 DOA-BASED LOCALIZATION

The second approach to target localization in passive radar is to use direction of arrival (DOA) measurements. By combining the DOA with the bistatic range, the target position can be calculated in Cartesian coordinates. In the case when only the azimuth angle is available, the intersection of DOA and the bistatic ellipse results in a 2D target position estimate. If azimuth and elevation angles are available, a 3D position can be calculated.

8.3.1 Algorithms for Target Localization

The geometry considered during target localization with the use of DOA is shown in Figure 8.12. It is assumed that the receiver is placed at the origin of the coordinate system. The elevation and azimuth angles are measured with respect to the receiver position. The elevation angle θ is defined as the angle between the receiver-target vector and the x-y plane. Elevation angle $\theta = 0°$ corresponds to target at ground level. Positive values of θ indicate a target above the x-y plane. The azimuth angle ϕ is the angle between the projection of the target on the x-y plane and the positive y semi-axis. Thus, $\phi = 0°$ corresponds to the positive y semi-axis direction, which also coincides with the geographical North.

The concept of using angle measurement for target localization is shown in Figure 8.13. If only the azimuth angle is available, 2D position estimation is possible. The intersection of the bistatic ellipse and the ray corresponding to the azimuth angle indicates the estimated target position, as shown in Figure 8.13(a). When both azimuth and elevation angles are measured, 3D position estimation can be carried out. In such a case, the ray determined by the azimuth and elevation angles is intersected with the 3D bistatic ellipsoid, as visualized in Figure 8.13(b).

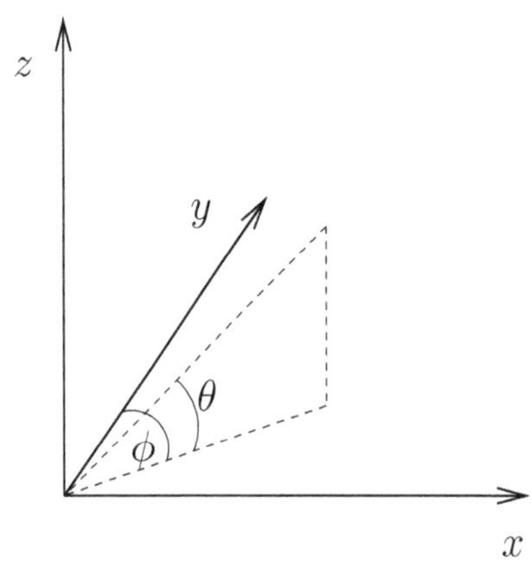

Figure 8.12 Azimuth (ϕ) and elevation (θ) angle definitions.

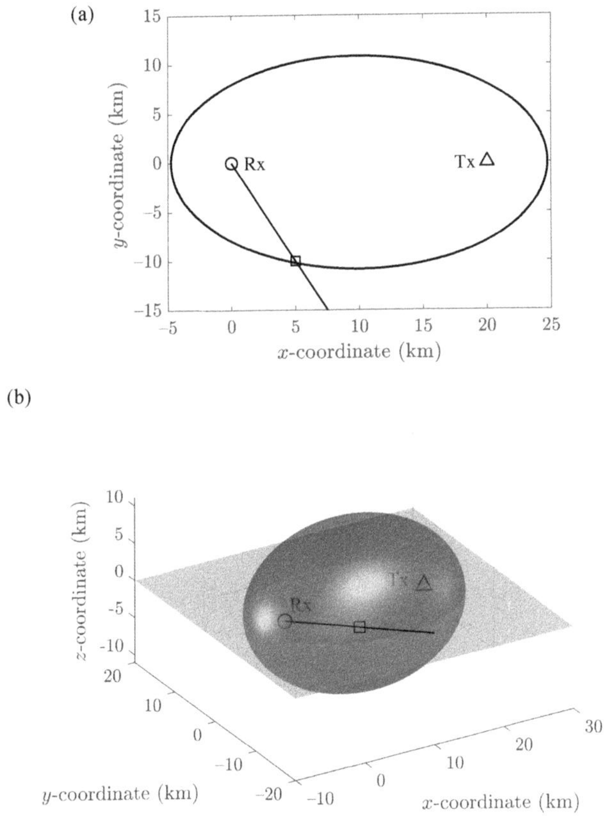

Figure 8.13 Concept of target localization with the use of bistatic range and angle measurements. (a) 2D case with azimuth measurement only. (b) 3D case with azimuth and elevation measurements available.

In the 2D case, the target position $\hat{\mathbf{x}} = [\tilde{x}, \tilde{y}]'$ can be calculated using closed-form formulas as follows [26]:

$$\tilde{x} = \frac{\left((R + R_b)^2 - x_t^2 - y_t^2\right)\cos(\phi)}{2\left(R + R_b - x_t\cos(\phi) - y_t\sin(\phi)\right)} \tag{8.35}$$

$$\tilde{y} = \frac{\left((R + R_b)^2 - x_t^2 - y_t^2\right)\sin(\phi)}{2\left(R + R_b - x_t\cos(\phi) - y_t\sin(\phi)\right)} \tag{8.36}$$

where $[x_t, y_t]$ is the transmitter position, ϕ is the azimuth angle, R is the bistatic range, and R_b is the baseline length (transmitter-receiver range).

If the azimuth (ϕ) and elevation (θ) angles are available, the 3D target position $\hat{\mathbf{x}} = [\tilde{x}, \tilde{y}, \tilde{z}]'$ can be calculated as follows:

$$\tilde{x} = \frac{\left((R+R_b)^2 - x_t^2 - y_t^2 - z_t^2\right)\cos(\phi)\cos(\theta)}{2\left(R+R_b - x_t\cos(\phi)\cos(\theta) - y_t\sin(\phi)\cos(\theta) - z_t\sin(\theta)\right)} \tag{8.37}$$

$$\tilde{y} = \frac{\left((R+R_b)^2 - x_t^2 - y_t^2 - z_t^2\right)\sin(\phi)\cos(\theta)}{2\left(R+R_b - x_t\cos(\phi)\cos(\theta) - y_t\sin(\phi)\cos(\theta) - z_t\sin(\theta)\right)} \tag{8.38}$$

$$\tilde{z} = \frac{\left((R+R_b)^2 - x_t^2 - y_t^2 - z_t^2\right)\sin(\theta)}{2\left(R+R_b - x_t\cos(\phi)\cos(\theta) - y_t\sin(\phi)\cos(\theta) - z_t\sin(\theta)\right)} \tag{8.39}$$

8.3.2 Accuracy Analysis of Target Localization

Similar to the localization based on ellipsoid intersection, the localization accuracy will be investigated for the DOA-based approach. First, the 2D case with azimuth angle measurement only is analyzed. Let us define the vector of measurements $\mathbf{r}_{R\phi} = [R, \phi]'$ and its covariance matrix:

$$\mathbf{R}_{R\phi} = \text{cov}\left(\mathbf{r}_{R\phi}\right) = \text{diag}\left([\sigma_R^2, \sigma^2{}_\phi]\right) \tag{8.40}$$

Unlike the bistatic ellipsoid intersection, here there is a straightforward closed-form relationship between the measurements and target position ((8.35)–(8.36)). For this reason, the covariance of the target position estimate can be calculated as:

$$\mathbf{P} = \text{cov}(\hat{\mathbf{x}}) \approx \left(\frac{\partial \hat{\mathbf{x}}}{\partial \mathbf{r}_{R\phi}}\right) \mathbf{R}_{R\phi} \left(\frac{\partial \hat{\mathbf{x}}}{\partial \mathbf{r}_{R\phi}}\right)', \tag{8.41}$$

where the Jacobian is expressed as:

$$\frac{\partial \hat{\mathbf{x}}}{\partial \mathbf{r}_{R\phi}} = \begin{bmatrix} \partial\tilde{x}/\partial R & \partial\tilde{x}/\partial\phi \\ \partial\tilde{y}/\partial R & \partial\tilde{y}/\partial\phi \end{bmatrix} \tag{8.42}$$

where $\tilde{x}$ and $\tilde{y}$ are expressed using (8.35) and (8.36), respectively.

In the 3D case, the measurement covariance matrix has the following form:

$$\mathbf{R}_{R\phi\theta} = \text{cov}\left(\mathbf{r}_{R\phi\theta}\right) = \text{diag}\left([\sigma_R^2, \sigma_\phi^2, \sigma_\theta^2]\right) \tag{8.43}$$

where $\sigma_R^2, \sigma_\phi^2, \sigma_\theta^2$ are the variance of bistatic range, azimuth angle, and elevation angle, respectively.

The covariance matrix of the target position estimate in the Cartesian coordinates can be calculated as:

$$\mathbf{P} = \text{cov}(\hat{\mathbf{x}}) \approx \left(\frac{\partial \hat{\mathbf{x}}}{\partial \mathbf{r}_{R\phi\theta}}\right) \mathbf{R}_{R\phi\theta} \left(\frac{\partial \hat{\mathbf{x}}}{\partial \mathbf{r}_{R\phi\theta}}\right)', \tag{8.44}$$

where the Jacobian is expressed as:

$$\frac{\partial \hat{\mathbf{x}}}{\partial \mathbf{r}_{R\phi\theta}} = \begin{bmatrix} \partial\tilde{x}/\partial R & \partial\tilde{x}/\partial\phi & \partial\tilde{x}/\partial\theta \\ \partial\tilde{y}/\partial R & \partial\tilde{y}/\partial\phi & \partial\tilde{y}/\partial\theta \\ \partial\tilde{z}/\partial R & \partial\tilde{z}/\partial\phi & \partial\tilde{z}/\partial\theta \end{bmatrix} \tag{8.45}$$

The values of $\tilde{x}$, $\tilde{y}$, and $\tilde{z}$ are calculated using (8.37), (8.38) and (8.39), respectively.

8.3.3 Numerical Results

8.3.3.1 Simulation Results

The calculation of the accuracy for the case of 2D localization will be illustrated with an example. The scenario's geometry is shown in Figure 8.14(a). The receiver is located at (0,0) km and the transmitter is located at (50,0) km. The target is located at the ground level so that the bistatic range is 70 km and azimuth angle is 45° (with respect to the receiver). Figures 8.14(b, c) show the error ellipse calculated based on the covariance matrix (8.41). The dots show target position estimates. In both cases the standard deviation was equal to 300m. The difference between Figures 8.14(b, c) is the standard deviation of the azimuth angle measurement. In the first case, the standard deviation is 1° and in the second case 5°. As can be seen, in the first case the error ellipse corresponds well to the obtained cloud of calculated target positions. In the second case, however, the discrepancy starts to be visible. The point cloud is bent along the ellipse, whereas the error ellipse is tangential to the bistatic ellipse. This is a result of approximating the nonlinear transformation with the first derivative used in (8.44).

Similar to the case of the intersection of bistatic ellipsoids, accuracy over a wide range is also analyzed for the localization with the use of bistatic range and DOA. The target was located at the ground level. The standard deviation of the bistatic range was 300m and azimuth angle 1°. The experiment was repeated 1,000 times for each position on the grid. The experimental accuracy was calculated as the standard deviation of the error between the estimated and true parameters. The theoretical accuracy was calculated according to (8.41). The standard deviation of

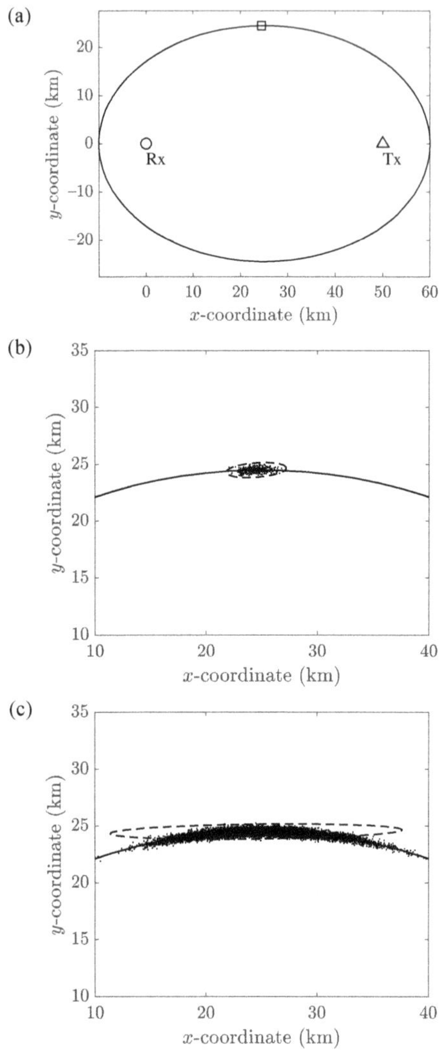

Figure 8.14 (a) Localization scenario geometry. Distribution of localized target positions (dots) and error ellipse (dashed line) for (b) 1° standard deviation of azimuth measurement, and (c) 5° standard deviation of azimuth measurement.

the x and y coordinates are shown in Figure 8.15 (a, b), respectively. As can be seen, the theoretical accuracy calculated using (8.41) coincides with the simulated results. The accuracy near the baseline deteriorates substantially, which is associated with the decrease of range accuracy for large bistatic angles.

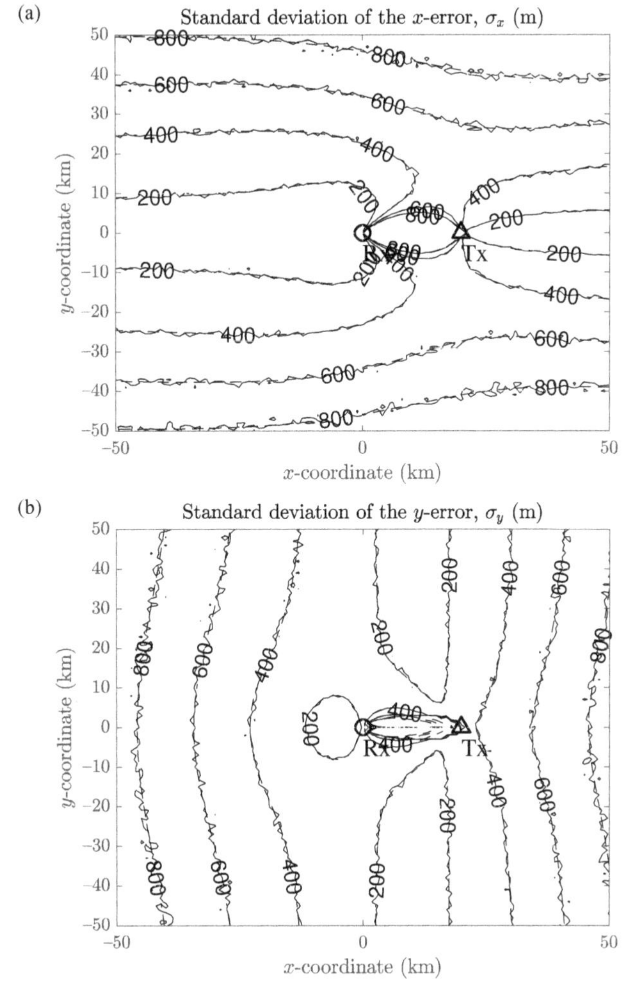

Figure 8.15 Simulated (solid line) and theoretical (dashed line) standard deviation of the position error using bistatic range and DOA measurement in the (a) x and (b) y directions (the values of the contour lines, marked only for the simulated data, are also the same for the theoretical data).

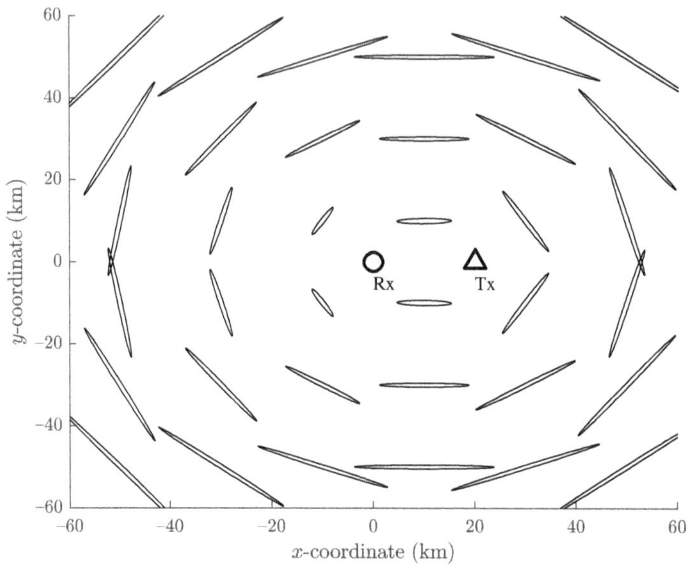

Figure 8.16 Error ellipses for different positions of the target for localization using bistatic range and DOA measurements.

Observing standard deviation in x and y coordinates separately does not provide direct information on the directivity of the error; directions in which the accuracy is better or worse than in others. This information can be obtained by observing error ellipses for different positions of the hypothetical target. Figure 8.16 shows error ellipses for different positions of the target on the x-y plane with respect to the transmitter and receiver. It was assumed that the bistatic range standard deviation is 300m and azimuth angle standard deviation is 5°. Each of the ellipses, placed on 10×10 km grid, shows the error ellipse for a target located at this position. A characteristic pattern can be observed formed from the error ellipses, which follows the shape of the bistatic ellipses.

8.3.3.2 Real-Data Results

Here, an example of measured data will be shown, where the azimuth measurement together with the bistatic range was used to localize a target in 2D Cartesian coordinates. The target under observation was a small drone. The passive radar was using DVB-T signals. The distance from the receiver to the transmitter was 42 km. The azimuth measurement was realized by phase interferometry (phase monopulse);

two antennas were displaced horizontally by approximately $\lambda/4$. The phase difference between the channels was measured in the following way: the cross-ambiguity function was calculated in a traditional way between the reference and both echo signals. As the two surveillance antennas were very close to each other and pointed in the same direction, the cross-ambiguity functions were very similar in amplitude. The phase difference conveyed information on the geometrical relationship of the received echo with respect to the two antennas. Target detection was performed on one of the cross-ambiguity functions by thresholding its absolute value. When a detection was declared for a particular range-velocity cell (R_0,V_0), the phase difference α for this echo was calculated as a phase difference between this particular cell for both cross-ambiguity functions ($\psi_L(R,V)$ for the left antenna and $\psi_R(R,V)$ for the right antenna):

$$\alpha = \arg\{\psi_L(R_0, V_0)\psi_R^*(R_0, V_0)\} \tag{8.46}$$

The azimuth angle ϕ was calculated as follows:

$$\phi = \arcsin\left(\frac{\alpha\lambda}{2\pi d}\right) \tag{8.47}$$

where d is the spacing between the antennas.

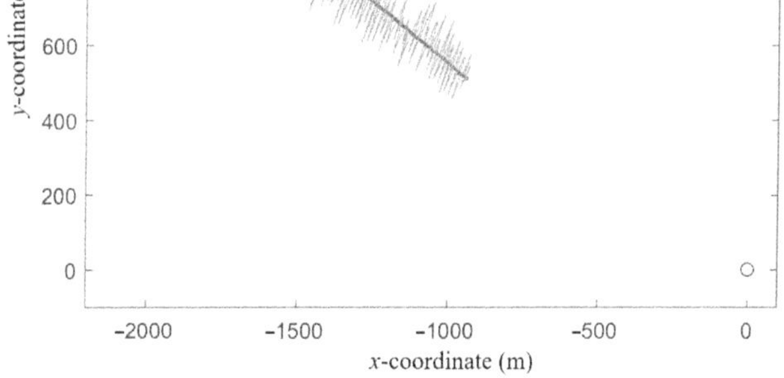

Figure 8.17 The 2D drone position estimated from bistatic range and DOA measurement.

Target localization was carried out by intersecting the bistatic ellipse with a ray indicated by the azimuth measurement using (8.35) and (8.36). The results of target localization are shown in Figure 8.17 with a gray curve. For comparison, the global positioning system (GPS) track from the logger mounted onboard the drone is shown with black dots. The GPS position was treated as a ground truth. As can be seen, the position provided by the radar follows the GPS trajectory. The calculated standard deviation of the azimuth error was close to 5°.

8.4 SUMMARY

In this chapter, two approaches to target localization have been analyzed. In the first case, bistatic measurements are used to calculate the intersection of bistatic ellipsoids. In the second approach, bistatic measurements are used together with the angle measurements.

The target localization in passive radar based solely on bistatic measurements is potentially a challenging task. The estimation of the target position could be obtained by numerical optimization of the appropriate cost function. However, such a solution would be time-consuming (due to the iterative nature of the optimization algorithms) and not very robust (due to the possibility of falling into a local minimum of the cost function). Closed-form equations are much more desirable. By appropriate rearrangement of the equations, an analytical solution can be found to a problem that is very close to the original localization problem [2]. This approach, originally proposed for the TDOA systems, can be adapted to the passive radar scenario. Two versions of the algorithm were investigated: spherical-interpolation and spherical-intersection. It has been demonstrated that, for the analyzed scenario, the latter outperforms the former. This result is contrary to that of the TDOA case, where the spherical-interpolation method performed better.

Target localization with the use of DOA is much simpler; closed-form equations are readily available. In the case when only the azimuth measurement is available, 2D position estimation can be obtained. When azimuth and elevation angles are measured, the target can be localized in 3D space. The accuracy of both approaches has been analyzed.

The fundamental practical difference between the multistatic localization and DOA-based localization is the ghost target phenomenon. In the case of intersecting the bistatic ellipsoids, the origin of each ellipsoid is unknown. For this reason, intersecting ellipsoids originating from different targets leads to false localizations, called ghost targets. The more measurements, the more severe the ghost target problem is. An additional difficulty is dealing with the multitude of combinations that have to be tested, which results from the fact that this is a combinatorial-explosion type of problem. The DOA-based localization is free from these problems, as the bistatic range and angle measurements are well associated with a single target.

References

[1] M. Malanowski, *Signal Processing Algorithms and Target Localisation Methods for Continuous-Wave Passive Bistatic Radar*. No. 179 in Prace Naukowe Politechniki Warszawskiej. Elektronika, Warszawa, ul. Polna 50: Oficyna Wydawnicza Politechniki Warszawskiej, 2012.

[2] M. Malanowski and K. Kulpa, "Two Methods for Target Localization in Multistatic Passive Radar," *IEEE Trans. Aerospace and Electronic Systems*, vol. 48, pp. 572–580, January 2012.

[3] J. Smith and J. Abel, "The Spherical Interpolation Method of Source Localization," *IEEE Journal Oceanic Eng.*, vol. OE–12, pp. 246–252, January 1987.

[4] B. Friedlander, "A Passive Localization Algorithm and Its Accuracy Analysis," *IEEE Journal Oceanic Eng.*, vol. OE–12, pp. 234–245, January 1987.

[5] H. Schau and A. Robinson, "Passive Source Localization Employing Intersecting Spherical Surfaces from Time-of-Arrival Differences," *IEEE Trans. Acoustics, Speech and Signal Processing*, vol. ASSP–35, pp. 1223–1225, August 1987.

[6] G. Mellen II, M. Pachter, and J. Raquet, "Closed-Form Solution for Determining Emitter Location Using Time Difference of Arrival Measurements," *IEEE Trans. Aerospace and Electronic Systems*, vol. 39, pp. 1056–1058, July 2003.

[7] J. Smith and J. Abel, "Closed-Form Least-Squares Source Location Estimation from Range-Difference Measurements," *IEEE Trans. Acoustics, Speech and Signal Processing*, vol. ASSP–35, pp. 1661–1669, December 1987.

[8] M. Malanowski, K. Kulpa, and J. Misiurewicz, "PaRaDe – Passive Radar Demonstrator Family Development at Warsaw University of Technology," in *Proc. Microwaves, Radar and Remote Sensing Symposium – MRRS*, (Kiev, Ukraine), pp. 75–78, September 2008.

[9] M. Malanowski, G. Mazurek, K. Kulpa, and J. Misiurewicz, "FM Based PCL Radar Demonstrator," in *Proc. International Radar Symposium 2007*, (Cologne, Germany), pp. 431–435, September 2007.

[10] M. Nalecz and K. Kulpa, "Range and Azimuth Estimation Using Raw Data in DSP-Based Radar System," in *Proc. International Conference on Microwaves and Radar – MIKON*, (Cracow, Poland), pp. 871–875, May 1998.

[11] M. Malanowski and K. Kulpa, "Experimental Analysis of Passive Radar Accuracy," in *Proc. International Radar Symposium 2009*, (Hamburg, Germany), pp. 331–335, September 2009.

[12] M. Tobias and A. D. Lanterman, "A Probability Hypothesis Density-Based Multitarget Tracker Using Multiple Bistatic Range and Velocity Measurements," in *Thirty-Sixth Southeastern Symposium on System Theory, 2004. Proceedings of the*, pp. 205–209, March 2004.

[13] M. Tobias and A. D. Lanterman, "Probability Hypothesis Density-Based Multitarget Tracking with Bistatic Range and Doppler Observations," *IEE Proceedings – Radar, Sonar and Navigation*, vol. 152, pp. 195–205, June 2005.

[14] S. Tian, Y. He, and G. Wang, "PHD Filter of Multi-Target Tracking with Passive Radar Observations," in *2006 8th International Conference on Signal Processing*, vol. 4, November 2006.

[15] D. E. Hack, L. K. Patton, A. D. Kerrick, and M. A. Saville, "Direct Cartesian Detection, Localization, and De-Ghosting for Passive Multistatic Radar," in *2012 IEEE 7th Sensor Array and Multichannel Signal Processing Workshop (SAM)*, pp. 45–48, June 2012.

[16] L. Lamentowski, T. Brenner, and M. Nieszporski, "De-Ghosting and Target Extraction Algorithm Analysis for the PCL-PET Fusion System," in *2015 16th International Radar Symposium (IRS)*, pp. 48–53, June 2015.

[17] M. A. Hadi, R. Umar, M. Shoaib, M. Bilal, and K. Jamil, "Effectiveness of Deghosting Process for Multi-Target Localization in Multistatic Passive Radar," in *2018 15th European Radar Conference (EuRAD)*, pp. 142–145, September 2018.

[18] M. Daun and C. R. Berger, "Track Initialization in a Multistatic DAB/DVB-T Network," in *2008 11th International Conference on Information Fusion*, pp. 1–8, June 2008.

[19] M. Daun, "Deghosting in Passive Air Surveillance Systems," in *11-th International Radar Symposium*, pp. 1–8, June 2010.

[20] M. Adjrad and K. Woodbridge, "Multistatic Target Tracking in DAB/DVB-T Networks," in *Proceedings of 2011 IEEE CIE International Conference on Radar*, vol. 1, pp. 42–45, October 2011.

[21] J. Yi, X. Wan, Z. Zhao, and H. Ke, "Evaluation of Velocity Threshold Constraint to Deghosting in Single Frequency Network Based Passive Radar," in *2013 14th International Radar Symposium (IRS)*, vol. 2, pp. 1043–1048, June 2013.

[22] Y. D. Zhang and B. Himed, "Moving Target Parameter Estimation and SFN Ghost Rejection in Multistatic Passive Radar," in *2013 IEEE Radar Conference (RadarCon13)*, pp. 1–5, April 2013.

[23] A. El Gonnouni and F. Lehmann, "Deghosting Method for Multiple Target Tracking in a Single Frequency Network," in *2014 International Radar Conference*, pp. 1–6, October 2014.

[24] J. Yi, X. Wan, F. Cheng, Z. Zhao, and H. Ke, "Deghosting for Target Tracking in Single Frequency Network Based Passive Radar," *IEEE Trans. Aerospace and Electronic Systems*, vol. 51, pp. 2655–2668, October 2015.

[25] P. Gomez-del-Hoyo, J. L. Bàrcena-Humanes, N. del-Rey-Maestre, J. Rosado-Sanz, and M. P. Jarabo-Amores, "Study of the Ghost Target Phenomenon on a Real DVB-T Passive Radar Scenario," in *2017 Signal Processing Symposium (SPSympo)*, pp. 1–6, September 2017.

[26] L. Ortenzi, L. Timmoneri, and D. Vigilante, "Unscented Kalman Filter (UKF) Applied to FM Band Passive Radar," in *2009 International Radar Conference "Surveillance for a Safer World" (RADAR 2009)*, pp. 1–6, October 2009.

Chapter 9

Cartesian Tracking

9.1 INTRODUCTION

The estimation of the Cartesian parameters of a target can be facilitated by the tracking algorithm, similar as in the bistatic coordinates. However, the problem of target tracking in Cartesian coordinates is more complicated than in the bistatic domain. One of the reasons is that the relationship between the measurement space and the target state space is nonlinear. Therefore, the standard linear Kalman filter cannot be used. Other tracking filters that can be applied include the extended Kalman filter (EKF) [1], the particle filter [1, 2], or the probability hypothesis density filter [3]. Another reason for the increased difficulty in tracking with passive radar is that an initial estimate of the target parameters is needed to perform tracking. The initial estimate of the target position in the Cartesian coordinates has to be obtained from the bistatic coordinates in the localization process, which was the focus of the previous chapter.

In this chapter, an example of a complete tracking algorithm that aims at solving the aforementioned problems is presented [4]. The algorithm is based on two stages: bistatic tracking and Cartesian tracking [5, 6]. First, the bistatic measurements are used by the bistatic tracker, described in Chapter 7. This eliminates false alarms to a great extent, as random detections that do not form probable trajectories are rejected. Moreover, precise bistatic parameters of targets are estimated, as the history of measurements is used in the tracking algorithm. The bistatic tracks are used by the localization algorithm, presented in Chapter 8. The estimated Cartesian positions are used by the Cartesian tracker, presented in this chapter. The considered tracker is based on the EKF.

The chapter begins with an analysis of the EKF for Cartesian target tracking. Parallel and sequential updating schemes are analyzed for the EKF. Next, the two-stage tracker is proposed. The theoretical considerations are followed by the simulation results and a real data example.

9.2 EXTENDED KALMAN FILTER

Assume the following classical target motion model in the Cartesian coordinates:

$$\mathbf{x}_c(k+1) = \mathbf{F}_c\mathbf{x}_c(k) + \mathbf{u}_c(k), \tag{9.1}$$

where $\mathbf{x}_c(k)$ is the state vector, $\mathbf{F}_c$ is the transition matrix, and $\mathbf{u}_c(k)$ is the process noise (the subscript "c" stands for Cartesian).

The measurement in passive radar is modeled as:

$$\mathbf{z}_{bi}(k) = h_i(\mathbf{x}_c(k)) + \mathbf{w}_i(k), \tag{9.2}$$

where $\mathbf{z}_{bi}(k)$ is the vector of bistatic measurements corresponding to the ith transmitter, $h_i(\cdot)$ is the nonlinear function transforming the Cartesian parameters into the bistatic parameters for the ith transmitter, and $\mathbf{w}_i(k)$ is the vector of the zero-mean Gaussian-distributed bistatic errors with the covariance matrix:

$$\tilde{\mathbf{R}}_{ci} = E\left[\mathbf{w}_i(k)\mathbf{w}_i'(k)\right] = \begin{bmatrix} \sigma_R^2 & 0 \\ 0 & \sigma_V^2 \end{bmatrix}. \tag{9.3}$$

As the relationship between the state vector $\mathbf{x}_c(k)$ and the measurement vector $\mathbf{z}_{bi}(k)$ is nonlinear, the classical Kalman filter cannot be applied. A standard approach is to linearize the nonlinear relationship by the calculation of the partial derivatives, and use the classical Kalman filter equations. This approach is called the EKF [7, 8].

The prediction stage of the filtering is the same as in the classical Kalman filter due to the linear relationship in the motion model:

$$\hat{\mathbf{x}}_c(k+1|k) = \mathbf{F}_c\hat{\mathbf{x}}_c(k|k), \tag{9.4}$$

$$\mathbf{P}_c(k+1|k) = \mathbf{F}_c\mathbf{P}_c(k|k)\mathbf{F}_c' + \mathbf{Q}_c, \tag{9.5}$$

where $\mathbf{P}_c(k+1|k)$ is the a priori state covariance matrix, $\mathbf{P}_c(k|k)$ is the a posteriori state covariance matrix, and $\mathbf{Q}_c$ is the process noise covariance matrix.

The nonlinear function $h_i(\cdot)$ is linearized in the EKF by calculating the Jacobian matrix:

$$\mathbf{H}_{ci}(k) = \left. \frac{\partial h_i(\mathbf{x}_c)}{\partial \mathbf{x}_c} \right|_{\hat{\mathbf{x}}_c(k|k-1)}, \tag{9.6}$$

which corresponds to calculating the value of the partial derivatives for the predicted state vector $\hat{\mathbf{x}}_c(k|k-1)$. The $\mathbf{H}_{ci}(k)$ matrix is then used in the standard Kalman filtering equations. There are two approaches that can be applied to the update stage of filtering, parallel and sequential [7, 8], which will be analyzed next.

9.2.1 Parallel Updating

First, consider the parallel updating scheme. In this approach, the measurements corresponding to all transmitters are used in a single step to update the state estimate and its covariance.

The measurement vectors corresponding to all individual transmitters $\mathbf{z}_{bi}(k)$ are stacked to create a combined measurement vector $\mathbf{z}_b(k)$:

$$\mathbf{z}_b(k) = \begin{bmatrix} \mathbf{z}_{b1}(k) \\ \mathbf{z}_{b2}(k) \\ \mathbf{z}_{b3}(k) \\ \vdots \end{bmatrix}. \tag{9.7}$$

The measurement model matrix $\mathbf{H}_c(k)$ is created in a similar way, by stacking the matrices $\mathbf{H}_{ci}(k)$ corresponding to individual transmitters:

$$\mathbf{H}_c(k) = \begin{bmatrix} \mathbf{H}_{c1}(k) \\ \mathbf{H}_{c2}(k) \\ \mathbf{H}_{c3}(k) \\ \vdots \end{bmatrix}. \tag{9.8}$$

The covariance matrix of the measurement error $\mathbf{R}_c$ is created as a block diagonal matrix of the individual covariance matrices $\tilde{\mathbf{R}}_{ci}$:

$$\mathbf{R}_c = \operatorname{diag}\left(\tilde{\mathbf{R}}_{c1}, \tilde{\mathbf{R}}_{c2}, \tilde{\mathbf{R}}_{c3}, \ldots\right). \tag{9.9}$$

With the combined vector $\mathbf{z}_b(k)$ and matrices $\mathbf{H}_c(k)$ and $\mathbf{R}_c$, the classical Kalman filtering equations are applied. The covariance of the innovation is calculated as:

$$\mathbf{S}_c(k+1) = \mathbf{H}_c(k+1)\mathbf{P}_c(k+1|k)\mathbf{H}'_c(k+1) + \mathbf{R}_c. \tag{9.10}$$

The Kalman gain is calculated according to:

$$\mathbf{K}_c(k+1) = \mathbf{P}_c(k+1|k)\mathbf{H}'_c(k+1)\mathbf{S}_c^{-1}(k+1). \tag{9.11}$$

The innovation (i.e., the difference between the prediction and measurement) is calculated as:

$$\mathbf{v}_c(k+1) = \begin{bmatrix} \mathbf{z}_{b1}(k+1) - h_1(\hat{\mathbf{x}}_c(k+1|k) \\ \mathbf{z}_{b2}(k+1) - h_2(\hat{\mathbf{x}}_c(k+1|k) \\ \mathbf{z}_{b3}(k+1) - h_3(\hat{\mathbf{x}}_c(k+1|k) \\ \vdots \end{bmatrix}. \tag{9.12}$$

The a posteriori state estimate is calculated by updating it with the use of the Kalman gain and innovation:

$$\hat{\mathbf{x}}_c(k+1|k+1) = \hat{\mathbf{x}}_c(k+1|k) + \mathbf{K}_c(k+1)\mathbf{v}_c(k+1). \tag{9.13}$$

The a posteriori covariance matrix is calculated in a standard way:

$$\mathbf{P}_c(k+1|k+1) = \mathbf{P}_c(k+1|k) - \mathbf{K}_c(k+1)\mathbf{H}_c(k+1)\mathbf{P}_c(k+1|k). \tag{9.14}$$

The bistatic measurements are assigned to the Cartesian track based on an association gate. The association gate is defined as:

$$\mathbf{v}'_c(k)\mathbf{S}_c^{-1}(k)\mathbf{v}_c(k) \leq \gamma. \tag{9.15}$$

If numerous plots are present in the association gate, a decision has to be made as to which plots are used in the track update. Here, a simple nearest neighbor (NN) approach is used, in which the plot that is the closest to the predicted position (in the normalized error sense) is assigned to a track. Other approaches, such as probabilistic data association (PDA) or joint probabilistic data association (JPDA) can also be applied [8].

An important feature of the parallel updating strategy is that first a set of bistatic measurements has to be tested using the condition (9.15). If multiple bistatic measurements corresponding to different targets are available, it can lead to an unnecessary increase in computational burden, as numerous tests have to be carried out before the actual updating can take place. This drawback can be avoided in the sequential updating scheme, presented in the next section.

In the considered parallel updating approach, a fixed structure of the filter is assumed: data from all transmitters are expected. In the case of a lack of target detection corresponding to one of the transmitters, the measurement covariance matrix $\tilde{\mathbf{R}}_c$ corresponding to the appropriate transmitter is set to a large value. In this way, information that the measurement is not available is conveyed to the filter.

9.2.2 Sequential Updating

In the sequential updating scheme, the state estimate and its covariance are updated partially, with bistatic measurements corresponding to one transmitter at a time.

First, temporary variables corresponding to the state estimate and its covariance are created:

$$\hat{\mathbf{x}}_{c0}(k+1|k) = \hat{\mathbf{x}}_c(k+1|k), \tag{9.16}$$

$$\mathbf{P}_{c0}(k+1|k) = \mathbf{P}_c(k+1|k). \tag{9.17}$$

Next, the updating is performed separately, for each transmitter according to:

for $i = 1, \ldots, N_{\text{Tx}}$

$$\mathbf{S}_{ci}(k+1) = \mathbf{H}_{ci}(k+1)\mathbf{P}_{c(i-1)}(k+1|k)\mathbf{H}_{ci}{}'(k+1) + \tilde{\mathbf{R}}_{ci}, \tag{9.18}$$

$$\mathbf{K}_{ci}(k+1) = \mathbf{P}_{c(i-1)}(k+1|k)\mathbf{H}_{ci}{}'(k+1)\mathbf{S}_{ci}^{-1}(k+1), \tag{9.19}$$

$$\hat{\mathbf{x}}_{ci}(k+1|k) = \hat{\mathbf{x}}_{c(i-1)}(k+1|k) + \mathbf{K}_{ci}(k+1)\mathbf{v}_{ci}(k+1), \tag{9.20}$$

$$\mathbf{P}_{ci}(k+1|k) = \mathbf{P}_{c(i-1)}(k+1|k) - \mathbf{K}_c(k+1)\mathbf{H}_{ci}(k+1)\mathbf{P}_{c(i-1)}(k+1|k), \tag{9.21}$$

end

where N_{Tx} is the number of transmitters used. At the end, the updated state estimate and its covariance are assigned as follows:

$$\hat{\mathbf{x}}_c(k+1|k+1) = \hat{\mathbf{x}}_{cN_{\text{Tx}}}(k+1|k), \tag{9.22}$$

$$\mathbf{P}_c(k+1|k+1) = \mathbf{P}_{cN_{\text{Tx}}}(k+1|k). \tag{9.23}$$

The bistatic measurements are assigned to the Cartesian track based on the association gate; however, in contrast with the parallel updating scheme, the association gate is formed for the measurements corresponding to each transmitter separately:

$$\mathbf{v}_{ci}'(k)\mathbf{S}_{ci}^{-1}(k)\mathbf{v}_{ci}(k) \leq \gamma, \tag{9.24}$$

where $\mathbf{v}_{ci}(k)$ is the innovation for the ith transmitter. In this way, the computational burden connected with the validation of the bistatic measurements is decreased significantly, compared with the parallel updating approach. Moreover, in practice, it will often happen that not all bistatic measurements corresponding to all transmitters are available, due to a probability of detection less than 1. This is not a problem when the sequential approach is used, as the track can be successfully updated only with bistatic measurements corresponding to selected transmitters.

9.3 TWO-STAGE TRACKING ALGORITHM

In order to simplify the tracking in passive radar, a two-stage tracking algorithm can be applied consisting of the bistatic and Cartesian trackers [5, 6]. In this way, several problems associated with target tracking in passive radar are alleviated.

The algorithm, whose block diagram is shown in Figure 9.1, operates in the following way: the bistatic plots corresponding to the individual transmitter-receiver pairs are fed to the Cartesian tracker. If Cartesian tracks exist, they are mapped to the bistatic coordinates, where the plots from individual pairs are assigned to tracks based on the validation gates. Next, the Cartesian tracks are updated using assigned plots with the sequential updating. All unassigned plots are processed by the bistatic trackers, such as the one described in Chapter 7, working independently for each pair. The bistatic tracks can be initialized using the M/N logic. This means that a tentative track that is created for each unassigned plot has to be confirmed at least M times during an observation window of length N. By using appropriate values of M and N, the required suppression of false tracks can be

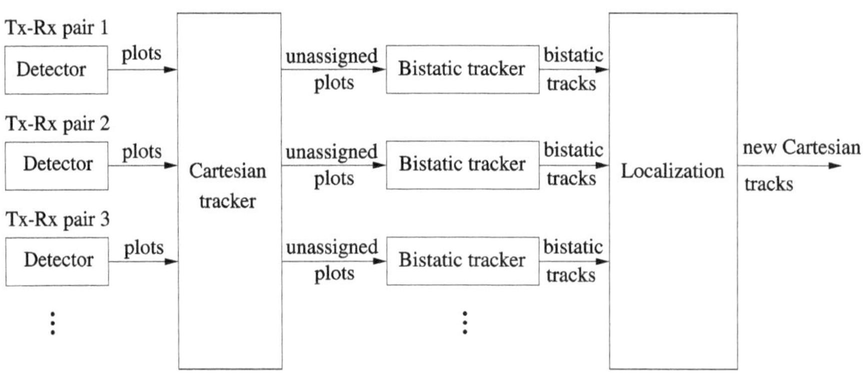

Figure 9.1 Block diagram of the two-stage tracker.

obtained. The confirmed bistatic tracks are used by the localization algorithm, presented in Chapter 8. All possible combinations of the bistatic measurements are tested for the intersection of the bistatic ellipsoids. The obtained intersections are potential candidates for the Cartesian tracks. It is worth pointing out that the transformation from the Cartesian coordinates (state vector space) to the bistatic coordinates (measurement space) is easy and unambiguous. For this reason, once the Cartesian track is formed, its updating with the bistatic measurements is easy. This is in contrast to the transformation from the bistatic coordinates to the Cartesian coordinates, which is problematic.

The order in which the consecutive stages of the processing from Figure 9.1 are used is slightly different than the figure suggests. First, all plots that are fed to the Cartesian tracker are unassigned, as there are no Cartesian tracks. Therefore, the first operation is forming the bistatic tracks from bistatic plots. Next, the bistatic tracks are used to form the Cartesian tracks in the localization algorithm. Finally, the Cartesian tracker maintains the Cartesian tracks, using the plots for updating.

The proposed tracking algorithm is advantageous for the following reasons: the bistatic trackers filter the bistatic measurements in terms of the elimination of false alarms. Due to the application of the track initialization procedure, only the measurements which form a probable trajectory in the consecutive observations are used to create a bistatic track. As a result, only the measurements that, with a high probability, originate from a real target are used in the localization algorithm. This leads to a decrease in the computational burden associated with the localization process, as well as to a mitigation of the ghost-target phenomenon. Moreover, the bistatic parameters of a target are estimated with a higher accuracy with the tracking algorithm than directly from a single measurement. As a consequence, an increased precision of the Cartesian localization can be expected, in comparison with the direct use of the bistatic measurements.

A disadvantage of the proposed two-stage algorithm is the delay associated with the initialization of the bistatic tracks. Before the bistatic tracks are established, the localization cannot take place. Therefore, the minimization of the initialization time of the bistatic tracks becomes important. This has been the topic of [9]. The research led to the development of an optimization procedure that minimizes the track confirmation time, taking the false track initialization rate as a constraint. As a result of the procedure, a set of processing parameters is obtained, including the integration time and the detection threshold, providing optimal performance for a selected value of the target echo signal-to-noise ratio.

9.4 NUMERICAL RESULTS

9.4.1 Simulations

It was assumed that a nearly constant velocity (NCV) motion model was used in x, y, and z coordinates. The state vector was:

$$\mathbf{x}_c(k) = \begin{bmatrix} x(k) \\ v_x(k) \\ y(k) \\ v_y(k) \\ z(k) \\ v_z(k) \end{bmatrix}, \tag{9.25}$$

where $[x(k), y(k), z(k)]'$ is the position vector of a target and $[v_x(k), v_y(k), v_z(k)]'$ is the velocity vector of a target at time instant k. The transition matrix had the following form:

$$\mathbf{F}_c = \begin{bmatrix} \tilde{\mathbf{F}}_c & \mathbf{0} & \mathbf{0} \\ \mathbf{0} & \tilde{\mathbf{F}}_c & \mathbf{0} \\ \mathbf{0} & \mathbf{0} & \tilde{\mathbf{F}}_c \end{bmatrix}, \tag{9.26}$$

where

$$\tilde{\mathbf{F}}_c = \begin{bmatrix} 1 & T \\ 0 & 1 \end{bmatrix}, \tag{9.27}$$

where T is the refresh interval, equal to the coherent processing interval (CPI).

The process noise was defined by the covariance matrix:

$$E\left[\mathbf{u}_c(k)\mathbf{u}_c'(j)\right] = \delta(k,j)\mathbf{Q}_c, \tag{9.28}$$

where

$$\mathbf{Q}_c = \begin{bmatrix} \tilde{\mathbf{Q}}_{cxy} & \mathbf{0} & \mathbf{0} \\ \mathbf{0} & \tilde{\mathbf{Q}}_{cxy} & \mathbf{0} \\ \mathbf{0} & \mathbf{0} & \tilde{\mathbf{Q}}_{cz} \end{bmatrix}, \tag{9.29}$$

$$\tilde{\mathbf{Q}}_{cxy} = q_{cxy} \begin{bmatrix} T^3/3 & T^2/2 \\ T^2/2 & T \end{bmatrix}, \tag{9.30}$$

$$\tilde{\mathbf{Q}}_{cz} = q_{cz} \begin{bmatrix} T^3/3 & T^2/2 \\ T^2/2 & T \end{bmatrix}. \tag{9.31}$$

The covariance matrix of the process noise in the xy dimensions $\tilde{\mathbf{Q}}_{cxy}$ was separated from the one in the z dimension $\tilde{\mathbf{Q}}_{cz}$ to reflect typical target maneuverability characteristics (higher in the x and y than in z dimension).

The nonlinear function $h_i(\cdot)$ converting the parameters of a target from the Cartesian to bistatic coordinates for the ith transmitter had the following form:

$$\begin{bmatrix} R_i \\ V_i \end{bmatrix} = h_i\left(\begin{bmatrix} x \\ v_x \\ y \\ v_y \\ z \\ v_z \end{bmatrix}\right), \tag{9.32}$$

where the bistatic range R_i is calculated as:

$$\begin{aligned} R_i = & \sqrt{(x - x_{ti})^2 + (y - y_{ti})^2 + (z - z_{ti})^2} + \\ & + \sqrt{x^2 + y^2 + z^2} - \sqrt{x_{ti}^2 + y_{ti}^2 + z_{ti}^2}, \end{aligned} \tag{9.33}$$

where $[x_{ti}, y_{ti}, z_{ti}]'$ is the position of the ith transmitter. In (9.33) it was assumed that the receiver position is $[0, 0, 0]'$. The bistatic velocity V_i is calculated according to:

$$V_i = \frac{(x - x_{ti})v_x + (y - y_{ti})v_y + (z - z_{ti})v_z}{\sqrt{(x - x_{ti})^2 + (y - y_{ti})^2 + (z - z_{ti})^2}} + \frac{xv_x + yv_y + zv_z}{\sqrt{x^2 + y^2 + z^2}}. \tag{9.34}$$

The two-stage tracking algorithm has been tested by simulating a scenario with three transmitters, one receiver, and three targets. The scenario is shown in Figure 9.2. The relative positions of the transmitters with respect to the receivers are listed in Table 9.1. Three targets were simulated, whose trajectories were generated according to the assumed motion model (9.1) with power spectral densities of process noise $q_{cxy} = 1$ m^2/s^3 and $q_{cz} = 0.1$ m^2/s^3. The Cartesian parameters, that is, target position and velocity components (9.25), were converted to the bistatic ones using (9.33) and (9.34). The bistatic measurements were distorted by additive Gaussian error with a standard deviation $\sigma_R = 300$m for the bistatic range and $\sigma_V = 0.5$ m/s for the bistatic velocity. The probability of detection was set to 0.7 for all transmitters. The refresh interval was equal to $T = 1s$.

The results of tracking in bistatic coordinates are shown in Figure 9.3(a–c) for the three transmitters, respectively. The results for one of the tracked targets are shown (the results for other targets are of similar nature). The plots show the

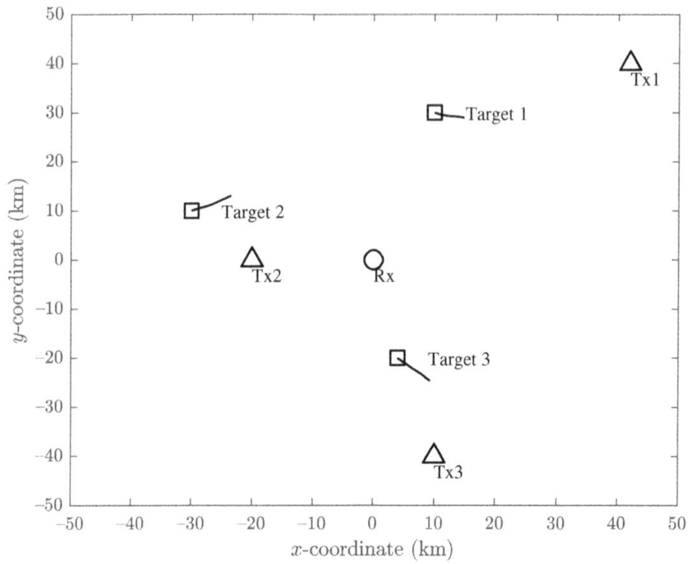

Figure 9.2 Simulation scenario and target trajectories used for testing the two-stage tracking algorithm.

Table 9.1
Relative Positions of the Transmitters Used for Testing of the Two-Stage Tracking Algorithm

Transmitter	*x-offset (km)*	*y-offset (km)*	*z-offset (km)*
Tx1	42.00	40.00	0.00
Tx2	−20.00	0.00	0.20
Tx3	10.00	−40.00	0.15

raw measurements and the tracked and true values of the bistatic range and bistatic velocity. Initially, the tracked bistatic range and velocity values are obtained from the bistatic trackers. After the Cartesian track is established, the tracked values are obtained by mapping the tracked Cartesian position to the bistatic coordinates. It can be seen that in the case of the bistatic range, the accuracy of the tracked value is substantially better than of the raw measurement; this effect was already presented in Chapter 7. In the case of the bistatic velocity, the improvement of the accuracy is not so significant.

In Figure 9.4 a comparison of the true and tracked position for the Cartesian tracking for one of the simulated targets in the x, y, and z directions is shown (sequential updating has been used in the experiment). The Cartesian tracking algorithm operates correctly, as it follows the true target position. The convergence

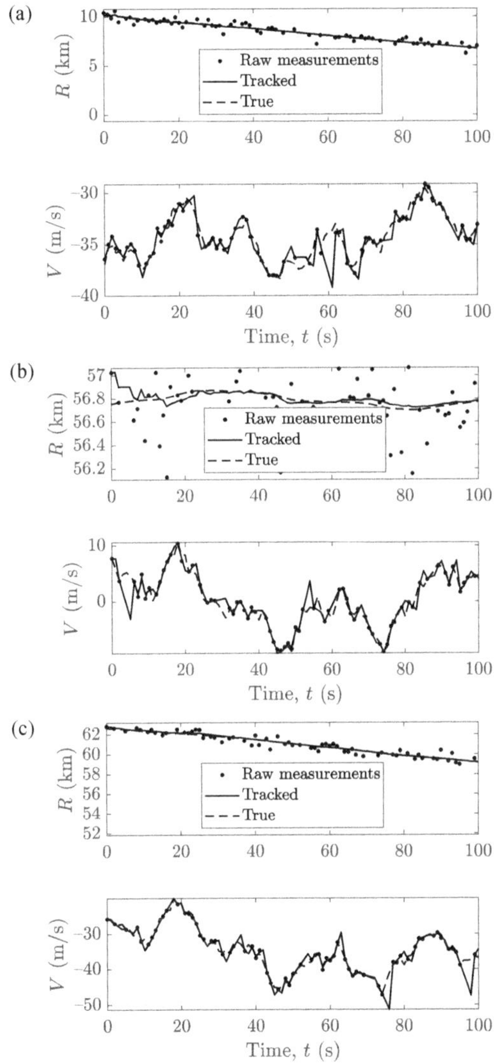

Figure 9.3 Comparison of the measured, tracked, and true bistatic parameters of a target, corresponding to: (a) Tx1, (b) Tx2, and (c) Tx3.

of the filter can be seen, as the initial error decreases with time. From the results presented, it can be concluded that the error in the z direction decreases slower than

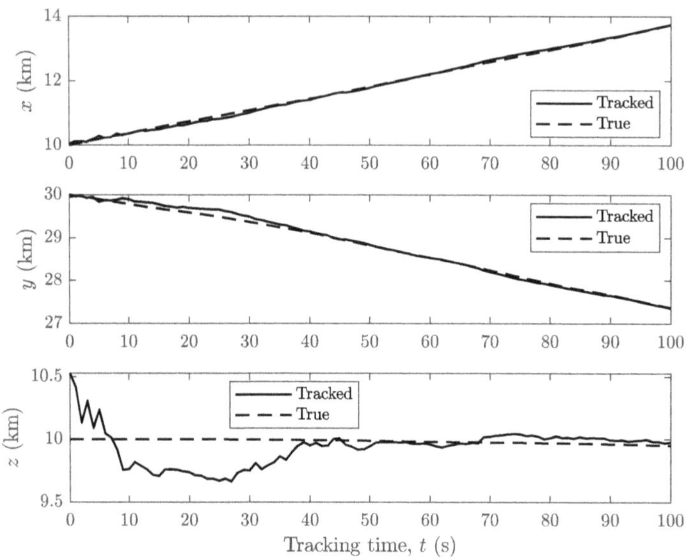

Figure 9.4 Comparison of the tracked and true position of a simulated target (sequential updating).

in the x and y directions. This is a typical feature of the passive radar due to the scene geometry.

An important feature of the Kalman filter is that in parallel with the state estimation, its covariance matrix is calculated. The covariance matrix provides information on the accuracy of the estimation, which can be very useful in practice. In theory, the filter-calculated covariance matrix should describe the actual accuracy in a statistical sense. In Figure 9.5 a comparison of the actual and filter-calculated position accuracy is shown for the sequential updating scheme (Figure 9.5(a)) and parallel (Figure 9.5(b)) updating scheme. The actual accuracy was calculated by performing a series of tests, and calculating the standard deviation of the error between the tracked and true position. The filter-calculated value was obtained by averaging the appropriate element of the state covariance matrix $\mathbf{P}_c(k|k)$ (e.g., element at position 1,1 in the case of the position error in the x direction). The accuracy calculated by the filter is very close to the actual accuracy in a statistical sense. The obtained results show that in the simulated scenario the accuracy in the order of 100m in all x, y, z directions was obtained after the steady state was reached. Initially, the error in the z direction is larger than in the case of the x and y directions; however, when the tracking filter converges, comparable values of accuracy are obtained in all directions. The results for the sequential updating (Figure 9.5(a)) and the parallel updating (Figure 9.5(b)) are very similar. As mentioned before, the sequential updating is much more practical, as the track can be updated independently by the measurements corresponding to individual transmitters. The sequential updating will therefore be used in the subsequent experiments.

A corresponding result of the filter-calculated and actual accuracy of the velocity estimation is shown in Figure 9.6 (for the sequential updating). As can be seen, the accuracy calculated by the filter is also very close to the actual accuracy

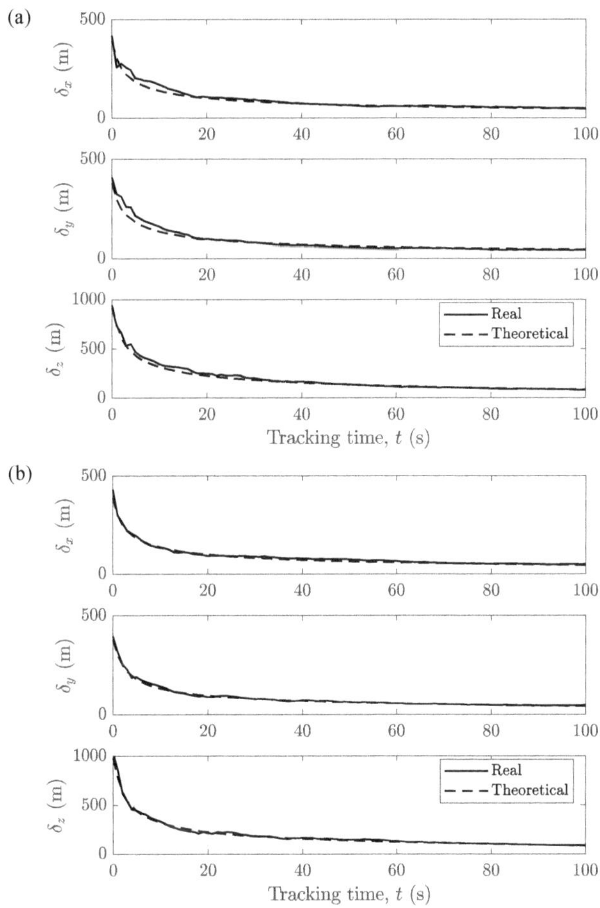

Figure 9.5 Comparison of the measured and theoretical errors of the position estimate of a target for the (a) sequential and (b) parallel updating schemes.

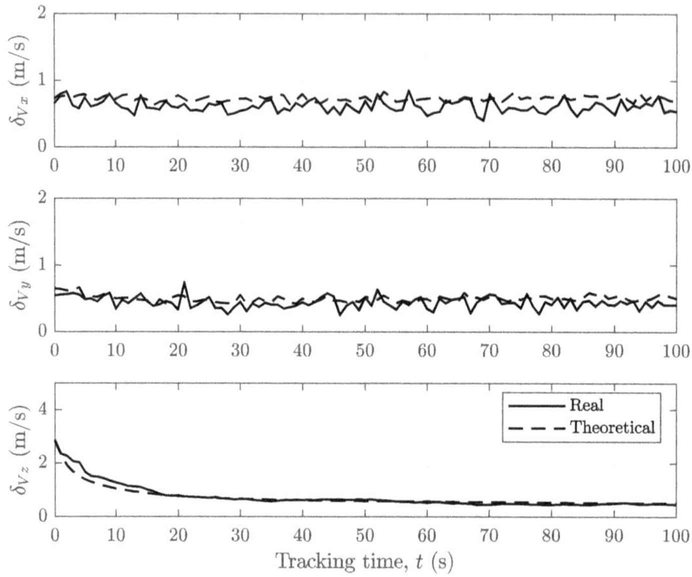

Figure 9.6 Comparison of the measured and theoretical errors of the velocity estimate of a target (sequential updating).

in this case. In the simulated scenario, the accuracy below 1 m/s was obtained in all directions. However, much slower convergence in the z direction can be observed.

9.4.2 Real-Data Results

Here an example of Cartesian target tracking will be investigated based on measured data. The target of interest was an amateur supersonic rocket, which was constructed by a rocket aficionado Damian Mayer [10–13]. The rocket is equipped with a global positioning system (GPS) receiver and inertial measurement unit, which allow the flight parameters to be read. However, as these kinds of tests are very experimental in nature, failure of the equipment installed onboard the rocket is very probable. For this reason, it is desirable to have an external system that could provide information on the target trajectory, especially maximum altitude and velocity. Passive radar seems to be a very good solution, as it does not require a permit to emit electromagnetic energy; the illumination is already provided by existing transmitters. Moreover, passive radar can provide a high refresh rate (several times per second), as well as 3D tracking capability thanks to the multistatic operation.

The target under investigation is shown in Figure 9.7. The name of the rocket is Strega, and it was constructed by Damian Mayer. The length of the rocket is 2.3m, and the diameter is 80–84 mm. It has three stabilizers. The rocket (except the nose) is almost entirely made from aluminum 6060 alloy. The engine of the rocket was powered by 5.2 kg of KNO3 and Sorbitol propellant. The mass of the rocket without the propellant is 8.5 kg. The rocket is equipped with a simple flight parameter recorder named Altimax. Altimax features an accelerometer and a pressure sensor which operates at the frequency of 200 Hz for 0.5s after launch and then switches to 40 Hz to save space on the microSD card used for data recording. The precision of the acceleration measurement is ± 0.04g. The Altimax G2 SD also has 3 pyro outputs and 2 outputs for servomechanisms. The whole flight recording unit is controlled by an Atmega1284 CPU.

The measurement scenario geometry is shown in Figure 9.8. Three DVB-T transmitters were used. The distance to the transmitters was 73 km, 44 km, and 90 km, respectively. The launch site is close to the passive radar receiver; 200m west of the radar.

In Figure 9.9 the recorded spectrum is shown, with four channels of the DVB-T transmission used in the experiment. The signal at the carrier frequency of 634 MHz originates from transmitter Tx1. Signals at 642 MHz and 650 MHz originate from transmitter Tx2. The signal at the frequency of 658 MHz is emitted

Figure 9.7 Launch of an experimental rocket that was tracked with the passive radar. Photograph by Andrzej Chwastek from the Polish Rocket Society. (©2018 IEEE Reprinted, with permission, from [12].)

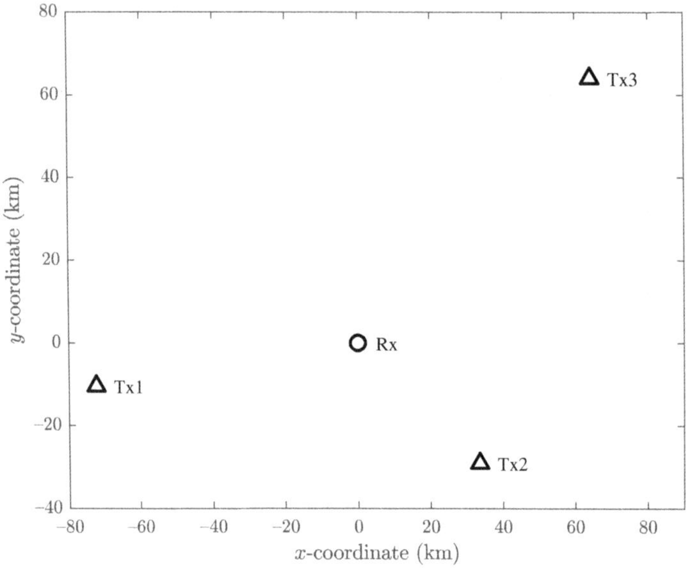

Figure 9.8 Geometry of the measurement scenario. Three transmitters used in the experiment were 73 km, 44 km, and 90 km from the receiver, respectively. (©2018 IEEE Reprinted, with permission, from [12].)

by transmitter Tx3. As the two adjacent DVB-T channels (642 MHz and 650 MHz) originate from the same transmitter, it was possible to treat them as one signal, with the bandwidth equal to approximately 16 MHz. This allowed the range resolution to be improved twofold in comparison to a single DVB-T channel.

An example of the cross-ambiguity functions corresponding to three transmitters is presented in Figure 9.10. The coherent processing interval (CPI) was equal to 25 ms. The rocket echo, marked with a circle, has different range and velocity values for each of the transmitters, as the transmitter-target-receiver geometry is different in each case. A finer noise structure can be observed for Tx2, as two DVB-T channels were used from this transmitter. This increases the range resolution, which is the effect observed on the cross-ambiguity function.

In Figure 9.11 the history of detections is shown for each of the transmitters. The rocket's trajectory in the bistatic coordinates can be clearly seen in each case. However, at a certain point, the trajectory ends abruptly. This was caused by the premature activation of the parachute installed in the rocket, whose aim was to ensure the safe landing of the part of the rocket with electronic equipment. When the parachute was activated, the rocket had a substantial velocity (approximately

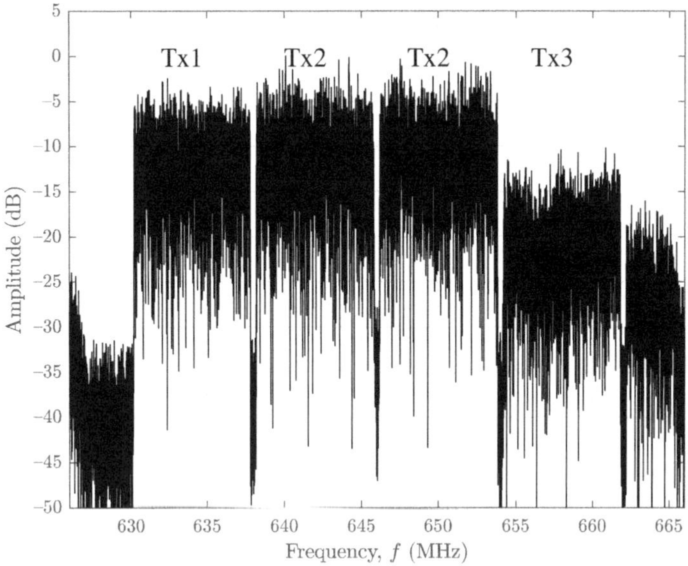

Figure 9.9 Spectrum of the recorded signal. Four DVB-T channels were used in the experiment. (©2018 IEEE Reprinted, with permission, from [12].)

1,500 km/h), which caused its disintegration. Some of the parts of the rocket decelerated very quickly and started a free fall. The radar cross-section of individual parts of the rocket was reduced in comparison to the whole rocket. For this reason, the reliable detection and tracking of the remains of the rocket could not be obtained.

In the tracking algorithm applied, nearly constant acceleration (NCA) model in x, y, and z coordinates was used, with the state vector:

$$\mathbf{x}_c(k) = \begin{bmatrix} x(k) \\ v_x(k) \\ a_x(k) \\ y(k) \\ v_y(k) \\ a_y(k) \\ z(k) \\ v_z(k) \\ a_z(k) \end{bmatrix}, \tag{9.35}$$

where $x(k), y(k), z(k)$ correspond to the target position, $v_x(k), v_y(k), v_z(k)$ correspond to the target velocity, and $a_x(k), a_y(k), a_z(k)$ correspond to the target accel-

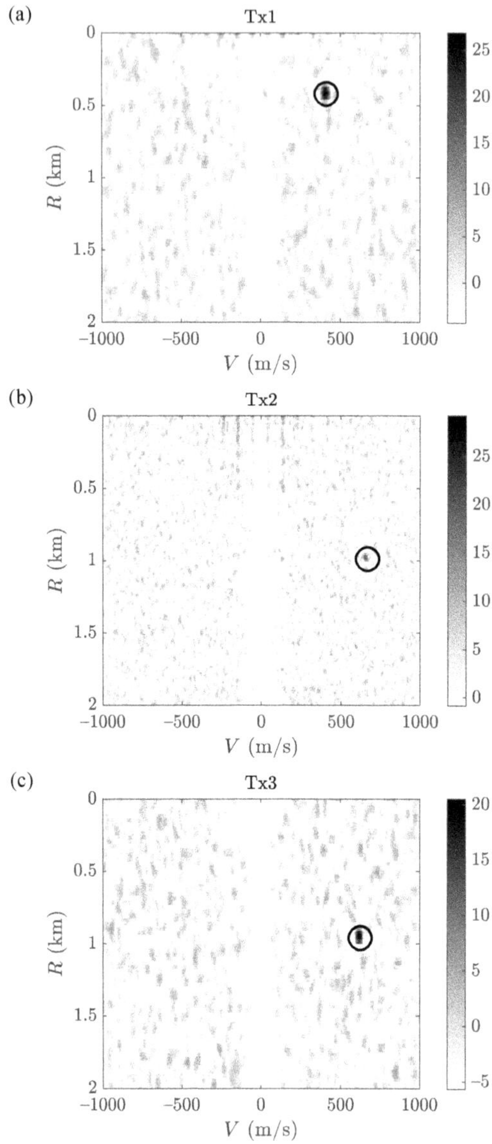

Figure 9.10 Cross-ambiguity functions corresponding to transmitters (a) Tx1, (b) Tx2, and (c) Tx3 with the rocket echo visible. (©2018 IEEE Reprinted, with permission, from [12].)

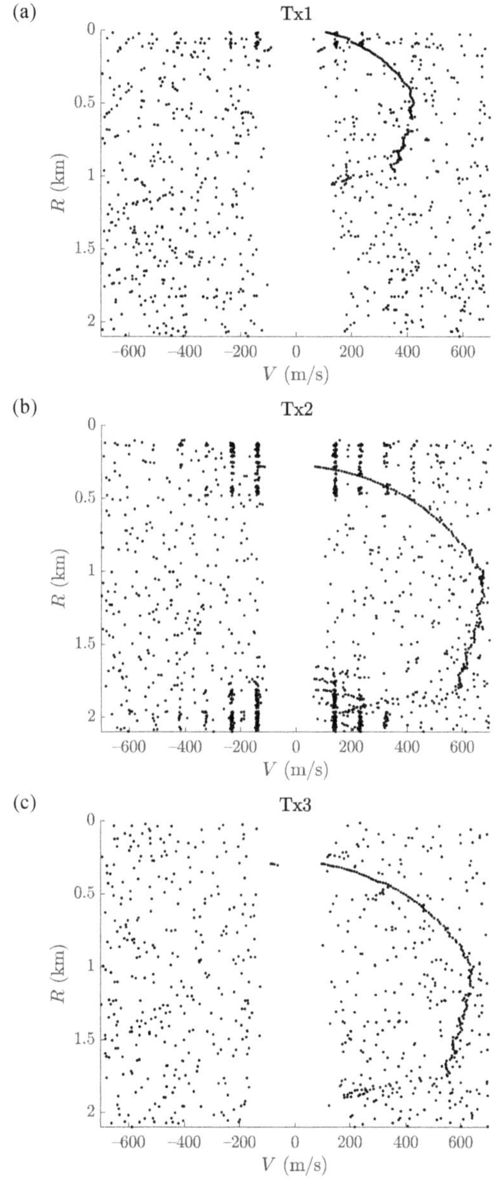

Figure 9.11 Cumulated history of detection corresponding to transmitters (a) Tx1, (b) Tx2, and (c) Tx3 with the rocket trajectory visible. (©2018 IEEE Reprinted, with permission, from [12].)

eration. The transition matrix had the following form:

$$\mathbf{F}_c = \begin{bmatrix} \tilde{\mathbf{F}}_c & \mathbf{0} & \mathbf{0} \\ \mathbf{0} & \tilde{\mathbf{F}}_c & \mathbf{0} \\ \mathbf{0} & \mathbf{0} & \tilde{\mathbf{F}}_c \end{bmatrix}, \tag{9.36}$$

where

$$\tilde{\mathbf{F}}_c = \begin{bmatrix} 1 & T & T^2/2 \\ 0 & 1 & T \\ 0 & 0 & 1 \end{bmatrix}, \tag{9.37}$$

where T is the refresh interval.

The process noise covariance matrix was defined as:

$$\mathbf{Q}_c = \begin{bmatrix} \tilde{\mathbf{Q}}_{cxy} & \mathbf{0} & \mathbf{0} \\ \mathbf{0} & \tilde{\mathbf{Q}}_{cxy} & \mathbf{0} \\ \mathbf{0} & \mathbf{0} & \tilde{\mathbf{Q}}_{cz} \end{bmatrix}, \tag{9.38}$$

where:

$$\tilde{\mathbf{Q}}_{cxy} = q_{cxy} \begin{bmatrix} T^5/3 & T^4/8 & T^3/6 \\ T^4/8 & T^3/3 & T^2/2 \\ T^3/6 & T^2/2 & T \end{bmatrix}, \tag{9.39}$$

$$\tilde{\mathbf{Q}}_{cz} = q_{cz} \begin{bmatrix} T^5/3 & T^4/8 & T^3/6 \\ T^4/8 & T^3/3 & T^2/2 \\ T^3/6 & T^2/2 & T \end{bmatrix}. \tag{9.40}$$

The values of q_{cxy} and q_{cz} were chosen to be sufficiently large in order to accommodate the high acceleration of the target. Depending on the target to be tracked, different motion models can be chosen in the Cartesian tracker. An approach using several motion models in parallel, such as interacting multiple models (IMM), can also be applied [8].

The results of the rocket tracking can be observed in Figure 9.12, 9.13, and 9.14. Figure 9.12 shows the target position in the x, y, and z coordinates. As can be seen, the initial target position is approximately $(-200, 0, 0)$m, which is consistent with the placement of the launch site of the rocket, 200m west of the radar. During the target flight, its position in the x and y coordinates does not change substantially, as the rocket was flying almost straight up. The most significant change can be observed in the z coordinate, as the rocket was climbing from 0 to $1,500$m. The

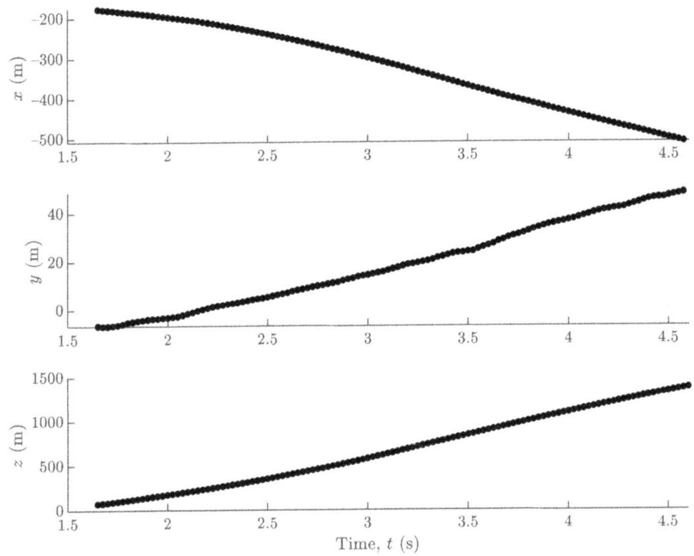

Figure 9.12 Rocket position in the x, y, and z coordinates estimated as a result of the tracking.

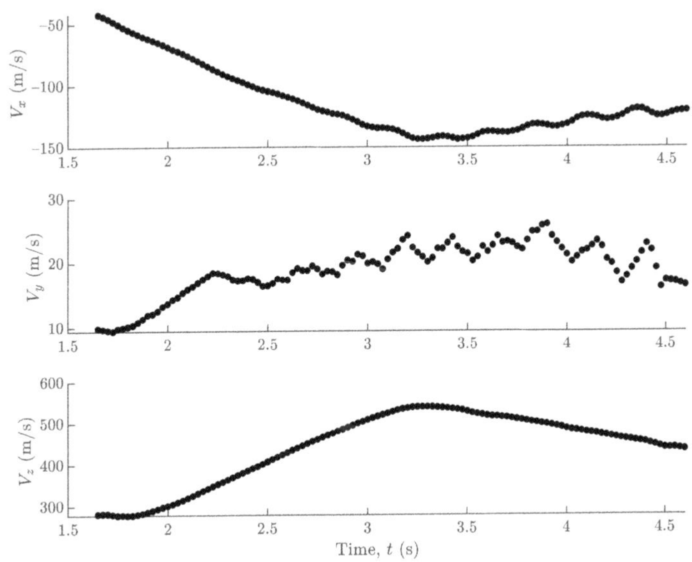

Figure 9.13 Rocket velocity in the x, y, and z coordinates estimated as a result of the tracking. (©2018 IEEE Reprinted, with permission, from [12].)

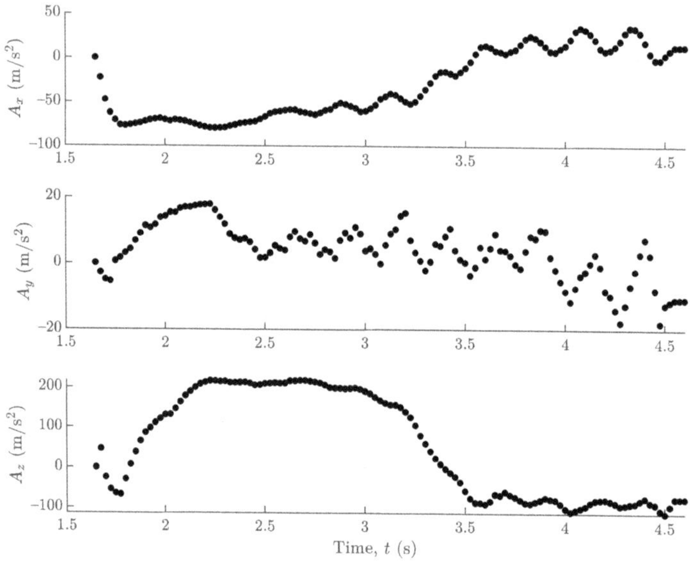

Figure 9.14 Rocket acceleration in the x, y, and z coordinates estimated as a result of the tracking. (©2018 IEEE Reprinted, with permission, from [12].)

target was tracked for 4.5s, at which point the parachute was activated, which resulted in track loss. The velocity components of the tracked rocket are shown in Figure 9.13. The tracking started when the rocket already had a substantial velocity. This is the result of a delay associated with the process of establishing the track. As can be seen, the highest values of the velocity were obtained for the z coordinate, reaching approximately 600 m/s (2,160 km/s). This clearly confirms that the rocket achieved supersonic speed. The components of the acceleration of the rocket can be observed in Figure 9.14. The acceleration in the z coordinate exceeds 200 m/s^2, which is a very high value, making a rocket a challenging target to be tracked.

9.5 SUMMARY

In this chapter, the problem of target tracking in Cartesian coordinates with passive radar was considered. The Cartesian tracker based on the extended Kalman filter was presented. When the track is updated with multiple simultaneous measurements, as in the case of passive radar, two approaches can be used: parallel updating and sequential updating. In the former one, all measurements are used to update the track in a single step. In the latter approach, the track is updated with a measurement corresponding to a single transmitter at a time. It has been shown that both methods yield similar results in the considered scenario. However, as the sequential approach is much more convenient to use in practice, it is more favorable.

A two-stage tracking algorithm with the Cartesian tracker used in conjunction with the bistatic tracker was shown. In this algorithm, the measurements are initially used to form bistatic tracks. When the tracks are established, they are used by the localization algorithm. As a result, target parameters in the Cartesian space are found, which are used to create a Cartesian track. From this point, the tracking is taken over by the Cartesian tracker; the Cartesian track is updated with the bistatic measurements. The aim of the intermediate bistatic tracking stage is the elimination of false alarms. In this way, target localization operates (almost) only on true measurements, which alleviates the problem of ghost targets. Moreover, the increased accuracy of tracked bistatic parameters (in comparison with the raw bistatic measurements) leads to more accurate Cartesian position estimates.

The results of tracking a real target show that passive radar can be successfully used to track even such challenging targets as rockets. The 3D tracking was obtained, with position, velocity, and acceleration components. A fast update rate allows the parameters of the target to be tracked precisely, even if they change rapidly.

References

[1] S. Herman and P. Moulin, "A Particle Filtering Approach to FM-Band Passive Radar Tracking and Automatic Target Recognition," in *Proc. IEEE Aerospace Conference 2002*, (Big Sky, Montana, USA), pp. 4–1789–4–1808, March 2002. vol. 4.

[2] A. Benavoli and A. D. Lallo, "Why Should We Use Particle Filtering in FM Band Passive Radars?," in *Proc. European Radar Conference – EURAD*, (Amsterdam, The Netherlands), pp. 344–347, October 2008.

[3] M. Tobias and A. D. Lanterman, "Probability Hypothesis Density-Based Multitarget Tracking with Bistatic Range and Doppler Observations," *IEE Proc. Radar, Sonar and Navigation*, vol. 152, pp. 195–205, June 2005.

[4] M. Malanowski, *Signal Processing Algorithms and Target Localisation Methods for Continuous-Wave Passive Bistatic Radar.* No. 179 in Prace Naukowe Politechniki Warszawskiej. Elektronika, Warszawa, ul. Polna 50: Oficyna Wydawnicza Politechniki Warszawskiej, 2012.

[5] M. Malanowski, K. Kulpa, and R. Suchożebrski, "Two-Stage Tracking Algorithm for Passive Radar," in *Proc. 12th International Conference on Information Fusion – FUSION*, (Seattle, Washington, USA), pp. 1800–1806, July 2009.

[6] M. Malanowski, "Target Tracking Using FM-Based Passive Radar," in *Proc. Signal Processing Symposium*, (Jachranka, Poland), p. CD, June 2011.

[7] Y. Bar-Shalom and X. R. Li, *Estimation and Tracking: Principles, Techniques, and Software.* Norwood, MA: Artech House, 1993.

[8] Y. Bar-Shalom and X. R. Li, *Multitarget-Multisensor Tracking: Principles and Techniques.* Storrs, CT: Yaakov Bar-Shalom, 1995.

[9] M. Malanowski and K. Kulpa, "Optimization of Confirmation Time of Bistatic Tracks in Passive Radar," *IEEE Trans. Aerospace and Electronic Systems*, vol. 47, pp. 1060–1072, April 2011.

[10] S. Rzewuski, K. Borowiec, K. Kulpa, M. Malanowski, M. K. Bączyk, K. Klincewicz, M. Wielgo, and A. Kurowska, "Supersonic Target Detection in Passive Radar," in *2015 16th International Radar Symposium (IRS)*, pp. 89–94, June 2015.

[11] K. Borowiec and M. Malanowski, "Accelerating Rocket Detection Using Passive Bistatic Radar," in *2016 17th International Radar Symposium (IRS)*, pp. 1–5, May 2016.

[12] M. Malanowski, K. Borowiec, S. Rzewuski, and K. Kulpa, "Detection of Supersonic Rockets Using Passive Bistatic Radar," *IEEE Aerospace and Electronic Systems Magazine*, vol. 33, pp. 24–33, January 2018.

[13] M. Malanowski, K. Borowiec, and S. Rzewuski, "Rocket Detection Using Passive Radar – Challenges and Solutions," in *2018 International Conference on Radar (RADAR)*, pp. 1–5, August 2018.

Chapter 10

Examples of Passive Radar Systems

10.1 INTRODUCTION

One of the most famous passive radars of the modern era was the Silent Sentry developed by the Lockheed Martin Corporation in the 1990s [1]. It operated using FM radio as its signal source. The system had several versions. One of them had a linear array mounted on the side of a building, which served as the surveillance channel. A separate directional antenna was used to receive the reference signal. In another version of the system, a transportable linear array was mounted on the side of a trailer. A third version of the system used uniform circular array to provide 360° of coverage. The development of the system was stopped some time ago, and no further generations were released. Therefore, the system can now be treated as historical.

In this chapter, several examples of modern passive radar systems and their applications will be presented. The number of passive radar systems developed in recent years is substantial. The selection of the systems is by definition not complete; therefore, no conclusion should be drawn regarding the importance or maturity of the system from the fact that it is or is not presented here. The institutions responsible for developing passive radar systems are introduced in alphabetical order.

10.2 ERA, CZECH REPUBLIC

Era is a Czech company which has decades of experience in constructing civilian and military multilateration (MLAT) systems, that is, systems that use TDOA (time-difference-of-arrival) measurements for target localization. Some of the well-known products of the company include RAMONA, TAMARA, and VERA (fifth generation), which are military multilateration systems. The company has also developed a proof-of-concept passive radar demonstrator called the Silent Guard [2–4] and new a generation of passive radar called SICORRA. The Silent Guard system, shown in Figure 10.1, is based on FM radio transmitters. It is equipped with a 6-element circular antenna array, with horizontally polarized array elements. The antennas are placed on a mast, the height of which can be extended up to 8m. The receivers and signal processor are placed in a van. The reception of the signal is based on direct or RF (radio frequency) sampling, which means that the analog signal is filtered, amplified, and then sampled directly, without analog downconversion. Such a solution, typically applied in FM-based passive radar, provides minimum distortion and maximum interchannel coherency. The basic system configuration (extendable) is capable of processing signals from up to 8 transmitters; therefore, 8 bistatic pairs can be created. This is more than enough for 3D target tracking. As Era has significant experience with multilateration systems, the combination of TDOA and passive radar techniques is also pursued by the company [3]. The direction of arrival, which can be very helpful in target localization, is also implemented in the system [2, 4].

Signal processing in the passive radar is realized using efficient GP-GPU (general-purpose graphics processing unit) cards [2]. The processing starts with ground clutter removal and cross-ambiguity function calculation. Next, a sequential target elimination algorithm is applied, which is similar to the CLEAN algorithm described in Chapter 5. As a result of the processing of a pair of reference and

Figure 10.1 A picture of Era's Silent Guard passive radar showing the mast with a 6-element antenna array and van with the receivers and processing units. (Courtesy of Era.)

echo signals, bistatic plots are obtained, along with the bistatic range and Doppler measurements, as well as other parameters such as power and phase. The bistatic plots are then used in range-Doppler trackers (trackers operating in the bistatic coordinates). Next, the bistatic tracks from different pairs are combined in order to create Cartesian tracks. Notice that this is the same two-stage tracking approach as the one presented in Chapter 9.

An example of the result obtained with the Silent Guard system is shown in Figure 10.2. The image shows a history of detections from the passive radar that were confirmed with a multilateration system (with an approximate detection

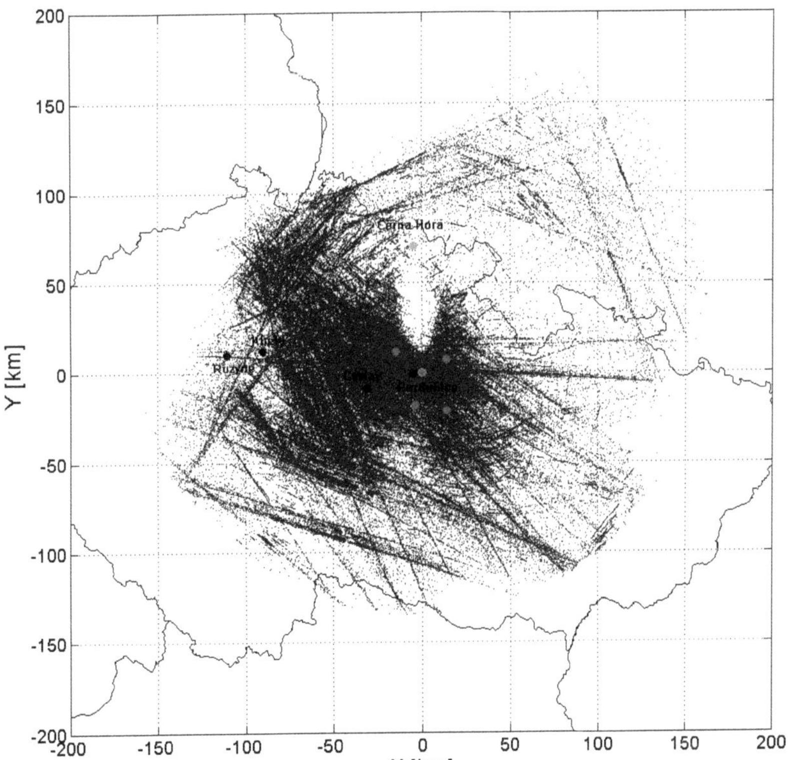

Figure 10.2 An example of FM-based passive radar detections confirmed with the multilateration system. A detection range of 150–180 km can be observed. (©2011 IEEE Reprinted, with permission, from [2].)

range of 400 km). The confirmation was done by transforming the geographic target coordinates to the bistatic coordinates and verifying if a passive radar detection is within a certain gate. The observed targets were mainly noncooperative airliners. As can be seen, a detection range of 150–180 km is obtained with an FM-based passive radar. The detection range is not the same in every direction, which results from terrain masking and the mutual relationship of the receiver and the transmitters.

Experiences with the Silent Guard as a proof-of-concept system allowed the company to develop a multistatic system called SICORRA (Silent Correlation Radar). SICORRA, as a further evolution of passive radar in ERA, consists of several nodes networked into one system. There are basically two types of nodes – a receiving station and an evaluation station. As well as its predecessor, SICORRA is based on FM transmitters. While the signal processing remains the same from the hardware point of view (i.e., the concept of direct sampling is applied), there are new software algorithms implemented in GP-GPUs. Stretch correlation processing for range and Doppler walk compensation is one of the most important improvements that allows the coherent processing time to be increased and therefore improves the system sensitivity. A trailer-mounted receiving and evaluation station of the system is shown in Figure 10.3. The rack of the antenna system contains two four-elements circular arrays placed one above the other at the height of 15m. The station operates unattended and allows full remote control, if required.

Figure 10.3 Trailer mounted receiving and evaluation station of SICORRA system. (Courtesy of Era.)

An interesting project that Era was involved in was the active MSPSR (Multistatic Primary Surveillance Radar) [5–8, 8]. The active MSPSR is not exactly a passive radar that is the topic of this book, but it is closely related to it. The main difference between the MSPSR under consideration and a passive radar is that a passive radar relies on noncooperative transmitters, whereas the MSPSR uses dedicated transmitters. The similarities between the two systems include bistatic geometry and problems with clutter rejection. Era developed an MSPSR demonstrator operating in the L-band. An example of the results obtained with the system is presented in Figure 10.4 [8]. In the plots, an example of a 3D target trajectory tracked with a high-precision multilateration system is shown. In the experiment, only one bistatic pair was used in the MSPSR, therefore no 3D localization could be obtained. For this reason, the 3D measurements from the multilateration system were transformed to the bistatic coordinates and compared with the detections obtained with the MSPSR system. The markers on the trajectory show the capabilities of MSPSR system detection.

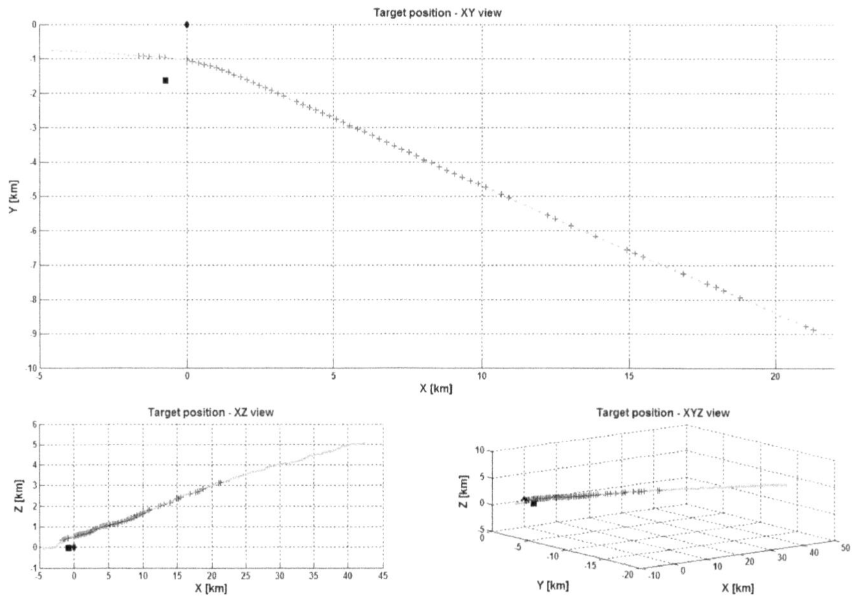

Figure 10.4 An example of 3D target trajectory tracked with the MLAT system with marked detections from the L-band MSPSR system. (©2016 IEEE Reprinted, with permission, from [8].)

10.3 FRAUNHOFER FHR, GERMANY

Fraunhofer is a German research organization with numerous divisions that deal with different branches of science and technology. One of the branches is the Fraunhofer Institute for High Frequency Physics and Radar Techniques (FHR), located in Wachtberg, near Bonn. The institute has substantial experience with passive radar technology, especially with DVB-T-based systems.

One interesting project that the Fraunhofer FHR realized is called PARASOL [9–11], *Passiv Radar basierte Schaltung der Objektbefeuerung für die Luftfahrt* in German, which translates to: passive radar-based switching of object illumination for air traffic. The aim of the project was to design a low-cost passive radar system mounted on wind turbine masts that would detect the presence of an aircraft in the vicinity. When the target was detected, signaling lights on the mast would be

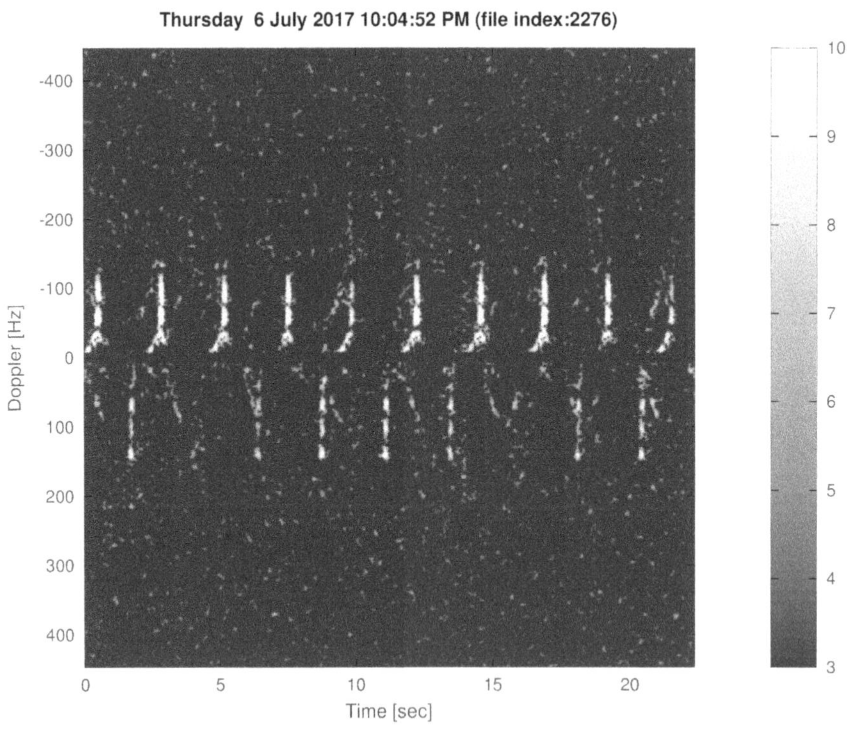

Figure 10.5 Spectrogram of the received signal showing the influence of the rotor blades. (Courtesy of Fraunhofer FHR.)

activated to warn the aircraft. In the case of no target near the mast, the lights are deactivated as they disturb people living in the area and attract birds, which can be hit by the blades of the turbine. Another related application of passive radar is monitoring of the wind turbines status and operation [12, 13].

The designed passive radar is based on DVB-T transmission. Three nodes of the passive radar enable the 3D target position to be estimated. One of the challenges was the elimination of the strong echoes (flashes) from the rotating blades of the wind turbine, which can seriously degrade the air target detection performance. An example of a spectrogram showing the influence of the rotating blades is shown in Figure 10.5. The characteristics of the signal originating from the blade depend on the type of the turbine, wind speed, and many other factors. Typically, the blade flash period is in the order of 1s. An algorithm has been developed whose aim is to estimate the instants of the flashes, and use clean time intervals between the flashes for passive radar processing. In the first stage of the algorithm, the frequency of blade flash is estimated based on the cepstral analysis. Next, the phase offset of the flash is calculated using short-time Fourier transformation (STFT).

The radar has to measure the elevation angle in order to be able to distinguish ground moving targets, such as cars, from aircraft. This was achieved by using two antennas displaced vertically, which enabled interferometric elevation measurement. In Figure 10.6 the antenna system of the radar mounted on a pole is shown. Two antennas used for the elevation measurement are visible. One of the problems of a DVB-T-based radar, as mentioned in Chapter 3, is the relatively wide bandwidth of the received signals, roughly from 450 to 900 MHz. In order to provide good

Figure 10.6 Two-element discone antenna array of the passive radar mounted on the pole of the wind turbine. (Courtesy of Fraunhofer FHR.)

reception of the signal in the whole band, the discone antennas were used. These antennas are built with a cone covered with a disk, hence their name.

The result of target tracking on the range-Doppler plane is shown in Figure 10.7. An air target at the distance of more than 10 km of bistatic range is visible.

Another research topic investigated at Fraunhofer FHR is a passive radar on a mobile platform. Two areas can be distinguished within this topic: the detection of ground moving targets (ground moving target indication (GMTI)) [14–17] and radar imaging using the synthetic aperture radar (SAR) technique [18, 19]. Here an example of the detection of ground moving targets is presented [15]. In Figure 10.8 a picture of the van used in a measurement campaign is presented. A single directional antenna is used for receiving the reference signal. A 7-element antenna array is used for receiving the echo signals. The elements of the array are the discone antennas. Five channels from the middle of the array are used for receiving the signal. One of the outer elements was used for array calibration, and the other one was terminated.

During the experiment, the van was moving at a velocity of 11 m/s. This caused the clutter to be spread in the Doppler dimension. An example of the range-Doppler map from the experiment is shown in Figure 10.9. As can be seen, the clutter occupies the spectrum of $-30, \ldots, +30$ Hz.

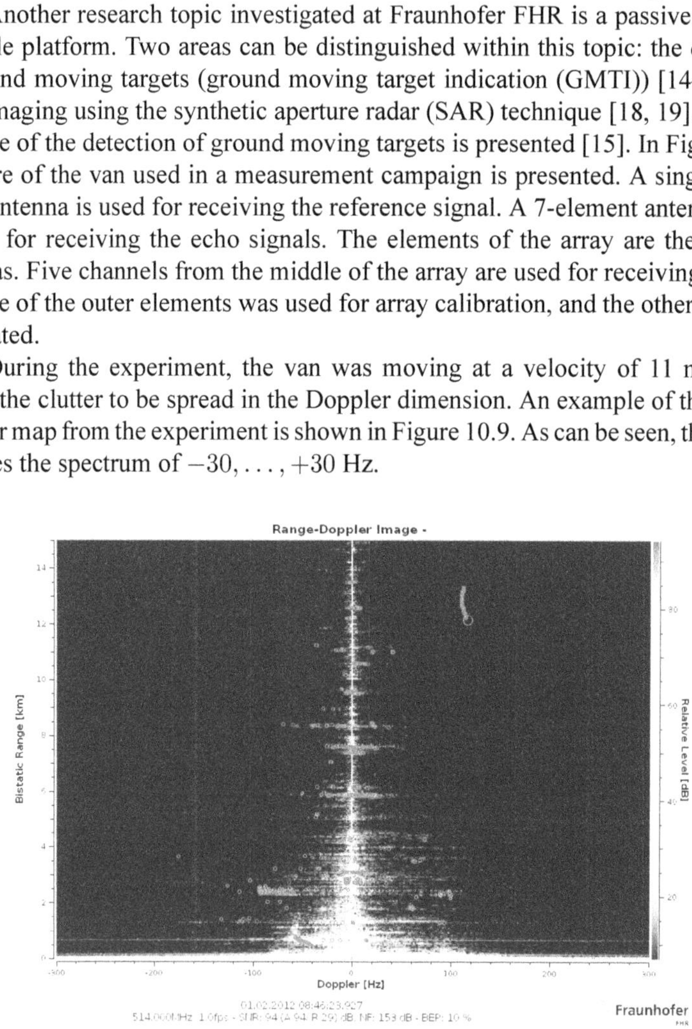

Figure 10.7 Range-Doppler plane showing an aerial tracked target at (12 km,120 Hz) with interference caused by rotor blades. (Courtesy of Fraunhofer FHR.)

Figure 10.8 The van used in the measurement campaign. A single directional antenna is used for receiving the reference signal. The 7-element antenna array is used for receiving the echo signal (5 elements are actually used for the signal reception). (©2018 IEEE Reprinted, with permission, from [15].)

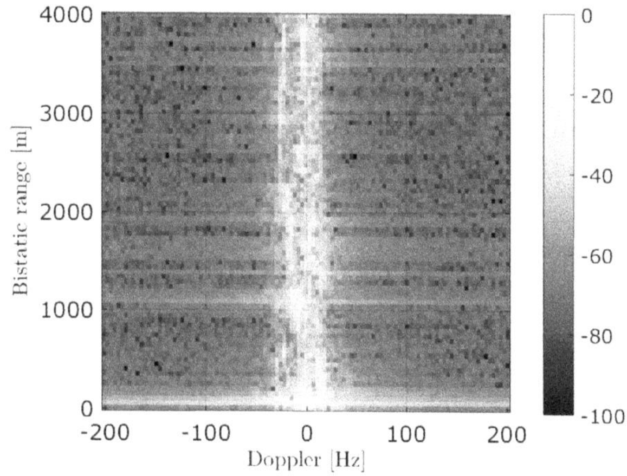

Figure 10.9 Range-Doppler plane before the application of STAP clutter cancellation. Doppler-spread clutter due to the van motion is visible. (©2018 IEEE Reprinted, with permission, from [15].)

The classical approach to Doppler-spread clutter filtering, which originates from airborne monostatic radar, is space-time adaptive processing (STAP). In this approach, an antenna array is used, which allows selective filtering to be performed in the space domain. One of the main problems of STAP processing is the estimation of the interference covariance matrix. In [15] the adjacent-bin post-Doppler (ABPD) approach has been applied. The result of the application of the STAP approach to

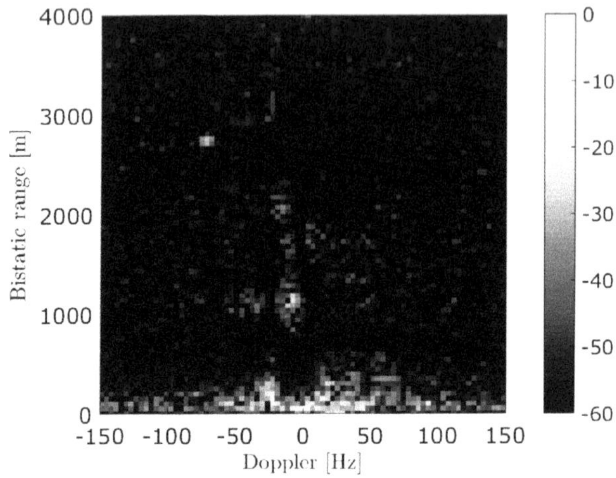

Figure 10.10 Range-Doppler plane after the application of STAP clutter cancellation. Clutter is reduced substantially. A moving target is now visible at (2, 700m, −71 Hz). (©2018 IEEE Reprinted, with permission, from [15].)

the data is shown in Figure 10.10. As can be seen, the clutter has been substantially attenuated. This allowed for a moving target to be easily visible (with the bistatic range of approximately 2.7 km and Doppler shift of −71 Hz).

The conducted experiments show that passive radar can be successfully used on a mobile platform. The ultimate aim in this area of research seems to be installing an operational passive radar onboard an aircraft, which would enable GMTI and SAR capabilities without the need to emit energy. Both capabilities were already successfully tested (see [14–17] for GMTI and [20–27] for SAR), however, this is still an active research area where significant improvement is to be achieved.

10.4 HENSOLDT, GERMANY

Hensoldt, formerly known as Airbus, Cassidian, and EADS, is a multinational company with its headquarters in Germany, specializing in radar, optoelectronics, and avionics. The company has been developing passive radars for several years [28–37]. As a result, a multiband passive system was created.

The multiband antenna system of the Hensoldt passive radar is shown on the left side of Figure 10.11. The top-level antenna operates in the UHF band (470–840 MHz) and it is used for receiving the DVB-T signal. The antenna in the middle

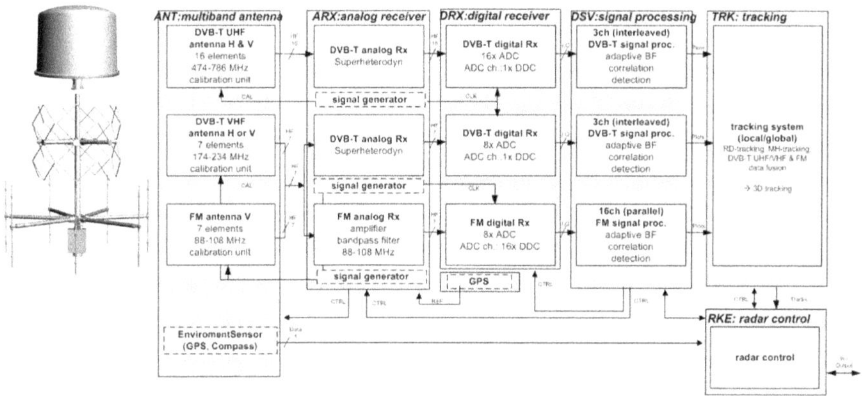

Figure 10.11 Block diagram of the system architecture of the Hensoldt passive radar. (Courtesy of Hensoldt.)

works in the VHF band (178–240 MHz) and can be used for the reception of the DAB and DVB-T signals. The antenna array at the bottom is used for FM radio (88–108 MHz). The UHF and VHF antennas are dual-polarized, and the FM antennas are vertically polarized.

A block diagram of the system architecture is presented on the right side of Figure 10.11. The first stage is the 3-band antenna system (ANT). The signals are then fed to the analog receiver section (ARX). In the case of the FM signals, only filters and amplifiers are used (without analog downconversion). In the case of the VHF and UHF bands, superheterodyne receivers are used. The analog signals are converted to the digital domain in the digital receiver section (DRX). In the case of the UHF signals, a 16-channel ADC is used, as there are 16 elements in the antenna array. For VHF and FM signals, 8-channel ADC is used (the FM and VHF antennas have 7 elements). The signal processing unit (DSV) deals with adaptive beamforming, correlation, and detection. For FM radio, 16 channels can be processed simultaneously. In the case of the VHF and UHF signals, 3 channels for each band can be processed in an interleaved manner. The processing is realized using Intel 64Bit XEON Server Hardware equipped with Graphics Processing Unit (GPU) cards. The bistatic plots are then fed to the tracker (TRK). The tracker performs bistatic range-velocity tracking, target localization and Cartesian tracking. A multihypothesis tracker is used. The developed passive radar was used to track various types of targets. The performance of the advanced tracking algorithm was verified in challenging scenarios, such as the observation of highly maneuvering military targets. An example of the tracking results is shown in Figure 10.12. In

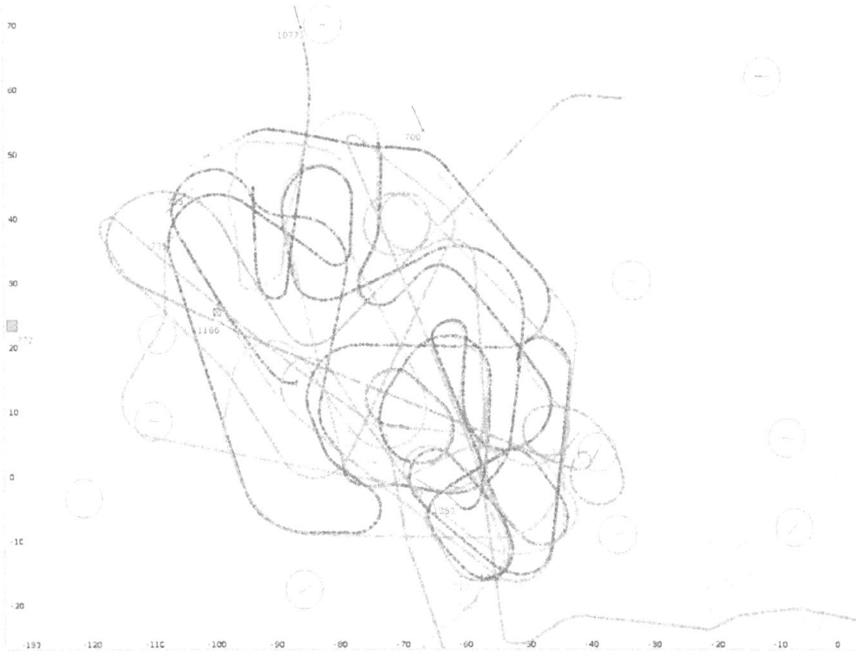

Figure 10.12 Example of trajectories of tracked targets. Several Eurofighter 2000 Typhoon fighters in a simulated dogfight. (Courtesy of Hensoldt.)

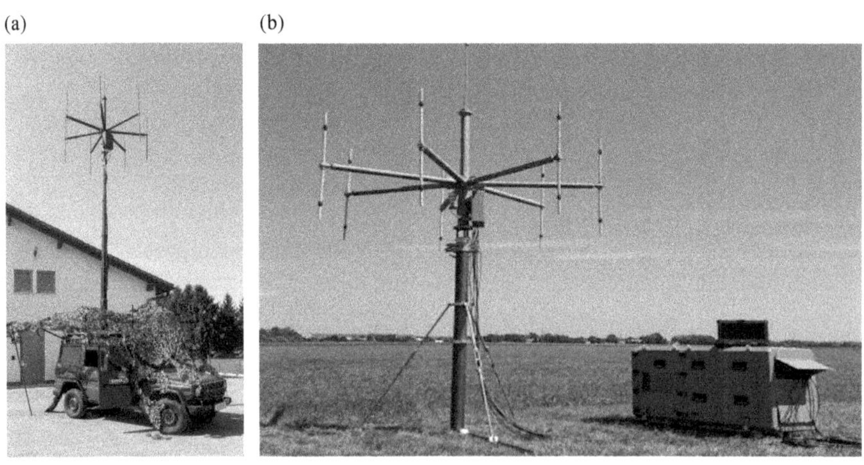

Figure 10.13 Hensoldt's PCL system: (a) military mobile version and (b) portable version. (Courtesy of Hensoldt.)

this case, several Eurofighter 2000 Typhoon fighters were performing complicated maneuvers, including high-g turns and trajectory crossings. The passive radar was able to reliably track the targets even in such a challenging scenario.

In Figure 10.13 two versions of the Hensoldt system are presented. These systems use FM and DAB transmitters. The antenna system is composed of 7 vertically polarized dual-band dipoles. This solution shows one of the main advantages of the passive radar; a small and portable system can be built, providing large capabilities at a relatively low cost, when compared with active radar solutions.

10.5 PIT-RADWAR, POLAND

PIT-Radwar is a Polish radar manufacturer, focusing on ground-based radars for military applications. In 2012, in response to Polish Ministry of Defense (MOD) requirements a consortium was formed, consisting of PIT-Radwar as the leader, the Warsaw University of Technology (WUT), and AM Technologies, a private Polish company. The consortium is involved in a project whose aim is to develop a passive location system combining two technologies: PCL (passive coherent location) and PET (passive emitter tracking, or passive ESM [electronic support measures] tracking) [38]. In the case of PCL, three types of transmitters are used:

- FM radio;
- DVB-T television;
- GSM telephony.

The PET part of the system consists of:

- ELINT (Electronic Intelligence): radar signals transmitted by aircraft;
- COMINT (Communications Intelligence): communication signals transmitted by aircraft;
- IFF (Identification Friend-or-Foe): signals emitted by the aircraft transponders;
- NAVIG: navigation signals transmitted by aircraft.

Within the project, four passive radar stations were created. One of the stations (a prototype vehicle) is realized at TRL 9 (Technology Readiness Level), and 3 stations (the laboratory stations) are realized at TRL 6. The system is by definition multistatic and multiband. In the case of the PCL system, each station can operate with several illuminators of opportunity, producing bistatic measurements. In the case of the PET system, each pair of the receiving stations produces a TDOA measurement. In addition, direction of arrival, here referred to as direction finding (DF), measurements can be obtained for the ELINT system. The measurements from all four stations from the PCL and PET subsystems are processed in a fusion subsystem [39–43, 43, 44, 44].

A block diagram of the fusion and tracking system is shown in Figure 10.14 [43]. The PCL subsystem produces bistatic plots and tracks. The PET subsystem, in turn, generates TDOA and DF measurements. The data from the two subsystems are processed in the MEASUREMENT DATA ASSOCIATION module, which assigns the measurements to the currently tracked targets. The measurements which are assigned to the tracks are used for the updating of the tracks in the TRACKER ALGORITHM block. The unassigned measurements can potentially create new Cartesian tracks, which is verified in the INTERSECTION CALCULATION module, where triplets of measurements are intersected. Once the intersections are calculated, they are processed in the CLUSTERING & OBJECT EXTRACTION block. The aim of this module is to combine individual intersections into clusters that represent potential targets. Next, the Cartesian parameters of the target (position and velocity) are calculated in the ESTIMATE GENERATION module. The estimate is next passed to the TRACK MANAGEMENT module, which initiates new Cartesian tracks.

An example of clustering of intersections performed in the CLUSTERING & OBJECT EXTRACTION block is shown in Figure 10.15. The image on the left side represents the input data: intersection points of the measurements from PCL and PET subsystems. The bistatic range measurements represent ellipsoids in the Cartesian space. The TDOA measurements correspond to hyperboloids in the Cartesian space. The DF measurement defines a ray showing the direction of arrival

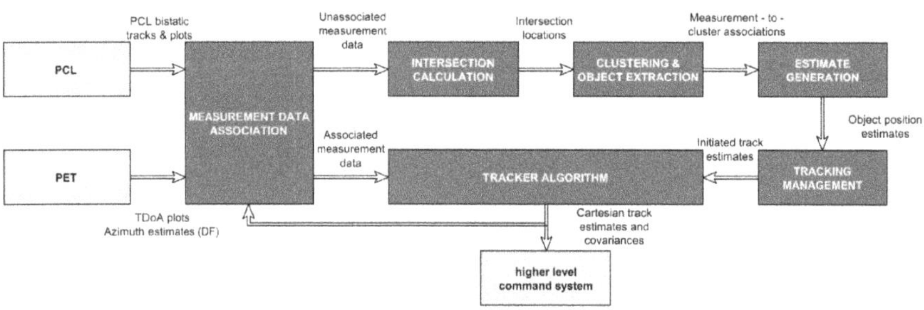

Figure 10.14 Block diagram of the fusion and tracking algorithm used in the PCL-PET system. (©2018 IEEE Reprinted, with permission, from [43].)

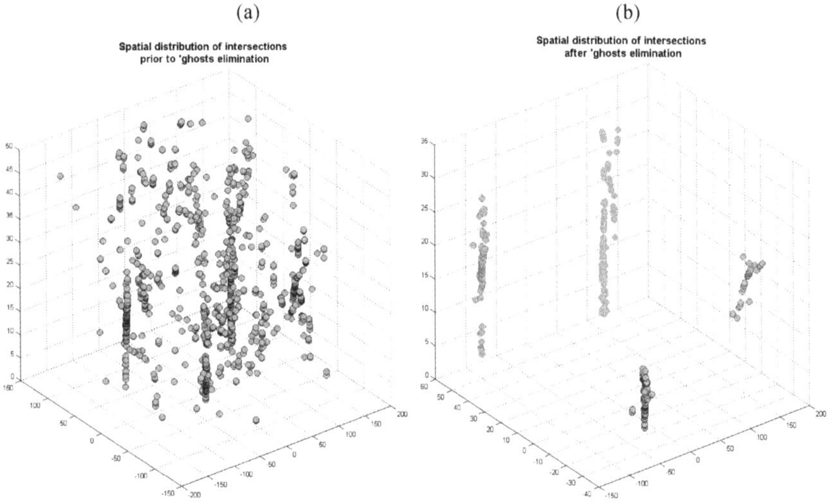

Figure 10.15 Intersections of triplets of the PCL and PET measurements (a) before clustering and (b) after clustering using the DBSCAN algorithm. Note different axes ranges on both images. (©2018 IEEE Reprinted, with permission, from [43].)

of the target. When triplets of those measurements are intersected, numerous ghost targets can appear. In order to combine the intersection points into clusters, the DBSCAN algorithm is applied. The result of the application of clustering is shown on the right side of Figure 10.15. As can be seen, four clusters were created, which corresponds to the true number of targets. The clusters are spread in the altitude dimension, which results from relatively poor altitude estimation accuracy due to the geometrical setup of the system.

A picture of the prototype vehicle (TRL 9) is shown in Figure 10.16. The vehicle is equipped with two masts. On the front mast ELINT, IFF and NAVIG antennas are installed. The rear mast is equipped with the PCL antenna arrays for FM (bottom level) and DVB-T/GSM (middle level). The COMINT antenna is at the top of the rear mast. It is worth noting that the PCL antennas are dual-polarized, which enables greater flexibility in using illuminators of opportunity.

The system was designed from the beginning as a military product. This entails complying with strict regulations on military equipment. This results in a significant increase in development time, as well as in cost. The system, however, can be used as a fully-fledged military radar when it is completed. At the time of this writing, the system has been demonstrated to military customers several times and is currently being prepared for factory and then Polish MOD qualification tests.

Figure 10.16 Picture of the PCL-PET system developed by PIT-Radwar, WUT, and AM Technologies, Poland. (Courtesy of PIT-Radwar.)

10.6 THALES, ONERA, FRANCE

Thales, a French manufacturer of aerospace, space, and defense systems, and ONERA (Office National d'Etudes et de Recherches Aérospatiales), a French research laboratory, have been cooperating on passive radar technology, focusing on FM, DAB, and DVB-T illuminators of opportunity [45–50]. One interesting aspect of passive radar pursued by the two institutions was elevation angle measurement [51]. Elevation angle measurement in passive radar is not very common. When it is performed, it is usually carried out by phase comparison monopulse, with the antennas displaced vertically. By comparing the phases of signals from antennas at different heights, the elevation angle can be calculated. The information on the elevation angle of the target is very valuable in passive radar, as it allows one to discriminate between ground moving targets, low-flying targets, and high-flying targets. It is even more important when the typical geometry of passive radar is considered; the transmitters and the radar are placed very close to the ground plane, and the altitude estimation resulting from intersecting bistatic ellipsoids is relatively poor. For this reason, direct elevation measurement can improve the localization accuracy significantly. Moreover, ghost targets with a calculated altitude inconsistent with the elevation measurement can be easily eliminated.

In Figure 10.17 the antenna array used for elevation measurement is shown. The antennas are used to receive the DVB-T signal in the UHF band. The 16 elements of the antenna allow for azimuth and elevation beam steering and angle estimation capabilities. The elements are placed on 4 panels. In two of the panels (top and bottom), 5 elements are arranged horizontally. In the two middle panels, 3 elements are arranged vertically. In the initial phase of the project, the multipath effect was expected during the elevation angle measurement. This effect, however, was not observed in practice.

An example of an elevation estimation result is shown in Figure 10.18. The solid line shows the ground truth obtained with a global positioning system (GPS) receiver mounted onboard the target. The crosses show the elevation measurements obtained with the passive radar. As can be seen, the measurements are very close to the actual elevation angles. In this experiment, accuracy in the order of 1° was obtained, which can improve target localization significantly.

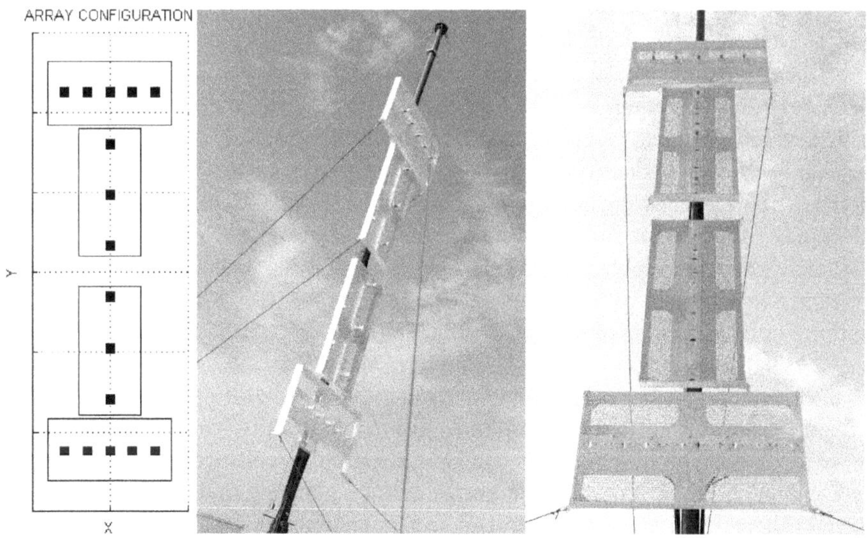

Figure 10.17 A 16-element DVB-T antenna array used for elevation estimation. (©2012 IEEE Reprinted, with permission, from [51].)

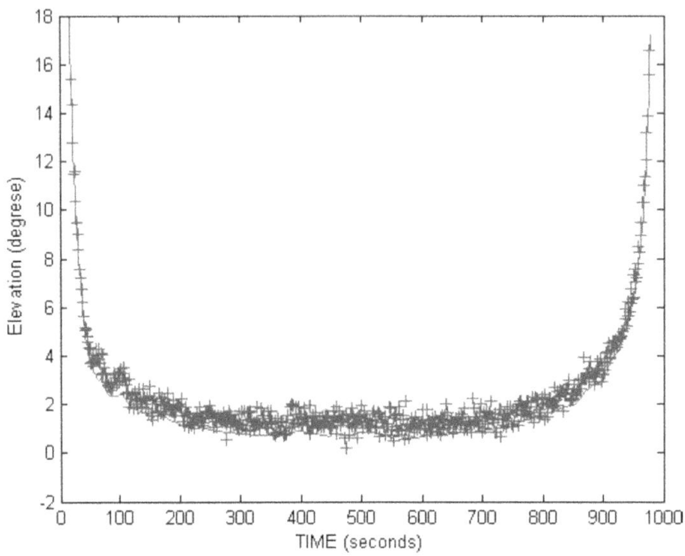

Figure 10.18 Target elevation measured with the GPS receiver (solid line) and passive radar (crosses). (©2012 IEEE Reprinted, with permission, from [51].)

Thales developed an operational passive radar system called the Ground Alerter 100. The system uses FM radio and DVB-T television transmitters. It provides 360° azimuth coverage and 60° elevation coverage. It is advertised as especially efficient in detecting low-flying small targets with low velocities, such as drones. An image showing the antenna arrays of the Ground Alerter 100 is presented in Figure 10.19. In the middle, the FM antenna array is shown. It consists of 8 vertically polarized dipoles. The DVB-T antennas, visible on each side of the FM antenna, are equipped with 16 dipole elements. In both cases, FM and DVB-T, the uniform circular arrays provide full 360° visibility.

As indicated earlier in the book, passive radar performance is highly dependent on the operating scenario, especially transmitter-target-receiver geometry, radiation patterns of the transmitters, target altitude, and terrain profile. For this reason, a tool for predicting the performance of the radar in a particular scenario is very important. In Figure 10.20 a screenshot of the Thales mission planning tool is presented. The top part shows the probability of target detection overlaid on the map. The detection range can have an irregular shape depending on the direction, which is a typical case for passive radar. The bottom part shows the power of the received signal

Figure 10.19 Antenna arrays of Thales' Ground Alerter 100 system. FM array in the middle, two DVB-T arrays at both sides. (Courtesy of Thales.)

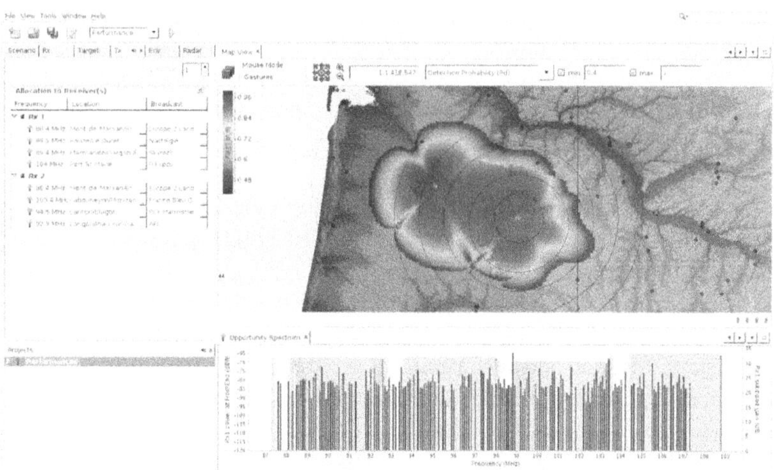

Figure 10.20 Screenshot of the mission planning tool for passive radar. (Courtesy of Thales.)

from different FM transmitters, which allows one to verify which transmitters are available in a particular setup of the passive radar.

10.7 UNIVERSITY OF PISA, CNIT, ITALY

The University of Pisa, and the Radar and Surveillance System National Laboratory, National Interuniversity Consortium for Telecommunications (CNIT) have significant experience in the area of passive radar and radar imaging [52–59]. One of the topics investigated by these closely cooperating institutions is passive inverse synthetic aperture radar (P-ISAR). The idea of this technique is to create a radar image of the target using illuminators of opportunity.

The resolution in the range dimension is obtained using a wide-bandwidth signal, whereas the cross-range resolution is obtained thanks to the fine Doppler resolution, which depends on the integration time. The integration time can be increased almost arbitrarily (the limit is usually imposed by the ability to model target motion); however, the signal bandwidth is determined by the type of illuminator used. For example, the DVB-T signal bandwidth of 7.6 MHz is wide from the point of view of target detection, but from the target imaging point of view it is not sufficient. For this reason, multiple adjacent DVB-T channels are often employed in order to increase the range resolution. In [57] signals from 3 adjacent DVB-T channels were used, which provide approximately 24 MHz of bandwidth. However, there are gaps in the spectrum of approximately 0.8 MHz between the channels. These gaps distort the range profile response from a point scatterer, increasing the sidelobe level. An algorithm was used by the authors, called grating lobe cancellation (GLC), which mitigates the distortion.

An example of the P-ISAR imaging of a ship from [57] will be shown here. The target under investigation is shown in Figure 10.21. It is a container ship called LIWIA P with the size of 212m by 32m.

The target observed in the range-slow time domain is shown in Figure 10.22. The data were obtained by extracting the target echo from the two-dimensional cross-ambiguity function and performing an inverse Fourier transform in the velocity dimension. As can be seen, the target changes the bistatic range over time. Two time windows are marked in the image that were analyzed in [57]. Here results from Frame #1 will be shown.

At the top of Figure 10.23, the data in the range-slow time domain after autofocusing are shown. For the autofocusing of the image, the image-contrast-based autofocusing (ICBA) algorithm has been applied [60]. In the algorithm,

Figure 10.21 Picture of the target under investigation – LIWIA P container ship. (©2013 IEEE Reprinted, with permission, from [57].)

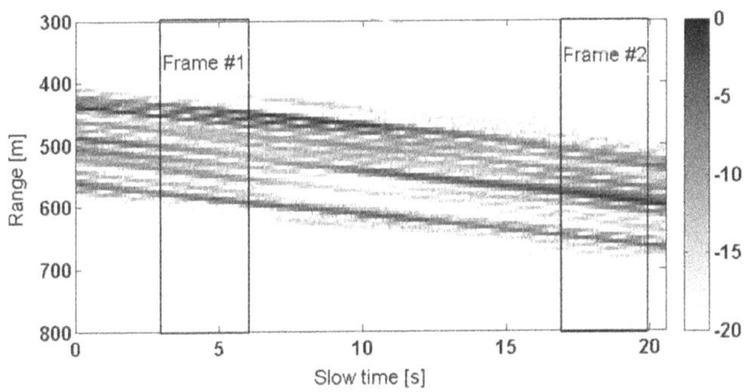

Figure 10.22 Target observed in range-slow time domain. Two analyzed time windows are marked. (©2013 IEEE Reprinted, with permission, from [57].)

a parametric target motion model is used, and the parameters of the model are tuned in an optimization process. The optimized cost function is based on the image contrast; the better the image contrast, the better the fit of the motion model to reality. As can be seen, the target echo migration through the range resolution cells has been eliminated. The ISAR image is then obtained by performing a Fourier transform in the slow-time dimension. The resulting ISAR image is shown at the bottom of Figure 10.23. As can be seen, the image resolution is not as good as is usually obtained with the active ISAR technique, which results primarily from the limited bandwidth of the applied signal. However, such images can be used for the determination of the size of the target and automatic target recognition. Improvement of the results can be expected by using signals with wider bandwidth (e.g., using

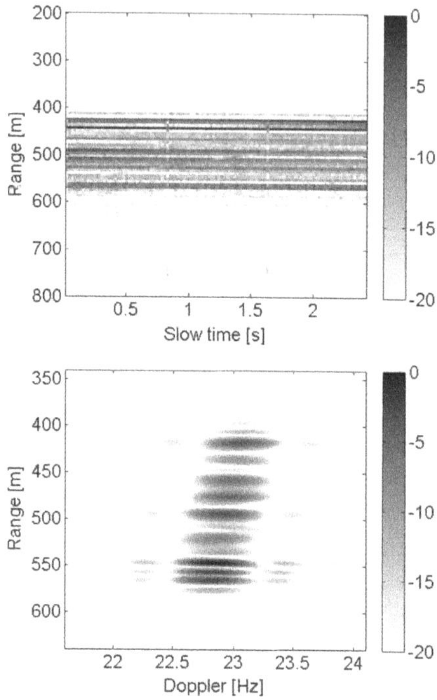

Figure 10.23 Target in the range-slow time domain after the application of the motion compensation algorithm (top). Obtained ISAR image (bottom). (©2013 IEEE Reprinted, with permission, from [57].)

more DVB-T channels) or by using multistatic configuration and combining the results from different transmitter-receiver pairs.

10.8 WARSAW UNIVERSITY OF TECHNOLOGY, POLAND

Work on passive radar technology at the Warsaw University of Technology (WUT) started in 2002. At the beginning, analog TV transmitters were considered as illuminators of opportunity. However, due to the poor characteristics of the signal, other types of transmissions were investigated. The most promising type of illuminator proved to be FM radio. Its drawbacks, such as low and content-dependent bandwidth and analog modulation, were outweighed by the advantages, which include relatively long detection range, the widespread availability of the transmitters, and the relative ease with which an FM-based system can be constructed.

The system constructed at WUT in 2007 is called *PaRaDe*. The system is modular and flexible, which makes it possible to use in many configurations: stationary ground-based, mounted on a ground moving vehicle, and airborne [61–68]. Depending on the version, the system can use FM or DVB-T illuminators of opportunity.

The receiving hardware (digital part) of the *PaRaDe* system is shown in Figure 10.24. The hardware includes four sampling modules, a clock distribution module, a clock signal generator, a calibration signal generator, a power supplier, a GPS receiver and an ADS-B (Automatic Dependent Surveillance-Broadcast) receiver (used as a source of the reference data).

The *PaRaDe* system has been tested during several measurement campaigns. In the campaigns, usual civilian air traffic, as well as military targets during exercises were observed [69]. In Figure 10.25 the antenna array of the *PaRaDe* system during one of the measurement campaigns is shown. The array is mounted on a 12m deployable mast. The mast is installed on a trailer, which enables system mobility. The analog part of the receiver system (amplifiers and filters) is placed at the top of the mast in the center of the antenna array. The digital part of the receiver system (shown in Figure 10.24) and the processing platforms are placed in a caravan (also visible in Figure 10.25).

An example of the bistatic display is shown in Figure 10.26. In the figure, apart from the passive radar detections, reference data obtained with the ADS-B receiver are shown. The ADS-B data indicate the aircraft's geographical position (longitude and latitude), heading, velocity, altitude and identifier. The aircraft parameters are converted to bistatic coordinates using the transmitter, receiver, and aircraft posi-

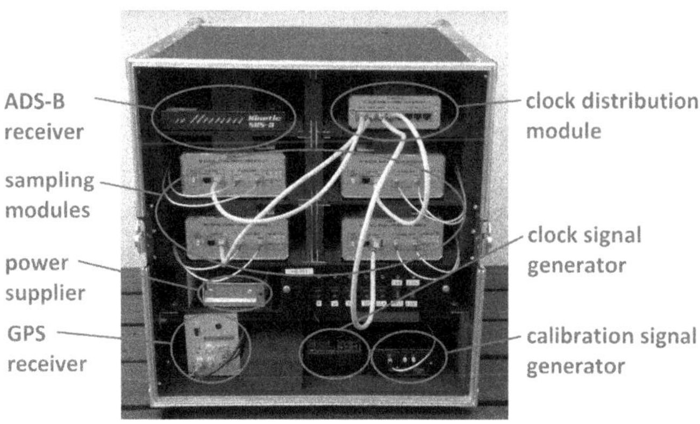

Figure 10.24 The digital part of the receiving system of the *PaRaDe* radar.

Figure 10.25 Antenna array of the *PaRaDe* system on a deployable 12-m mast.

tions. The results are shown on the bistatic display as squares. In Figure 10.26 a target track at the bistatic range close to 600 km can be seen. The track is confirmed with the ADS-B data. The distance between the aircraft and the radar receiver was close to 300 km (approximately half of the bistatic range). Such results are impressive, as it is comparable to the performance of an active long-range radar. Moreover, typical detection ranges for FM-based radar reported in the literature are in the order of 300 km in bistatic range [70, 71].

Apart from ground-based passive radars for airborne target detection, passive radar on mobile platforms was also researched at the Warsaw University of Technology. At the beginning, the *PaRaDe* radar was installed on a car, which resulted in the first trials with a mobile platform [72, 73]. The main aim of the experiment was to investigate the characteristics of clutter, as well as temporal characteristics of the received signal, which were highly influenced by the time-varying multipath propagation. The next step was to put the *PaRaDe* system on an aircraft [74–76]. The trials were conducted in 2008, which was one of the first airborne passive radar experiments in the world (possibly the first one published in open literature [63]). In the above-mentioned experiments, the aim was to detect moving targets hidden in the Doppler-spread clutter. Different approaches to clutter removal were used,

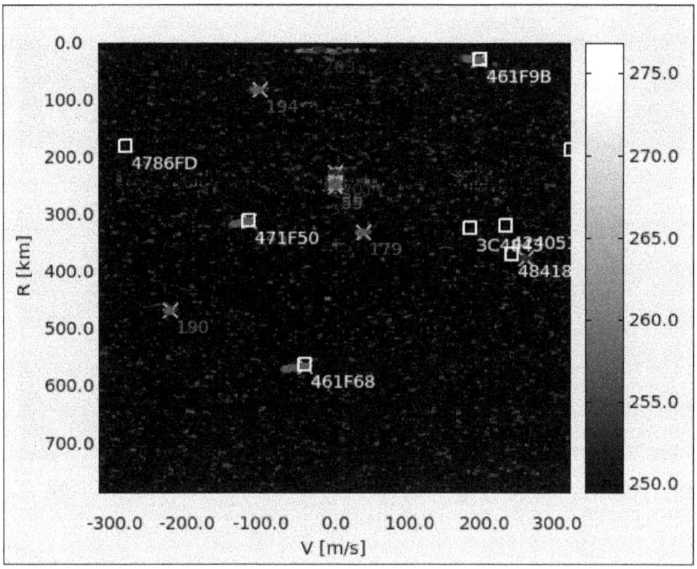

Figure 10.26 Screenshot of the *PaRaDe* bistatic display showing a confirmed detection at bistatic range close to 600 km (crosses – passive radar detections, squares – ADS-B data converted to bistatic coordinates).

including DPCA (displaced phase center antenna), STAP (space-time adaptive processing), and the CLEAN technique (see Chapter 5). The measurements were based on FM illuminators of opportunity. This provides long detection ranges, but poor range resolution.

Another area of research involving moving platforms was passive SAR imaging [23, 77, 78]. In this case, DVB-T transmitters were used, as they provide much finer range resolution than the FM transmitters (20m for DVB-T of quasi-monostatic range resolution versus 1.5 km for FM). In the first experiment, the receivers were mounted on a car [77]. Preliminary results were obtained with algorithms developed for image creation in the passive SAR mode. Next, the airborne trials were conducted [21, 23, 78]. A picture of an aircraft, called the PZL-04 Wilga, used in one of the measurement campaigns is shown in Figure 10.27. An antenna mounted on one side of the aircraft was used to receive the reference signal. Another antenna on the other side of the aircraft was used for echo signal reception. The distance to the transmitter was approximately equal to 11 km.

Three approaches were applied for bistatic SAR image creation: unfocused processing, focused processing, and back-projection [78]. In the first approach

Figure 10.27 Picture of the PZL-04 Wilga aircraft used in passive SAR experiment. (Courtesy of Damian Gromek.)

a low-pass filter is applied, which selects only a short part of the Doppler history of a point target, thus increasing the cross-range resolution. In the focused SAR processing, the matched filter is used to compress the signal in the azimuth direction. In both methods (unfocused and focused processing), range cell migration and Doppler centroid correction are applied before image creation. In the back-projection approach, the image is divided into pixels. For each pixel, a template signal is created with the parameters resulting from the specific position of the pixel in the image. The

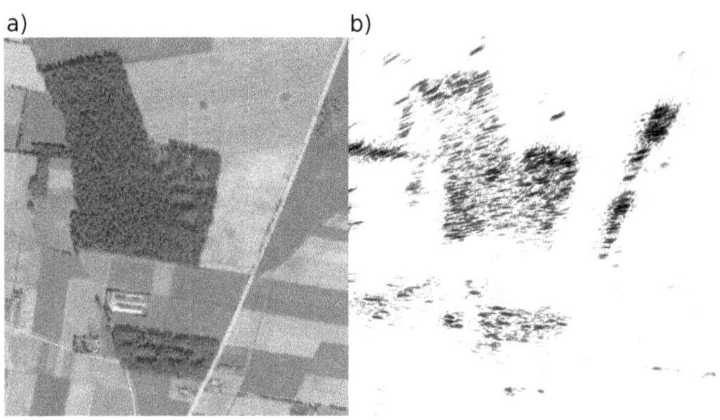

Figure 10.28 (a) Optical image of the mapped area and (b)passive SAR image obtained with the back-projection algorithm (b). (Courtesy of Damian Gromek.)

template is then compared with the received signal. When a match is found between the template and the received signal, a peak appears in the image. The advantage of the back-projection approach is that it takes into account geometrical distortions, such as range cell migration; therefore, no additional corrections are required. The disadvantage of this approach is the very high computational complexity and sensitivity to the precise data from the INS (inertial navigation system)/GPS unit. Fortunately, the calculations are performed independently for each pixel; therefore, the algorithm is easily parallelizable, for example, through the use of a graphics processing unit (GPU).

A comparison of an optical image and passive SAR image is shown in Figure 10.28. The SAR image has been obtained using the back-projection approach. As can be seen, the SAR image reflects the main features of the terrain. The obtained resolution, in the order of tens of meters in range and a few meters in cross-range, is far from what is typically achieved in active SAR systems. This is caused mainly by the limited bandwidth of the illuminator of opportunity (DVB-T signal with 7.6-MHz bandwidth). The chance to improve the image resolution lies in using transmitters with wider bandwidth or using several neighboring channels.

Despite its drawbacks, the undeniable advantage of using a passive SAR system is the ability to create radar imagery without revealing the presence of the radar. Moreover, passive SAR could be used on drones, as the radar consists of a lightweight receiver. Research in the area of passive SAR imaging is currently very active, and significant advances can be expected in the nearest future.

10.9 SUMMARY

In the chapter, several examples of passive radar systems and their applications were presented. As can be seen, passive radar can be successfully applied to airborne target tracking, even in the case of maneuvering targets, such as jet fighters. Mobile platforms, such as a car or an aircraft, can also be used as a carrier for the passive radar. Target detection in such a configuration requires advanced clutter filtering, usually based on the STAP-like approach. Moreover, passive radar can also be applied to radar imaging. In the case of ground imaging, the radar is placed onboard an aircraft, which enables SAR mapping to be performed. Target imaging in the ISAR mode is obtained by observing the target for a long time, so that the Doppler resolution allows individual parts of the target to be distinguished. The main limitation of the imaging application of passive radar is the relatively narrow bandwidth of the signal, which corresponds to the bandwidth of the illuminator of opportunity.

Improvement can be obtained when several channels are used, which effectively increases the signal bandwidth.

In the past, passive radars were usually constructed in the form of technology demonstrators. Many results were also obtained with the laboratory equipment put together just for the measurement campaigns, which can hardly be called a demonstrator. This is still done in order to test new ideas in passive radar; however, the situation is changing. More operational and military-grade systems are being built, which is a sign that the passive radar technology has reached a level of reliability and trust to enable operational use.

References

[1] B. D. Nordwall, "'Silent Sentry' A New Type of Radar," *Aviation Week and Space Technology*, vol. 149, pp. 70–71, November 1998.

[2] R. Plšek, V. Stejskal, M. Pelant, and M. Vojáček, "FM Based Passive Coherent Radar: From Detections to Tracks," in *2011 Tyrrhenian International Workshop on Digital Communications – Enhanced Surveillance of Aircraft and Vehicles*, pp. 123–127, September 2011.

[3] R. Plšek, V. Stejskal, M. Pelant, and M. Vojáček, "Passive Coherent Location and Passive ESM Tracker Systems Synergy," in *2013 14th International Radar Symposium (IRS)*, vol. 1, pp. 149–154, June 2013.

[4] T. Shejbal, R. Plšek, and A. Hermanek, "Comparison of Azimuth Estimation in PCL and MLAT Systems Applied on Measured Data," in *2017 18th International Radar Symposium (IRS)*, pp. 1–9, June 2017.

[5] P. Cabalkova, D. Kubal, and M. Pelant, "Comparison of Clutter Cancelation Methods," in *2014 24th International Conference Radioelektronika*, pp. 1–4, April 2014.

[6] P. Cabalkova, D. Kubal, M. Pelant, R. Plšek, V. Stejskal, and R. Balada, "Comparison of Clutter Cancelation Methods for Code Division Multiplex System in the Presence of Targets," in *2014 11th European Radar Conference*, pp. 467–470, October 2014.

[7] P. Cabalkova, "Statistical Comparison of Clutter Cancelation Techniques with Focus on Distortion Products," in *2015 16th International Radar Symposium (IRS)*, pp. 278–283, June 2015.

[8] P. Cabalkova and R. Plšek, "Comparison of Target Detections from Active MSPSR System with Outputs of MLAT System," in *2016 17th International Radar Symposium (IRS)*, pp. 1–6, May 2016.

[9] J. Heckenbach, H. Kuschel, J. Schell, and M. Ummenhofer, "Passive Radar Based Control of Wind Turbine Collision Warning for Air Traffic PARASOL," in *2015 16th International Radar Symposium (IRS)*, pp. 36–41, June 2015.

[10] H. Kuschel, J. Heckenbach, and M. Ummenhofer, "Mitigation of Rotor Blade-Flash Masking in Passive Radar Collision Warning System PARASOL," in *2015 IEEE Radar Conference*, pp. 185–190, October 2015.

[11] H. Kuschel, J. Heckenbach, and M. Ummenhofer, "Passive Radar Collision Warning System PARASOL," *IEEE Aerospace and Electronic Systems Magazine*, vol. 32, pp. 40–52, February 2017.

[12] M. Ummenhofer, C. Schwark, C. Kress, S. Mechlers, N. Denecke, A. Rohr, M. Fredebohm, M. Koch, and A. Friedmann, "Practical Investigation of Using Passive Radar for Structural Health Monitoring of Wind Farms," in *2018 IEEE International Conference on Environmental Engineering (EE)*, pp. 1–5, March 2018.

[13] C. Kress, S. Mechler, N. Denecke, A. Friedmann, C. Kühnert, R. Seyboldt, M. Ummenhofer, C. Schwark, S. Leupold, and R. Schelenz, "ISO.Wind – A Monitoring System for Wind Parks Using Passive Radar," *Journal of Physics: Conference Series*, vol. 1037, p. 032017, June 2018.

[14] P. Wojaczek, F. Colone, D. Cristallini, P. Lombardo, and H. Kuschel, "The Application of the Reciprocal Filter and DPCA for GMTI in DVB-T – PCL," in *International Conference on Radar Systems (Radar 2017)*, pp. 1–5, October 2017.

[15] P. Wojaczek, A. Summers, and D. Cristallini, "Preliminary Experimental Results of STAP for Passive Radar on a Moving Platform," in *2018 22nd International Microwave and Radar Conference (MIKON)*, pp. 589–592, May 2018.

[16] P. Wojaczek and D. Cristallini, "The Influence of Channel Errors in Mobile Passive Radar using DVB-T Illuminators of Opportunity," in *2018 19th International Radar Symposium (IRS)*, pp. 1–10, June 2018.

[17] P. Wojaczek, F. Colone, D. Cristallini, and P. Lombardo, "Reciprocal Filter-Based STAP for Passive Radar on Moving Platforms," *IEEE Trans. Aerospace and Electronic Systems*, vol. 55, pp. 967–988, April 2019.

[18] I. Walterscheid, P. Wojaczek, D. Cristallini, and A. Summers, "Challenges and First Results of an Airborne Passive SAR Experiment Using a DVB-T Transmitter," in *EUSAR 2018; 12th European Conference on Synthetic Aperture Radar*, pp. 1–4, June 2018.

[19] P. Wojaczek, A. Summers, D. Cristallini, I. Walterscheid, P. Lombardo, and F. Colone, "Results of Airborne PCL Under CCI Conditions Using DVB-T Illuminators of Opportunity," in *2018 International Conference on Radar (RADAR)*, pp. 1–6, August 2018.

[20] M. Antoniou and M. Cherniakov, "Experimental Demonstration of Passive GNSS-Based SAR Imaging Modes," in *IET International Radar Conference 2013*, pp. 1–5, April 2013.

[21] D. Gromek, P. Samczyński, K. Kulpa, P. Krysik, and M. Malanowski, "Initial Results of Passive SAR Imaging Using a DVB-T Based Airborne Radar Receiver," in *2014 11th European Radar Conference*, pp. 137–140, October 2014.

[22] F. Santi, D. Pastina, M. Bucciarelli, M. Antoniou, D. Tzagkas, and M. Cherniakov, "Passive Multistatic SAR with GNSS Transmitters: Preliminary Experimental Study," in *2014 11th European Radar Conference*, pp. 129–132, October 2014.

[23] D. Gromek, K. Kulpa, and P. Samczyński, "Experimental Results of Passive SAR Imaging Using DVB-T Illuminators of Opportunity," *IEEE Geoscience and Remote Sensing Letters*, vol. 13, pp. 1124–1128, August 2016.

[24] D. Tzagkas, M. Antoniou, and M. Cherniakov, "Coherent Change Detection Experiments with GNSS-Based Passive SAR," in *2016 European Radar Conference (EuRAD)*, pp. 262–265, October 2016.

[25] L. M. H. Ulander, P. Frölind, A. Gustavsson, R. Ragnarsson, and G. Stenström, "Airborne Passive SAR Imaging Based on DVB-T Signals," in *2017 IEEE International Geoscience and Remote Sensing Symposium (IGARSS)*, pp. 2408–2411, July 2017.

[26] T. K. Sjögren, P. Frölind, R. Ragnarsson, A. Haglund, A. Gustavsson, L. M. H. Ulander, V. T. Vu, and M. I. Pettersson, "Simultaneous Passive SAR Imaging and Detection of Airborne Targets," in *International Conference on Radar Systems (Radar 2017)*, pp. 1–5, October 2017.

[27] S. Brisken, M. Moscadelli, V. Seidel, and C. Schwark, "Passive Radar Imaging Using DVB-S2," in *2017 IEEE Radar Conference (RadarConf)*, pp. 0552–0556, May 2017.

[28] A. Schröder, M. Edrich, and F. Wolschendorf, "Multiband Experimental PCL System: Concept and Measurement Results," in *11th INTERNATIONAL RADAR SYMPOSIUM*, pp. 1–4, June 2010.

[29] A. Schroeder and M. Edrich, "CASSIDIAN Multiband Mobile Passive Radar System," in *2011 12th International Radar Symposium (IRS)*, pp. 286–291, September 2011.

[30] A. Schroeder, M. Edrich, and V. Winkler, "Multi-Illuminator Passive Radar Performance Evaluation," in *2012 13th International Radar Symposium*, pp. 61–64, May 2012.

[31] A. Schroeder, M. Edrich, and V. Winkler, "Stationary Cassidian FM Passive Radar Demonstrator for 24/7 Operation and Sensor Cluster Measurements," in *2013 14th International Radar Symposium (IRS)*, vol. 1, pp. 161–166, June 2013.

[32] M. Edrich and A. Schroeder, "Multiband Multistatic Passive Radar System for Airspace Surveillance: A Step Towards Mature PCL Implementations," in *2013 International Conference on Radar*, pp. 218–223, September 2013.

[33] M. Edrich, A. Schroeder, and F. Meyer, "Design and Performance Evaluation of a Mature FM/DAB/DVB-T Multi-illuminator Passive Radar System," *IET Radar, Sonar Navigation*, vol. 8, pp. 114–122, February 2014.

[34] A. Schroeder and M. Edrich, "Passive Radar Sensor Cluster Operation," in *2014 15th International Radar Symposium (IRS)*, pp. 1–4, June 2014.

[35] M. Edrich and A. Schroeder, "Design, Implementation and Test of a Multiband Multistatic Passive Radar System for Operational Use in Airspace Surveillance," in *2014 IEEE Radar Conference*, pp. 0012–0016, May 2014.

[36] M. Edrich and F. Hoffmann, "Transmitter Network Model and Diversity Experiments for a Multiband Passive Radar System," in *2017 IEEE Radar Conference (RadarConf)*, pp. 0140–0145, May 2017.

[37] S. Lutz, V. Winkler, R. Müller, and C. Klöck, "Multi Static Long Range Multi Band 3D Passive Radar – Latest Developments at Hensoldt Sensors," in *2018 19th International Radar Symposium (IRS)*, pp. 1–9, June 2018.

[38] https://www.defence24.com/polish-passive-location-system-response-to-active-jamming-and-stealth technologies, "DEFENCE24.COM," March 2019.

[39] T. Brenner, P. Kasprzak, and L. Lamentowski, "Position Estimation in the Multiband PCL-PET Fusion System," in *2014 15th International Radar Symposium (IRS)*, pp. 1–4, June 2014.

[40] L. Lamentowski, T. Brenner, and P. Kasprzak, "Simulation Scenario for Multi-Band Radar System," in *2014 15th International Radar Symposium (IRS)*, pp. 1–4, June 2014.

[41] L. Lamentowski, R. Mularzuk, T. Brenner, and M. Nieszporski, "Tracking Algorithm Analysis for the PCL-PET Fusion System," in *2015 Signal Processing Symposium (SPSympo)*, pp. 1–6, June 2015.

[42] L. Lamentowski, T. Brenner, and M. Nieszporski, "De-Ghosting and Target Extraction Algorithm Analysis for the PCL-PET Fusion System," in *2015 16th International Radar Symposium (IRS)*, pp. 48–53, June 2015.

[43] T. Brenner, W. Dyszynski, and L. Lamentowski, "Cluster Analysis for Multistatic Passive Combined PCL and PET Fusion System," in *2018 22nd International Microwave and Radar Conference (MIKON)*, pp. 595–596, May 2018.

[44] M. Konopko, L. Lamentowski, W. Dyszynski, and T. Brenner, "Analysis of Measurement Association Methods in PCL-PET Passive Location System," in *2018 19th International Radar Symposium (IRS)*, pp. 1–8, June 2018.

[45] D. Poullin, M. Flécheux, and M. Klein, "New Challenges for PCL System: 3D Requirements and 'Optimal' Resources Management for PCL Air Traffic Survey," in *11th International Radar Symposium*, pp. 1–3, June 2010.

[46] D. Poullin, M. Flécheux, and M. Klein, "New Capabilities for PCL System: 3D Measurement for Receiver in Multidonors Configuration," in *The 7th European Radar Conference*, pp. 344–347, September 2010.

[47] N. Millet and M. Klein, "Passive Radar Air Surveillance: Last Results with Multi-Receiver Systems," in *2011 12th International Radar Symposium (IRS)*, pp. 281 285, September 2011.

[48] M. Klein and N. Millet, "Multireceiver Passive Radar Tracking," *IEEE Aerospace and Electronic Systems Magazine*, vol. 27, pp. 26–36, October 2012.

[49] D. Poullin and M. Flécheux, "A Multistatic 3D Passive System Based on DVB-T," in *2014 International Radar Conference*, pp. 1–6, October 2014.

[50] N. Millet, M. Klein, J. Maintoux, and A. Hamon-Duquenne, "Lesson Learnt from Decades of Passive Radar Experiments," in *2014 International Radar Conference*, pp. 1–6, October 2014.

[51] D. Poullin, M. Flécheux, and M. Klein, "Elevation Angle Estimation for Low-Altitude Targets Using DVB (SFN Broadcasters)," *IEEE Aerospace and Electronic Systems Magazine*, vol. 27, pp. 27–35, November 2012.

[52] D. Petri, A. Capria, M. Conti, F. Berizzi, M. Martorella, and E. D. Mese, "High Range Resolution Multichannel DVB-T Passive Radar: Aerial Target Detection," in *2011 Tyrrhenian International Workshop on Digital Communications – Enhanced Surveillance of Aircraft and Vehicles*, pp. 129–132, September 2011.

[53] D. Olivadese, M. Martorella, E. Giusti, D. Petri, and F. Berizzi, "Passive ISAR with DVB-T Signal," in *EUSAR 2012; 9th European Conference on Synthetic Aperture Radar*, pp. 287–290, April 2012.

[54] L. Corucci, D. Petri, F. Berizzi, M. Martorella, and A. Capria, "An Integrated Space-Based Passive System for Maritime Operational Traffic Surveillance," in *2012 IEEE First AESS European Conference on Satellite Telecommunications (ESTEL)*, pp. 1–4, October 2012.

[55] D. Olivadese, M. Martorella, and F. Berizzi, "Multi-Channel P-ISAR Grating Lobes Cancellation," in *IET International Conference on Radar Systems (Radar 2012)*, pp. 1–5, October 2012.

[56] D. Olivadese, E. Giusti, D. Petri, M. Martorella, A. Capria, F. Berizzi, and R. Soleti, "Passive ISAR Imaging of Ships by Using DVB-T Signals," in *IET International Conference on Radar Systems (Radar 2012)*, pp. 1–4, October 2012.

[57] D. Olivadese, E. Giusti, D. Petri, M. Martorella, A. Capria, and F. Berizzi, "Passive ISAR with DVB-T Signals," *IEEE Trans. Geoscience and Remote Sensing*, vol. 51, pp. 4508–4517, August 2013.

[58] M. Martorella and E. Giusti, "Theoretical Foundation of Passive Bistatic ISAR Imaging," *IEEE Trans. Aerospace and Electronic Systems*, vol. 50, pp. 1647–1659, July 2014.

[59] E. Giusti and M. Martorella, "Passive 3D InISAR Using Target-Borne Illuminator of Opportunity," in *2018 International Conference on Radar (RADAR)*, pp. 1–6, August 2018.

[60] M. Martorella, F. Berizzi, and B. Haywood, "Contrast Maximisation Based Technique for 2-D ISAR Autofocusing," *IEE Proceedings – Radar, Sonar and Navigation*, vol. 152, pp. 253–262, August 2005.

[61] M. Malanowski, K. Kulpa, M. Mordzonek, and P. Samczyński, "PaRaDe – Reconfigurable Software Defined Passive Radar," in *Proc. NATO Specialist Meeting SET-136*, (Lisbon, Portugal), p. CD, June 2009.

[62] B. Dawidowicz, K. Kulpa, M. Malanowski, J. Misiurewicz, P. Samczyński, and M. Smolarczyk, "DPCA Detection of Moving Targets in Airborne Passive Radar," *IEEE Trans. Aerospace and Electronic Systems*, vol. 48, pp. 1347–1357, April 2012.

[63] K. Kulpa, M. Malanowski, J. Misiurewicz, M. Mordzonek, P. Samczyński, and M. Smolarczyk, "Airborne PCL Radar: the Concept and Primary Results," in *Proc. Military Radar 2008*, (Amsterdam, The Netherlands), p. CD, October 2008.

[64] B. Dawidowicz, K. Kulpa, and M. Malanowski, "Suppression of the Ground Clutter in Airborne PCL Radar Using DPCA Technique," in *Proc. European Radar Conference – EURAD*, (Rome, Italy), pp. 306–309, September – October 2009.

[65] K. Kulpa, M. Malanowski, P. Samczyński, and B. Dawidowicz, "The Concept of Airborne Passive Radar," in *Proc. Microwaves, Radar and Remote Sensing Symposium – MRRS*, (Kiev, Ukraine), pp. 267–270, August 2011.

[66] B. Dawidowicz, K. Kulpa, M. Malanowski, J. Misiurewicz, P. Samczyński, and M. Smolarczyk, "PCL Systems on Moving Platforms – Challenges and First Results," in *Proc. 3rd FHR Focus Days on PCL*, (Wachtberg, Germany), pp. 1–41, May 2011.

[67] K. Kulpa, M. Malanowski, P. Samczyński, and J. Misiurewicz, "On-Board PCL Systems for Airborne Platform Protection," in *Proc. the Tyrrhenian International Workshop on Digital Communications, Enhanced Surveillance of Aircraft and Vehicles (TIWDC/ESAV)*, (Capri, Italy), pp. 119–122, September 2011.

[68] M. Malanowski, *Signal Processing Algorithms and Target Localisation Methods for Continuous-Wave Passive Bistatic Radar*. No. 179 in Prace Naukowe Politechniki Warszawskiej. Elektronika, Warszawa, ul. Polna 50: Oficyna Wydawnicza Politechniki Warszawskiej, 2012.

[69] M. Malanowski, K. Kulpa, P. Samczyński, J. Misiurewicz, J. Kulpa, P. Roszkowski, and L. Podkalicki, "Experimental Results of the PaRaDe Passive Radar Field Trials," in *Proc. International Radar Symposium 2012*, (Warsaw, Poland), May 2012.

[70] P. E. Howland, D. Maksimiuk, and G. Reitsma, "FM Radio Based Bistatic Radar," *IEE Proc. Radar, Sonar and Navigation*, vol. 152, pp. 107–115, June 2005.

[71] A. D. Lallo, A. Farina, R. Fulcoli, P. Genovesi, R. Lalli, and R. Mancinelli, "Design, Development and Test on Real Data of an FM Based Prototypical Passive Radar," in *Proc. RadarCon 2008*, (Rome, Italy), p. CD, May 2008.

[72] B. Dawidowicz, K. S. Kulpa, and M. Malanowski, "Suppression of the Ground Clutter in Airborne PCL Radar Using DPCA Technique," in *2009 European Radar Conference (EuRAD)*, pp. 306–309, September 2009.

[73] K. Kulpa, M. Malanowski, P. Samczyński, and B. Dawidowicz, "The Concept of Airborne Passive Radar," in *2011 Microwaves, Radar and Remote Sensing Symposium*, pp. 267–270, August 2011.

[74] K. Kulpa, M. Malanowski, P. Samczyński, J. Misiurewicz, and M. Smolarczyk, "On-Board PCL Systems for Airborne Platform Protection," in *2011 Tyrrhenian International Workshop on Digital Communications – Enhanced Surveillance of Aircraft and Vehicles*, pp. 119–122, September 2011.

[75] B. Dawidowicz, K. S. Kulpa, M. Malanowski, J. Misiurewicz, P. Samczyński, and M. Smolarczyk, "DPCA Detection of Moving Targets in Airborne Passive Radar," *IEEE Trans. Aerospace and Electronic Systems*, vol. 48, pp. 1347–1357, April 2012.

[76] B. Dawidowicz, P. Samczyński, M. Malanowski, J. Misiurewicz, and K. S. Kulpa, "Detection of Moving Targets with Multichannel Airborne Passive Radar," *IEEE Aerospace and Electronic Systems Magazine*, vol. 27, pp. 42–49, November 2012.

[77] D. Gromek, P. Krysik, K. Kulpa, P. Samczyński, and M. Malanowski, "Ground-Based Mobile Passive Imagery Based on a DVB-T Signal of Opportunity," in *2014 International Radar Conference*, pp. 1–4, October 2014.

[78] D. Gromek, K. Radecki, J. Drozdowicz, P. Samczyński, and J. Szabatin, "Passive SAR Imaging Using DVB-T Illumination for Airborne Applications," *IET Radar, Sonar Navigation*, vol. 13, no. 2, pp. 213–221, 2019.

Chapter 11

Conclusions

In this book, an overview of signal processing algorithms and localization methods for passive radar was presented. The book describes the full processing chain, from the digital beamforming to the target tracking. Some of the methods used in passive radar, such as digital beamforming, have their counterparts in classical active radars. These methods are, however, adapted to specific conditions encountered in passive radar. For example, the requirements for digital beamforming in passive radar differ from those used in active radars. However, some of the methods used in passive radar are not usually applied in classical radar. An example is clutter cancelation with an adaptive filter, which is a standard procedure in passive radar not normally applied in active radars.

After reading the book the following question can be asked: how the passive radar can be used in practice and what are its weaknesses and strengths? Based on the presented content, the following disadvantages of passive radar can be formulated [1]:

- Dependency on external transmitters. The radar users, especially in the military, are used to the fact that they have full control over how the radar operates. In the case of the passive radar, the transmitter is beyond the control of the radar user. This may seem like a very serious disadvantage in both military and civilian applications. Some strategies are discussed between the radar manufacturers, transmitter operators and authorities, according to which the illuminators of opportunity could be treated as a part of the radar system infrastructure. This could include guaranteeing uninterrupted operation of the transmitter, which would provide a certain level of reliability of passive radar.

In the case of military applications, the use of auxiliary transmitters to be deployed if civilian transmitters are destroyed could be considered.

- Nonoptimal signal structure. The signals that are used by passive radars are not designed for target detection but for the transmission of information. Some of the features of the signals that are applied to facilitate data transmission, such as pilot frequencies or cyclic prefix, can be easily eliminated from the signal if they deteriorate the radar's performance. However, some of the signal characteristics, such as content-dependent signal bandwidth in FM radio, are difficult to overcome.
- Nonoptimal target illumination. Most of the transmitters utilized by the passive radar are used for transmitting the signal to the receivers that are placed on the ground surface. For this reason, the transmitted power is usually directed downwards, where the receivers are. This is nonoptimal from the point of view of airborne target detection. It often happens that the aircraft are illuminated by the transmitters through the sidelobe of the elevation pattern.
- High geometry dependency. As was discussed in Chapter 8 on target localization, the target positioning accuracy is highly dependent on the relative position of the target, receiver(s), and transmitter(s). For this reason, the performance of a passive radar can be satisfactory in an advantageous geometrical configuration, but it can also be unacceptable in an unfavorable configuration.
- Difficult performance prediction. The numerous factors that influence the passive radar's performance, such as geometric relationships, the influence of the transmitter azimuth and elevation radiation patterns, and the external noise level, make it difficult to predict the actual performance. For this reason, it is difficult to formulate requirements for the passive radar that would guarantee good performance of the system itself, disregarding the objective shortcomings of the passive radar in general.

The advantages of passive radar include:

- Lowcost and simplicity. As the passive radar does not require its own transmitter, it can be built relatively cheaply compared to the active radar. Moreover, a typical passive radar does not have moving parts, such as the rotating antenna in the classical active radar. In the simplest version, the passive radar can be built using ordinary radio/TV antennas, a two-channel SRD (software-defined radio) receiver, and a computer. The performance obtained using such a simple system may be very advantageous compared to the costs. This means that passive radar can be used in scenarios where until now radar was too expensive. An example is a small airfield, which normally would not be equipped with an active radar. However, the passive radar can be a feasible purchase, which would substantially increase the safety of the airfield.

- Fast refresh rate. Typically, passive radar outputs measurements with the period of 0.1–1.0s. This is much faster than a classical active ATC (air traffic control) radar with a mechanically scanned antenna, where the typical refresh rate is in the order of 5–10s. This feature of passive radar allows highly maneuvering targets to be tracked with high precision and without the effect of track loss.
- Fine Doppler resolution. In many cases, the bandwidth of the signal used in the passive radar is relatively low, such as in the case of the FM radio. However, the integration time in the order of 0.1–1s provides a fine Doppler resolution of 1–10 Hz. This usually corresponds to the velocity resolution in the order of 1 m/s, which provides very accurate measurement.
- Detection of stealth targets. The stealthiness of the targets is based mainly on two principles: target shaping aimed at deflecting incident energy in a different direction, and absorption of incident radiation. Both techniques can be effective in the higher frequency bands, at which active radars usually operate. The effectiveness of these approaches at lower frequencies is limited. In the case of target shaping, the idea is to deflect the radiation in a direction different than it came from. This is achievable when the wavelength is much smaller than the surface of the target. In the case of passive radar, where the typical wavelength is in the order of meters, the target shaping does not play such a significant role. Moreover, energy deflection is effective in the case of monostatic radar, but not necessarily bistatic radar, as the receiver is in a different location to the transmitter. The absorption of electromagnetic energy can be obtained when the thickness of the coating is comparable with the wavelength. Again, in the case of passive radar, the wavelength is usually longer than in active radars; therefore, energy absorption is much less effective.
- Detection of low-flying targets. As discussed in Chapter 2, the energy radiated by the illuminators of opportunity is usually directed towards the ground surface, as the receivers are usually placed there. This is not advantageous from the point of view of high-flying targets. However, low-flying targets should be illuminated relatively well. This provides the potential of detecting low-flying targets, such as drones and small private aircraft. Passive radar can

therefore be used as a gap-filler radar, whose aim is to detect targets in specific areas not accessible to active radars.

- Difficult to disable. Passive radar by its nature does not transmit electromagnetic energy in order to detect targets. Therefore, the traditional way of destroying active radars with antiradiation missiles (ARM) in a military conflict is not effective. One of the possibilities of disabling a passive radar would be to destroy the illuminators of opportunity. However, it is not possible to predict with 100% certainty which transmitters are used by a passive radar. Therefore, in order to be sure that a passive radar is not operating in a given area, destroying all transmitters would be required, which could be tens, hundreds, or even thousands of transmitters (for example cellular base stations). In such a case, the enemy would be forced to target numerous transmitters, which would be treated as strategic objects almost as important as active air-defense radars. This change of priorities would be itself a great advantage of possessing a passive radar, even if it was not actually used.
- Difficult to jam. There are two main approaches to jam a radar. One way is to generate a signal, usually a noise signal, which prevents a radar from detecting target echoes. The other way is to generate false echoes, so that the radar is not able to distinguish between the real and false target echoes. This technique is called deceptive jamming, and it is usually realized with DRFM (digital radio frequency memory) systems, which receive the radar signal, and repeat it with a delay and Doppler shift. In the case of a passive radar, the noise jamming would require generating noise signal continuously at multiple frequencies. This would be done usually without the prior knowledge of the frequencies used by the passive radar, nor the position of the passive radar. Moreover, the noise signal generated by the jammer does not integrate coherently in passive radar, whereas the target echoes do, with the integration gain in the order of 40–70 dB. For this reason, the effective jamming requires relatively high jammer resources (power and bandwidth). The deceptive jamming is possible; however, it is also not very effective. Because the position of the passive radar is usually not known, the jammer would not know where the energy of the jamming signal should be directed. Moreover, generating false target echoes on the range-Doppler is relatively easy; however, producing a credible target in the Cartesian coordinates would require taking into account all the geometrical dependencies of the multistatic geometry of a passive radar.

11.1 THE FUTURE OF PASSIVE RADAR

The rapid development of passive radar technology has been observed in the last decades. This, however, does not mean that new challenges and ideas in passive radar do not exist. The most important future trends in passive radar are:

- Multistatic operation. In its simplest form, a passive radar system is equipped with a single receiver. A more sophisticated setup involves the use of multiple receivers, communicating with one another [2–4]. An advantage of such an approach is increased detection coverage and localization accuracy. Moreover, such a scenario is more advantageous when a limited number of transmitters are available in a given area as the number of transmitter-receiver pairs is increased, which is important from the localization point of view. A problem which has to be dealt with is the appropriate synchronization and communication of the receiver nodes. The synchronization can be achieved using a GPS receiver with a reference frequency signal and time stamping. The communication between different passive radars can be based on a wired or wireless link. In the case of the wireless link, the fact that the receiver is no longer a passive device has to be taken into account.
- Multiband system. Each type of illuminator of opportunity that the passive radar uses has specific features, which may not be advantageous from the point of view of target detection [5–9]. For example, the bandwidth of the FM radio signal strongly depends on the program content, which decreases the reliability of an FM-based passive radar. However, DVB-T signals have a constant and time-invariant bandwidth, but the detection range is lower than in the case of the FM radio. By combining more than one type of signal source, their advantages can be used, and some drawbacks can be eliminated. Passive radars using more than one type of signal and, at the same time, more than one frequency band already exist. However, problems encountered when constructing such a system, such as the data fusion and wideband antenna arrays, need more research.
- Airborne systems. Usually, passive radars are ground-based, stationary systems. However, with the developments in the passive radar field, other system setups are starting to be considered. One of the possibilities is to install a passive radar on an airborne platform (manned aircraft, UAV, or helicopter) [10–13]. Due to the low complexity of the mechanical parts and the low power consumption of a passive radar, its application as an airborne radar has great potential. Numerous problems have to be solved, however, before

an operational airborne passive radar can be built. Some of the issues involve dealing with Doppler-spread clutter and constructing antenna arrays of a suitable size, in particular at low frequencies.

- Passive imaging. Apart from the target detection and localization considered in this work, target imaging can also be obtained with passive radar [14–18]. Classically, SAR (synthetic aperture radar) and ISAR (inverse SAR) techniques have been used for radar imaging. These techniques utilize large signal bandwidths to obtain high resolution in one dimension (called slant range). High resolution in the other dimension (called cross-range) is obtained by using the relative motion of the radar and target. In the case of passive radar, signals with a bandwidth in the order of 10 MHz are common (e.g., DVB-T), which results in slant-range resolutions of tens of meters. Better resolution can be obtained by combining signals from different frequency channels [19, 20]. Appropriate cross-range resolution in passive radar can be obtained by using the motion of the transmitter (e.g., satellite transmitter), target (e.g., flying aircraft), or receiver (e.g., radar placed on an airborne platform).

Thinking that the passive radar will replace the active radar in the future is not reasonable. A realistic scenario is that the passive radar will be used in specific scenarios as complementary technology to the active radar, for example, as a gap-filler, or low-cost alternative to active radar on small airfields.

On a very general level, it can be stated that passive radar will always be inferior to active radar, as the transmitter is beyond the control of the radar designer in the former case. However, once the drawbacks and limitations of passive radar are understood, the advantages can be used in suitable situations.

References

[1] M. Malanowski, *Signal Processing Algorithms and Target Localisation Methods for Continuous-Wave Passive Bistatic Radar*. No. 179 in Prace Naukowe Politechniki Warszawskiej. Elektronika, Warszawa, ul. Polna 50: Oficyna Wydawnicza Politechniki Warszawskiej, 2012.

[2] N. Millet and M. Klein, "Passive Radar Air Surveillance: Last Results with Multi-Receiver Systems," in *Proc. International Radar Symposium 2011*, (Leipzig, Germany), pp. 281–285, September 2011.

[3] M. Klein and N. Millet, "Multireceiver Passive Radar Tracking," *IEEE Aerospace and Electronic Systems Magazine*, vol. 27, pp. 26–36, October 2012.

[4] M. Edrich and F. Hoffmann, "Transmitter Network Model and Diversity Experiments for a Multiband Passive Radar System," in *2017 IEEE Radar Conference (RadarConf)*, pp. 0140–0145, May 2017.

[5] D. Gould, R. Pollard, C. Sarno, and P. Tittenso, "A Multiband Passive Radar Demonstrator," in *Proc. International Radar Symposium 2006*, (Cracow, Poland), pp. 657–660, May 2006.

[6] H. Kuschel, M. Glende, J. Heckenbach, S. Müller, J. Schell, and C. Schumacher, "Experimental Passive Radar Systems Using Digital Illuminators (DAB/DVB)," in *Proc. International Radar Symposium 2007*, (Cologne, Germany), pp. 411–417, September 2007.

[7] A. Schroeder, M. Edrich, and F. Wolschendorf, "Second-Generation Mobile Multiband Passive Radar Demonstrator," in *Proc. 3rd FHR Focus Days on PCL*, (Wachtberg, Germany), p. CD, May 2011.

[8] A. Schroeder and M. Edrich, "CASSIDIAN Multiband Mobile Passive Radar System," in *Proc. International Radar Symposium 2010*, (Vilnius, Lithuania), pp. 286–291, June 2010.

[9] S. Lutz, V. Winkler, R. Müller, and C. Klöck, "Multi Static Long Range Multi Band 3D Passive Radar – Latest Developments at Hensoldt Sensors," in *2018 19th International Radar Symposium (IRS)*, pp. 1–9, June 2018.

[10] K. Kulpa, M. Malanowski, J. Misiurewicz, M. Mordzonek, P. Samczyński, and M. Smolarczyk, "Airborne PCL Radar: the Concept and Primary Results," in *Proc. Military Radar 2008*, (Amsterdam, The Netherlands), p. CD, October 2008.

[11] J. Brown, K. Woodbridge, A. Stove, and S. Watts, "Air Target Detection Using Airborne Passive Bistatic Radar," *Electronics Letters*, vol. 46, pp. 1396–1397, September 2010.

[12] B. Dawidowicz, P. Samczyński, M. Malanowski, J. Misiurewicz, and K. S. Kulpa, "Detection of Moving Targets with Multichannel Airborne Passive Radar," *IEEE Aerospace and Electronic Systems Magazine*, vol. 27, pp. 42–49, November 2012.

[13] P. Wojaczek, F. Colone, D. Cristallini, and P. Lombardo, "Reciprocal Filter-Based STAP for Passive Radar on Moving Platforms," *IEEE Trans. Aerospace and Electronic Systems*, vol. 55, pp. 967–988, April 2019.

[14] K. Suwa, S. Nakamura, S. Morita, T. Wakayama, H. Maniwa, T. Oshima, R. Maekawa, S. Matsuda, and T. Tachihara, "ISAR Imaging of an Aircraft Target Using ISDB-T Digital TV Based Passive Bistatic Radar," in *Proc. Geoscience and Remote Sensing Symposium (IGARSS), 2010 IEEE International*, pp. 4103–4105, 2010.

[15] D. Pastina, M. Sedehi, and D. Cristallini, "Passive Bistatic ISAR Based on Geostationary Satellites for Coastal Surveillance," in *Proc. IEEE Radar Conference*, (Arlington, Virginia, USA), pp. 865–870, May 2010.

[16] F. Colone, P. Falcone, A. Macera, and P. Lombardo, "High Resolution Cross-Range Profiling with Passive Radar via ISAR Processing," in *Proc. International Radar Symposium 2011*, (Leipzig, Germany), pp. 301–306, September 2011.

[17] P. Wojaczek, A. Summers, D. Cristallini, I. Walterscheid, P. Lombardo, and F. Colone, "Results of Airborne PCL Under CCI Conditions Using DVB-T Illuminators of Opportunity," in *2018 International Conference on Radar (RADAR)*, pp. 1–6, August 2018.

[18] D. Gromek, K. Radecki, J. Drozdowicz, P. Samczyński, and J. Szabatin, "Passive SAR Imaging Using DVB-T Illumination for Airborne Applications," *IET Radar, Sonar Navigation*, vol. 13, no. 2, pp. 213–221, 2019.

[19] K. Olsen, *Investigation of Bandwidth Utilisation Methods to Optimise Performance in Passive Bistatic Radar*. PhD thesis, University College London, 2011.

[20] D. Petri, A. Capria, M. Conti, F. Berizzi, M. Martorella, and E. Mese, "High Range Resolution Multichannel DVB-T Passive Radar: Aerial Target Detection," in *Proc. the Tyrrhenian International Workshop on Digital Communications, Enhanced Surveillance of Aircraft and Vehicles (TI-WDC/ESAV)*, (Capri, Italy), pp. 129–132, September 2011.

Acronyms and Abbreviations

ADC	Analog-to-Digital Converter
ADS-B	Automatic Dependent Surveillance – Broadcast
ARM	Antiradiation Missile
CAF	Cross-Ambiguity Function
CFAR	Constant False Alarm Rate
CIC	Cascaded Integrator-Comb
COMINT	Communications Intelligence
CPI	Coherent Processing Interval
CUT	Cell Under Test
DBPSK	Differential Binary Phase-Shift Keying
DFT	Discrete Fourier Transform
DPCA	Displaced Phase Center Antenna
DPI	Direct Path Interference
DRFM	Digital Radio Frequency Memory
DVB-T	Digital Video Broadcasting – Terrestrial
EIRP	Equivalent Isotropic Radiated Power
EKF	Extended Kalman Filter
ELINT	Electronic Intelligence
ESM	Electronic Support Measures
FFT	Fast Fourier Transform
FIR	Finite Impulse Response
FM	Frequency Modulation

GMSK	Gaussian Minimum Shift Keying
GP-GPU	General Purpose – Graphics Processing Unit
GPS	Global Positioning System
GPU	Graphics Processing Unit
GSM	Global System for Mobile Communications
IFF	Identification, Friend or Foe
IFFT	Inverse Fast Fourier Transform
ISAR	Inverse Synthetic Aperture Radar
LMS	Least Mean Squares
LSL	Least Square Lattice
MIMO	Multiple-Input, Multiple-Output
MLAT	Multilateration
MSPSR	Multistatic Primary Surveillance Radar
MVDR	Minimum Variance Distortionless Response
NLMS	Normalized Least Mean Squares
NN	Nearest Neighbor
OFDM	Orthogonal Frequency-Division Multiplexing
PBR	Passive Bistatic Radar
PCL	Passive Coherent Location
PCR	Passive Covert Radar
PDF	Probability Density Function
PET	Passive Emitter Tracking, Passive ESM Tracking
PNFR	Peak-to-Noise Floor Ratio
PRF	Pulse Repetition Frequency
PSD	Power Spectral Density
QAM	Quadrature Amplitude Modulation
QPSK	Quadrature Phase-Shift Keying
RAM	Radiation-Absorbent Material
RCS	Radar Cross-Section
RDS	Radio Data Stream
RLS	Recursive Least Squares
SAR	Synthetic Aperture Radar
SDR	Software-Defined Radio

SFN	Single Frequency Network
SI	Spherical Interpolation
SN	Strongest Neighbor
SNR	Signal-to-Noise Ratio
STAP	Space-Time Adaptive Processing
SX	Spherical Intersection
TDOA	Time Difference of Arrival
UCA	Uniform Circular Array
UHF	Ultrahigh Frequency
ULA	Uniform Linear Array
VHF	Very High Frequency

About the Author

Mateusz Malanowski received his M.Sc., Ph.D., and D.Sc. (habilitation) degrees in Electrical Engineering from the Warsaw University of Technology, Warsaw, Poland, in 2004, 2009, and 2013 respectively. He was a Research Scientist with FGAN (Forschungsgesellschaft fuer Angewandte Naturwissenschaften), Germany, and an Engineer with Orpal, Poland. Currently, he is an Associate Professor at the Warsaw University of Technology. Professor Malanowski is the author/coauthor of over 180 conference and journal papers. His research interests are radar signal processing, passive radar, target tracking, synthetic aperture radar and noise radar. For the last 15 years he has been involved in numerous national and international projects, focusing on passive radar, synthetic aperture radar, and noise radar. He is currently involved in a project, whose aim is to develop the first Polish, and one of the first in the world, operational military passive radar system. Professor Malanowski has been a member of several NATO Science and Technology Organization groups. Professor Malanowski is a senior member of The Institute of Electrical and Electronics Engineers (IEEE) and a member of The Institution of Engineering and Technology (IET) and The European Microwave Association (EuMA).

Index

Recent Titles in the Artech House Radar Series

Dr. Joseph R. Guerci, Series Editor

Adaptive Antennas and Phased Arrays for Radar and Communications, Alan J. Fenn

Advanced Techniques for Digital Receivers, Phillip E. Pace

Advances in Direction-of-Arrival Estimation, Sathish Chandran, editor

Airborne Pulsed Doppler Radar, Second Edition, Guy V. Morris and Linda Harkness, editors

Basic Radar Analysis, Mervin C. Budge, Jr. and Shawn R. German

Basic Radar Tracking, Mervin C. Budge, Jr. and Shawn R. German

Bayesian Multiple Target Tracking, Second Edition , Lawrence D. Stone, Roy L. Streit, Thomas L. Corwin, and Kristine L Bell

Beyond the Kalman Filter: Particle Filters for Tracking Applications, Branko Ristic, Sanjeev Arulampalam, and Neil Gordon

Cognitive Radar: The Knowledge-Aided Fully Adaptive Approach, Joseph R. Guerci

Computer Simulation of Aerial Target Radar Scattering, Recognition, Detection, and Tracking, Yakov D. Shirman, editor

Control Engineering in Development Projects, Olis Rubin

Design and Analysis of Modern Tracking Systems, Samuel Blackman and Robert Popoli

Detecting and Classifying Low Probability of Intercept Radar, Second Edition, Phillip E. Pace

Digital Techniques for Wideband Receivers, Second Edition, James Tsui

Electronic Intelligence: The Analysis of Radar Signals, Second Edition, Richard G. Wiley

Electronic Warfare in the Information Age, D. Curtis Schleher

Electronic Warfare Target Location Methods, Second Edition, Richard A. Poisel

ELINT: The Interception and Analysis of Radar Signals, Richard G. Wiley

EW 101: A First Course in Electronic Warfare, David Adamy

EW 102: A Second Course in Electronic Warfare, David Adamy

EW 103: Tactical Battlefield Communications Electronic Warfare, David Adamy

FMCW Radar Design, M. Jankiraman

Fourier Transforms in Radar and Signal Processing, Second Edition, David Brandwood

Fundamentals of Electronic Warfare, Sergei A. Vakin, Lev N. Shustov, and Robert H. Dunwell

Fundamentals of Short-Range FM Radar, Igor V. Komarov and Sergey M. Smolskiy

Handbook of Computer Simulation in Radio Engineering, Communications, and Radar, Sergey A. Leonov and Alexander I. Leonov

High-Resolution Radar, Second Edition, Donald R. Wehner

Highly Integrated Low-Power Radars, Sergio Saponara, Maria Greco, Egidio Ragonese, Giuseppe Palmisano, and Bruno Neri

Introduction to Electronic Defense Systems, Second Edition, Filippo Neri

Introduction to Electronic Warfare, D. Curtis Schleher

Introduction to Electronic Warfare Modeling and Simulation, David L. Adamy

Introduction to RF Equipment and System Design, Pekka Eskelinen

Introduction to Modern EW Systems, Andrea De Martino

An Introduction to Passive Radar, Hugh D. Griffiths and Christopher J. Baker

Linear Systems and Signals: A Primer, JC Olivier

Meter-Wave Synthetic Aperture Radar for Concealed Object Detection, Hans Hellsten

The Micro-Doppler Effect in Radar, Second Edition, Victor C. Chen

Microwave Radar: Imaging and Advanced Concepts, Roger J. Sullivan

Millimeter-Wave Radar Targets and Clutter, Gennadiy P. Kulemin

MIMO Radar: Theory and Application, Jamie Bergin and Joseph R. Guerci

Modern Radar Systems, Second Edition, Hamish Meikle

Modern Radar System Analysis, David K. Barton

Modern Radar System Analysis Software and User's Manual, Version 3.0, David K. Barton

Monopulse Principles and Techniques, Second Edition, Samuel M. Sherman and David K. Barton

MTI and Pulsed Doppler Radar with MATLAB®, Second Edition, D. Curtis Schleher

Multitarget-Multisensor Tracking: Applications and Advances Volume III, Yaakov Bar-Shalom and William Dale Blair, editors

Non-Line-of-Sight Radar, Brian C. Watson and Joseph R. Guerci

Precision FMCW Short-Range Radar for Industrial Applications, Boris A. Atayants, Viacheslav M. Davydochkin, Victor V. Ezerskiy, Valery S. Parshin, and Sergey M. Smolskiy

Principles of High-Resolution Radar, August W. Rihaczek

Principles of Radar and Sonar Signal Processing, François Le Chevalier

Radar Cross Section, Second Edition, Eugene F. Knott, et al.

Radar Equations for Modern Radar, David K. Barton

Radar Evaluation Handbook, David K. Barton, et al.

Radar Meteorology, Henri Sauvageot

Radar Reflectivity of Land and Sea, Third Edition, Maurice W. Long

Radar Resolution and Complex-Image Analysis, August W. Rihaczek and Stephen J. Hershkowitz

Radar RF Circuit Design, Nickolas Kingsley and J. R. Guerci

Radar Signal Processing and Adaptive Systems, Ramon Nitzberg

Radar System Analysis, Design, and Simulation, Eyung W. Kang

Radar System Analysis and Modeling, David K. Barton

Radar System Performance Modeling, Second Edition, G. Richard Curry

Radar Technology Encyclopedia, David K. Barton and Sergey A. Leonov, editors

Radio Wave Propagation Fundamentals, Artem Saakian

Range-Doppler Radar Imaging and Motion Compensation, Jae Sok Son, et al.

Robotic Navigation and Mapping with Radar, Martin Adams, John Mullane, Ebi Jose, and Ba-Ngu Vo

Signal Detection and Estimation, Second Edition, Mourad Barkat

Signal Processing in Noise Waveform Radar, Krzysztof Kulpa

Signal Processing for Passive Bistatic Radar, Mateusz Malanowski

Space-Time Adaptive Processing for Radar, Second Edition, Joseph R. Guerci

Special Design Topics in Digital Wideband Receivers, James Tsui

Systems Engineering of Phased Arrays, Rick Sturdivant, Clifton Quan, and Enson Chang

Theory and Practice of Radar Target Identification, August W. Rihaczek and Stephen J. Hershkowitz

Time-Frequency Signal Analysis with Applications, Ljubiša Stanković, Miloš Daković, and Thayananthan Thayaparan

Time-Frequency Transforms for Radar Imaging and Signal Analysis, Victor C. Chen and Hao Ling

Transmit Receive Modules for Radar and Communication Systems, Rick Sturdivant and Mike Harris

For further information on these and other Artech House titles, including previously considered out-of-print books now available through our In-Print-Forever® (IPF®) program, contact:

Artech House
685 Canton Street
Norwood, MA 02062
Phone: 781-769-9750
Fax: 781-769-6334
e-mail: artech@artechhouse.com

Artech House
16 Sussex Street
London SW1V HRW UK
Phone: +44 (0)20 7596-8750
Fax: +44 (0)20 7630-0166
e-mail: artech-uk@artechhouse.com

Find us on the World Wide Web at: www.artechhouse.com

www.ingramcontent.com/pod-product-compliance
Ingram Content Group UK Ltd.
Pitfield, Milton Keynes, MK11 3LW, UK
UKHW021515250926
475443UK00015B/35

9 781630 816629